AF788747

Instructor's Solutions Manual

for

Evans/Rosenthal's

Probability and Statistics

The Science of Uncertainty

Michael J. Evans and Jeffrey S. Rosenthal
University of Toronto

W. H. Freeman and Company
New York

Printed in the United States of America

ISBN: 0-7167-4740-5

First printing 2003

W. H. Freeman and Company
41 Madison Avenue
New York, NY 10010
Houndmills, Basingstoke RG21 6XS, England

www.whfreeman.com

Contents

Preface

This is the Solutions Manual to accompany the text Probability and Statistics, The Science of Uncertainty by M.J. Evans and J.S. Rosenthal published by W. H. Freeman, 2003. The solutions to all **Exercises**, **Computer Exercises**, **Problems**, **Computer Problems** and **Challenges** are included.

The solutions to the computer based problems are presented using Minitab. This is not necessarily a recommendation that this software platform be used, but all these problems can be solved using this software. In general, we feel it will be fairly simple to modify the solutions so that they are suitable for other statistical software.

In the Appendix (also available in pdf format) we have included numerical answers to a selection of exercises and problems where we thought this suitable. Instructors can choose to provide these to students if they wish. Students are sometimes helped if they have an answer to work towards. In general, however, we do not feel that students should be simply provided with solutions. The true value in exercises and problems is the time the student spends thinking about the solution — even if they can't solve it. This is when learning takes place and it is the way to real understanding of the subject matter.

Many people helped in the preparation of this text. Not the least were the students who pointed out many things that lead to improvements. Also Aysha Hashim and Natalie Cheredeko, who taught from a preliminary version of the text, made many suggestions that lead to improvements. A special thanks goes to Hadas Moshonov who diligently looked at many of the problems in the text and helped with their solutions. Also many thanks to Heather and Rosemary Evans and Margaret Fulford, for their patience and support throughout this project.

Michael Evans and Jeffrey Rosenthal
Toronto, 2003

Chapter 1

Probability Models

1.2 Probability Models

Exercises

1.2.1
(a) $P(\{1,2\}) = P(\{1\}) + P(\{2\}) = 1/2 + 1/3 = 5/6$.
(b) $P(\{1,2,3\}) = P(\{1\}) + P(\{2\}) + P(\{3\}) = 1/2 + 1/3 + 1/6 = 1$.
(c) $P(\{1\}) = P(\{2,3\}) = 1/2$.

1.2.2
(a) $P(\{1,2\}) = P(\{1\}) + P(\{2\}) = 1/8 + 1/8 = 1/4$.
(b) $P(\{1,2,3\}) = P(\{1\}) + P(\{2\}) + P(\{3\}) = 1/8 + 1/8 + 1/8 = 3/8$.
(c) There are $\binom{8}{4} = 8!/4!4! = 70$ such events.

1.2.3 $P(\{2\}) = P(\{1,2\}) - P(\{1\}) = 2/3 - 1/2 = 1/6$.

1.2.4 No, since $P(\{2,3\}) \neq P(\{2\}) + P(\{3\})$.

1.2.5 Here $P(\{s\}) = P([s,s]) = s - s = 0$ for any $s \in [0,1]$.

1.2.6 We have that $a = A \cap B^c \cap C^c$, $b = A \cap B \cap C^c$, $c = A^c \cap B \cap C^c$, $d = A \cap B^c \cap C$, $e = A \cap B \cap C$, $f = A^c \cap B \cap C$, and $g = A^c \cap B^c \cap C$.

1.2.7 This is the subset $(A \cap B^c) \cup (A^c \cap B)$.

Problems

1.2.8 No, since $P([0,1]) = 1$, while $\sum_{s\in[0,1]} P(\{s\}) = \sum_{s\in[0,1]} 0 = 0$. Here additivity fails because $[0,1]$ is not countable.

1.2.9 No, since for countable S we would then have $P(S) = \sum_{s\in S} P(\{s\}) = \sum_{s\in[0,1]} 0 = 0$, contradicting the fact that $P(S) = 1$.

1.2.10 Yes. For example, this is true for the uniform distribution on $[0,1]$. Since $[0,1]$ is not countable, there is no contradiction.

1.3 Basic Results for Probability Models

Exercises

1.3.1
(a) $P(\{2,3,4,\ldots,100\}) = P(\{1,2,3,4,\ldots,100\}) - P(\{1\}) = 1 - 0.1 = 0.9$.
(b) $P(\{1,2,3\}) = P(\{1\}) + P(\{2\}) + P(\{3\}) \geq P(\{1\}) = 0.1$. And, $P(\{1,2,3\}) = 0.1$ if $P(\{2\}) = P(\{3\}) = 0$. So, 0.1 is the smallest possible value of $P(\{1\})$.

1.3.2 Let A be the event "Al watches the six o'clock news" and B be the event "Al watches the eleven o'clock news". Then $P(A) = 2/3, P(B) = 1/2$ and $P(A \cap B) = 1/3$. Therefore, the probability that Al only watches the six o'clock news is $P(A \backslash (A \cap B)) = P(A) - P(A \cap B) = 2/3 - 1/3 = 1/3$. The probability that Al watches neither news is given by $P((A \cup B)^c) = 1 - P(A \cup B) = 1 - P(A) - P(B) + P(A \cap B) = 1 - 2/3 - 1/2 + 1/3 = 1/6$.

1.3.3 $P(\text{late or early or both}) = P(\text{late}) + P(\text{early}) - P(\text{both}) = 10\% + 20\% - 5\% = 25\%$.

1.3.4 $P(\text{at least one knee sore}) = P(\text{right knee sore}) + P(\text{left knee sore}) - P(\text{both knees sore}) = 25\% - P(\text{both knees sore})$. The maximum is when $P(\text{both knees sore}) = 0$, where $P(\text{at least one knee sore}) = 25\%$. The minimum is when $P(\text{both knees sore}) = 10\%$ (so the right knee is always sore whenever the left one is), where $P(\text{at least one knee sore}) = 15\%$.

Problems

1.3.5 We see that $P(\{2,3,4,5\}) = P(\{1,2,3,4,5\}) - P(\{1\}) = 0.3 - 0.1 = 0.2$. Hence, the largest is $P(\{2\}) = 0.2$ (with $P(\{6\}) = 0.4$). The smallest is $P(\{2\}) = 0$ (with, e.g., $P(\{3\}) = 0.2$ and $P(\{6\}) = 0.4$).

Challenge

1.3.6

(a) Let $D = B \cup C$. Then $P(A \cup B \cup C) = P(A \cup D) = P(A) + P(D) - P(A \cap D) = P(A) + P(B \cup C) - P((A \cap B) \cup (A \cap C)) = P(A) + (P(B) + P(C) - P(B \cap C)) - (P(A \cap B) + P(A \cap C) - P((A \cap B) \cap (A \cap C)) = P(A) + (P(B) + P(C) - P(B \cap C)) - P(A \cap B) - P(A \cap C) - P(A \cap B \cap C)$, which gives the result.
(b) We use induction on n. We know the result is true for $n = 2$ from the text (and for $n = 3$ from part (a)). Assume it is true for $n - 1$, so that $P(B_1 \cup \ldots \cup B_{n-1}) = \sum_{i=1}^{n-1} P(B_i) - \sum_{\substack{i,j=1 \\ i<j}}^{n-1} P(B_i \cap B_j) + \sum_{\substack{i,j,k=1 \\ i<j<k}}^{n-1} P(B_i \cap B_j \cap B_k) - \ldots \pm P(B_1 \cap \ldots \cap B_{n-1})$ for any events $B_1, \ldots, B_{n-1}$. Let $D = A_1 \cup \ldots \cup A_{n-1}$. Then $P(A_1 \cup \ldots \cup A_n) = P(D \cup A_n) = P(D) + P(A_n) - P(D \cap A_n)$. Now, by the induction hypothesis, $P(D) = P(A_1 \cup \ldots \cup A_{n-1}) = \sum_{i=1}^{n-1} P(A_i) - \sum_{\substack{i,j=1 \\ i<j}}^{n-1} P(A_i \cap A_j) + \sum_{\substack{i,j,k=1 \\ i<j<k}}^{n-1} P(A_i \cap A_j \cap A_k) - \ldots \pm P(A_1 \cap \ldots \cap A_{n-1})$. Also, $P(D \cap A_n) = P((A_1 \cap A_n) \cup (A_2 \cap A_n) \cup \ldots \cup (A_{n-1} \cap A_n))$, so by the induction hypothesis this equals $\sum_{i=1}^{n-1} P(A_i \cap A_n) - \sum_{\substack{i,j=1 \\ i<j}}^{n-1} P(A_i \cap A_j \cap A_n) + \sum_{\substack{i,j,k=1 \\ i<j<k}}^{n-1} P(A_i \cap A_j \cap A_k \cap A_n) - \ldots \pm P(A_1 \cap \ldots \cap A_{n-1} \cap A_n)$. Putting this

all together, we see that $P(A_1 \cup \ldots \cup A_n) = \sum_{i=1}^n P(A_i) - \sum_{\substack{i,j=1 \\ i<j}}^n P(A_i \cap A_j) + \sum_{\substack{i,j,k=1 \\ i<j<k}}^n P(A_i \cap A_j \cap A_k) - \ldots \pm P(A_1 \cap \ldots \cap A_n)$. This proves the statement for this value of n. The general result then follows by induction.

1.4 Uniform Probability on Finite Spaces

Exercises

1.4.1
(a) By independence, $P(\text{all eight show six}) = (1/6)^8 = 1/1679616$.
(b) By additivity, $P(\text{all eight show same}) = \sum_{i=1}^6 P(\text{all eight show } i) = \sum_{i=1}^6 (1/6)^8 = 6\,(1/6)^8 = (1/6)^7 = 1/279936$.
(c) For the sum to equal 9, we need seven of the dice to show 1, and the eighth die to show 2. There are eight ways this can happen, each having probability $(1/6)^8$. So, $P(\text{sum equals nine}) = 8\,(1/6)^8 = 1/209952$.

1.4.2 There are $\binom{10}{2} = 45$ ways of choosing which two dice will have 2 showing. Then the probability that those two dice show 2, and the other eight do not, is $(1/6)^2(5/6)^8$. So, the answer is $45\,(1/6)^2(5/6)^8 = 1953125/6718464 \doteq 0.2907$.

1.4.3

$$\begin{aligned} P(\text{at least three heads}) &= 1 - P(\leq \text{two heads}) \\ &= 1 - P(0 \text{ heads}) - P(\text{one head}) - P(\text{two heads}) \\ &= 1 - (1/2)^{100} - \binom{100}{1}(1/2)^{100} - \binom{100}{2}(1/2)^{100} \\ &= 1 - (1 + 100 + 4950)(1/2)^{100} = 1 - 5051/2^{100} \end{aligned}$$

1.4.4
(a) There is only one way this can happen, so the probability is $1 / \binom{52}{5} = 1/2598960$.
(b) There are $\binom{13}{5}$ ways this can happen, so the probability is $\binom{13}{5} / \binom{52}{5} = 33/66640$.
(c) The number of ways this can happen is equal to $(52 \cdot 48 \cdot 44 \cdot 40 \cdot 36) / 5! = 1317888$, so the probability is $1317888 / \binom{52}{5} = 2112/4165$.
(d) The number of ways this can happen is equal to $(13)(12)\binom{4}{3}\binom{4}{2} = 3744$, so the probability is $3744 / \binom{52}{5} = 6/4165$.

1.4.5
(a) The number of ways this can happen is equal to $\binom{4}{1}\binom{13}{13}\binom{39}{13\ 13\ 13}$ $= 337912392291465600$, so the probability is

$$337912392291465600 / \binom{52}{13\ 13\ 13\ 13} = 1/158753389900.$$

(b) The number of ways this can happen is equal to $\binom{4}{1}\binom{4}{4}\binom{48}{9}\binom{39}{13\ 13\ 13}$, so the probability is

$$\binom{4}{1}\binom{4}{4}\binom{48}{9}\binom{39}{13\ 13\ 13} \Big/ \binom{52}{13\ 13\ 13\ 13} = 44/4165 \doteq 0.0106.$$

1.4.6 The complement of this event is the event that the sum is less than 4, which means we chose either two Aces, or one Ace and one 2. $P(\text{two Aces}) = \binom{4}{2} / \binom{52}{2} = 1/221$. $P(\text{one Ace and one 2}) = \binom{4}{1}\binom{4}{1} / \binom{52}{2} = 8/663$. So, $P(\text{sum} \geq 4) = 1 - 1/221 - 8/663 = 652/663$.

1.4.7 This is the probability that the first ten cards contain no Jack, so it equals $\binom{48}{10} / \binom{52}{10} = 246/595 \doteq 0.4134$.

1.4.8 Out of all 52! different orderings, the number where the Ace of Spades follows the Ace of Clubs is equal to 51! (since the two Aces can then be treated as a single card). The same number have the Ace of Clubs following the Ace of Spades. Hence, the probability that those two Aces are adjacent equals $2 \cdot 51! / 52! = 1/26$.

1.4.9 The probability of getting 7 on any one role equals $6/36 = 1/6$. Hence, the probability of *not* getting 7 on the first two roles, and then getting it on the third role, is equal to $(5/6)^2(1/6) = 25/216$.

1.4.10 There are $\binom{3}{2}$ ways of choosing which two dice are the same, and six ways of choosing which number comes up twice, and then 5 ways of choosing which number comes up once. Hence, the probability equals $\binom{3}{2}(6)(5)/(6{\cdot}6{\cdot}6) = 5/12$.

1.4.11 The probability they're all red equals $\left(\binom{5}{3} / \binom{12}{3}\right)\left(\binom{6}{3} / \binom{18}{3}\right) = 5/4488$. Similarly, the probability they're all blue equals $\left(\binom{7}{3} / \binom{12}{3}\right)\left(\binom{12}{3} / \binom{18}{3}\right) = 35/816$. Hence, the desired probability equals $5/4488 + 35/816 = 395/8976 \doteq 0.0440$.

Problems

1.4.12 If $A_1, A_2, \ldots$ are disjoint sets, then $|A_1 \cup A_2 \cup \ldots| = |A_1| + |A_2| + \ldots$. Hence, $P(A_1 \cup A_2 \cup \ldots) = |A_1 \cup A_2 \cup \ldots| / |S| = (|A_1| + |A_2| + \ldots) / |S| = |A_1|/|S| + |A_2|/|S| + \ldots = P(A_1) + P(A_2) + \ldots$. Hence, P is additive.

1.4.13 By considering all 8-tuples of numbers between 1 and 6, we see that 9 can occur if and only if one of the dice takes the value 2 and the remaining seven take the value 1. This occurs with probability $\binom{8}{1}\left(\frac{1}{6}\right)^8 = 4.7630 \times 10^{-6}$.

The value 10 can occur if and only if one of the dice takes the value 3 and the remaining seven take the value 1 or two of the dice take the value 2 and the remaining six take the value 1. This occurs with probability $\binom{8}{1}\left(\frac{1}{6}\right)^8 + \binom{8}{2}\left(\frac{1}{6}\right)^8 = 2.1433 \times 10^{-5}$.

The value 11 can occur if and only if one of the dice takes the value 4 and the remaining seven take the value 1 or one of the dice takes the value 3, one of the dice takes the value 2, and the remaining six take the value 1. This occurs with probability $\binom{8}{1}\left(\frac{1}{6}\right)^8 + \binom{8}{1}\binom{7}{1}\left(\frac{1}{6}\right)^8 = 3.8104 \times 10^{-5}$.

1.4.14 For $1 \le i \le 6$, the probability that the die equals i and the number of heads equals i is equal to $(1/6)\binom{6}{i}/2^6$. Hence, by additivity, the total probability is equal to $\sum_{i=1}^{6}(1/6)\binom{6}{i}/2^6 = (1/6)\sum_{i=1}^{6}\binom{6}{i}/2^6 = (1/6)(1) = 1/6$.

1.4.15 There are $\binom{10}{2\ 3\ 5} = 2520$ ways of choosing which two dice show 2, and which three dice show 3. For each such choice, the probability is $(1/6)^2(1/6)^3(4/6)^5$ that the dice show the proper combination of 2, 3, and other. Hence, the desired probability equals $2520(1/6)^2(1/6)^3(4/6)^5 = 280/6561$.

1.4.16 For $1 \le i \le 6$, the number of ways of apportioning the Spades to North, East, and Other (i.e. West and South combined), so that North and East each have i Spades, is equal to $\binom{13}{i\ i\ 13-2i}$. The number of ways of apportioning the non-Spades to North, East, and Other is then $\binom{39}{13-i\ 13-i\ 13+2i}$. Hence, the number of deals such that North and East each have i Spades is equal to

$$\binom{13}{i\ i\ 13-2i}\binom{39}{13-i\ 13-i\ 13+2i}.$$

On the other hand, the number of ways of apportioning all the cards to North, East, and Other is equal to $\binom{52}{13\ 13\ 26}$. It then follows by additivity that the desired probability is equal to

$$\sum_{i=1}^{6}\binom{13}{i\ i\ 13-2i}\binom{39}{13-i\ 13-i\ 13+2i}\Big/\binom{52}{13\ 13\ 26}$$
$$= 28033098249/158753389900 \doteq 0.1766/$$

1.4.17 For $1 \le i \le 9$ the probability that the card's value is i and that the number of heads equals i is equal to $(4/52)\binom{10}{i}/2^{10}$. For $i = 10$, $(4/52)$ is replaced by $(20/52)$ since any Ten, Jack, Queen, or King will do. Hence, by additivity, the total probability is equal to $\sum_{i=1}^{9}(4/52)\binom{10}{i}/2^{10} + (20/52)\binom{10}{10}/2^{10} = 79/1024 \doteq 0.0771$.

Challenges

1.4.18 For $2 \le i \le 7$, the probability that the sum of the numbers equals i is equal to $(i-1)/36$, while for $7 \le i \le 12$ it is equal to $(13-i)/36$. Hence, the desired probability is equal to $\sum_{i=2}^{7}((i-1)/36)\binom{12}{i}/2^{12} + \sum_{i=8}^{12}((13-i)/36)\binom{12}{i}/2^{12} = 18109/147456 \doteq 0.1228$.

1.4.19
(a) This equals $365/365^2 = 1/365$.
(b) This equals $365/365^C = 1/365^{C-1}$.
(c) This equals $1 - (365 \cdot 364 \cdots 366 - C)/365^C = 1 - 365!/(365-C)!\,365^C$.
(d) When $C = 23$, the probability equals 0.507297. That is, with 23 people in a room, there is more than a 50% chance that two share the same birthday. (If $C = 40$ this probability is 0.891232.) Many people find this surprising.

1.5 Conditional Probability and Independence

Exercises

1.5.1
(a) Here $P(\text{first die 6, and three dice 6}) = (1/6)\binom{3}{2}(1/6)^2(5/6) = 15/6^4$. Also, $P(\text{three dice 6}) = \binom{4}{3}(1/6)^3(5/6) = 20/6^4$. Hence, the conditional probability equals $(15/6^4) / (20/6^4) = 3/4 = 0.75$. (This also follows intuitively since any three of the four dice could have shown 6.)
(b) Here $P(\text{first die 6, and at least three dice 6}) = P(\text{first die 6, and three dice 6}) + P(\text{first die 6, and four dice 6}) = 15/6^4 + (1/6)^4 = 16/6^4$. Also, $P(\text{at least three dice 6}) = P(\text{three dice 6}) + P(\text{four dice 6}) = 20/6^4 + 1/6^4 = 21/6^4$. Hence, the conditional probability equals $(16/6^4) / (21/6^4) = 16/21 \doteq 0.762$.

1.5.2
(a) This probability equals $P(\text{one head, and die shows 1}) + P(\text{two heads, and die shows 2}) = \binom{2}{1}(1/2)^2(1/6) + \binom{2}{2}(1/2)^2(1/6) = 1/12 + 1/24 = 1/8$.
(b) This probability equals $P(\text{one head, and die shows 1}) / P(\text{die shows 1}) = (1/12) / (1/6) = 1/2$. (This makes sense since it is the same as the probability that the number of heads equals 1.)
(c) It is larger, since the die showing 1 makes it much easier for the number of heads to equal the number showing on the die.

1.5.3
(a) This probability equals $(1/2)^3 = 1/8$.
(b) Here $P(\text{number of heads odd}) = P(\text{one head}) + P(\text{three heads}) = \binom{3}{1}(1/2)^3 + \binom{3}{3}(1/2)^3 = 4/8 = 1/2$. Also $P(\text{number of heads odd, and all three coins heads}) = P(\text{all three coins heads}) = (1/2)^3 = 1/8$. Hence, desired conditional probability equals $(1/8) / (1/2) = 1/4$.
(c) Here $P(\text{number of heads even}) = P(\text{0 heads}) + P(\text{two heads}) = \binom{3}{0}(1/2)^3 + \binom{3}{2}(1/2)^3 = 4/8 = 1/2$. Also $P(\text{number of heads odd, and all three coins heads}) = 0$ since this is impossible. Hence, desired conditional probability equals $0/(1/2) = 0$.

1.5.4 $P(\text{five Spades}) = \binom{13}{5} / \binom{52}{5} = 33/66640$. Also, $P(\text{four Spades}) = \binom{13}{4}\binom{39}{1} / \binom{52}{5} = 143/13328$. Hence, $P(\text{five Spades} \mid \text{at least 4 Spades}) = P(\text{five Spades}) / P(\text{four Spades}) + P(\text{five Spades}) = (33/66640) / [(143/13328) + (33/66640)] = 3/68 \doteq 0.044$.

1.5.5 This probability equals $P(\text{four Aces}) / P(\text{four Aces}) = 1$.

1.5.6 This probability equals $\binom{13}{5}\binom{3}{1}^5 / \binom{39}{5} = 0.54318$.

1.5.7 This equals $P(\text{home run} \mid \text{fastball})\, P(\text{fastball}) + P(\text{home run} \mid \text{curve ball}) \times P(\text{curve ball}) = (8\%)(80\%) + (5\%)(20\%) = (0.08)(0.80) + (0.05)(0.20) = 0.074$.

1.5.8 By Bayes' Theorem, $P(\text{snow} \mid \text{accident}) = [P(\text{snow}) / P(\text{accident})] \times P(\text{accident} \mid \text{snow}) = [0.20/0.10]\,(0.40) = 0.80$.

1.5.9 Here $P(A) = 1/6$, $P(B) = 1/36$, $P(C) = 1/6$, and $P(D) = 1/6$.

(a) $P(A \cap B) = P(\text{both dice show 6}) = 1/36 \neq (1/6)(1/36) = P(A)\,P(B)$, so A and B are not independent.
(b) $P(A \cap C) = P(\text{both dice show 4}) = 1/36 = (1/6)(1/6) = P(A)\,P(C)$, so A and C are independent.
(c) $P(A \cap D) = P(\text{both dice show 4}) = 1/36 = (1/6)(1/6) = P(A)\,P(D)$, so A and D are independent.
(d) $P(C \cap D) = P(\text{both dice show 4}) = 1/36 = (1/6)(1/6) = P(C)\,P(D)$, so C and D are independent.
(e) $P(A \cap C \cap D) = P(\text{both dice show 4}) = 1/36 \neq (1/6)(1/6)(1/6) = P(A)\,P(C)\,P(D)$, so A and C and D are not all independent. (Thus, A and C and D are pairwise independent, but not independent.)

1.5.10 We have from the Exercise 1.4.11 solution that $P(\text{all red}) = 5/4488$, while $P(\text{all blue}) = 35/816$. Hence, $P(\text{all red} \mid \text{all same color}) = P(\text{all red}) \,/\, P(\text{all same color}) = (5/4488)/[(5/4488) + (35/816)] = 2/79 \doteq 0.025$.

Problems

1.5.11
(a) $P(\text{red}) = P(\text{card \#1})\,P(\text{red} \mid \text{card \#1}) + P(\text{card \#2})\,P(\text{red} \mid \text{card \#2}) + P(\text{card \#3})\,P(\text{red} \mid \text{card \#3}) = (1/3)(1) + (1/3)(0) + (1/3)(1/2) = 1/2$.
(b) $P(\text{card \#1} \mid \text{red}) = P(\text{card \#1, red}) \,/\, P(\text{red}) = P(\text{card \#1})\,P(\text{red} \mid \text{card \#1}) \,/\, P(\text{red}) = (1/3)(1) \,/\, (1/2) = 2/3$. (Many people think the answer will be 1/2.)
(c) Make three cards as specified, and repeatedly run the experiment. Discard all experiments where the one side showing is black. Of the experiments where the one side showing is red, count the fraction where the other side is also red. After many experiments, it should be close to 2/3.

1.5.12 Assume A and B are independent. Then since $A^C \cap B$ and $A \cap B$ are disjoint with union B, $P(A^C \cap B) + P(A \cap B) = P(B)$. Hence, $P(A^C \cap B) = P(B) - P(A \cap B) = P(B) - P(A)\,P(B) = P(B)[1 - P(A)] = P(B)\,P(A^C)$. So, A^C and B are independent. The converse then follows by interchanging A and A^C throughout.

1.5.13 If $P(A \mid B) > P(A)$, then $P(A \cap B) \,/\, P(B) > P(A)$, so $P(A \cap B) > P(A)\,P(B)$, so $P(A \cap B) \,/\, P(A) > P(B)$, so $P(B \mid A) > P(B)$. The converse follows by interchanging A and B throughout.

Challenges

1.5.14 Let q_i be the probability that the sum of the second and third dice (to be called the "other dice") equals i. Then the desired probability equals $P(\text{first die 4, sum of three dice 12}) \,/\, P(\text{sum of three dice 12}) = P(\text{first die 4, sum of other dice 8}) \,/\, \sum_{i=1}^{6} P(\text{first die } i\text{, sum of other dice } 12 - i) = (1/6)\,q_8 \,/\, \sum_{i=1}^{6}(1/6)\,q_{12-i} = q_8 \,/\, \sum_{j=7}^{12} q_j = (5/36) \,/\, [(6/36)+(5/36)+(4/36)+(3/36)+(2/36)+(1/36)] = 5 \,/\, [6+5+4+3+2+1] = 5/21$.

1.5.15

(a) This probability is equal to $P(\text{sum is 4} \mid \text{sum is 4 or 7}) = (3/36) / [(3/36) + (6/36)] = 3/9 = 1/3$.

(b) $p_2 = p_3 = p_{12} = 0$, and $p_7 = p_{11} = 1$. Also $p_4 = 1/3$ from part (a). For other i, let $q_i = P(\text{sum is } i)$ like in the previous solution. Then $p_i = P(\text{sum is } i \mid \text{sum is } i \text{ or } 7) = q_i / [q_i + (6/36)] = q_i / [q_i + (1/6)] = 1 / [1 + (1/6q_i)]$. Thus, $p_5 = 1 / [1 + (1/6(4/36))] = 2/5$, $p_6 = 1 / [1 + (1/6(5/36))] = 5/11$, $p_8 = 1 / [1 + (1/6(5/36))] = 5/11$, $p_9 = 1 / [1 + (1/6(4/36))] = 2/5$, and $p_10 = 1 / [1 + (1/6(3/36))] = 1/3$.

(c) By the law of total probability, the probability of winning at craps is $\sum_{i=2}^{12} P(\text{first sum } i)\, P(\text{win} \mid \text{first sum } i) = \sum_{i=2}^{12} q_i p_i = (1/36)(0) + (2/36)(0) + (3/36)(1/3) + (4/36)(2/5) + (5/36)(5/11) + (6/36)(1) + (5/36)(5/11) + (4/36)(2/5) + (3/36)(1/3) + (2/36)(1) + (1/36)(0) = 244/495 \doteq 0.492929$. This is just *barely* less than 50%; but that "barely less" is still enough to ensure that, if you play craps repeatedly, then eventually you will lose money (and the casino will get rich).

1.5.16

(a) Since you chose door A, the host will always open either door B or door C. Without any further information, those two events are equally likely, i.e., $P(\text{host opens B}) = P(\text{host opens C}) = 1/2$. Also, the car was originally equally likely to be behind any of the three doors, so $P(\text{car behind A}) = P(\text{car behind B}) = P(\text{car behind C}) = 1/3$. Also, if the car is actually behind A, then the host had a choice of opening door B or C, so $P(\text{host opens B} \mid \text{car behind A}) = 1/2$. Then by Bayes' Theorem $P(\text{win if don't switch} \mid \text{host opens B}) = P(\text{car behind A} \mid \text{host opens B}) = [P(\text{car behind A}) / P(\text{host opens B})]\, P(\text{host opens B} \mid \text{car behind A}) = [(1/3) / (1/2)]\,(1/2) = 1/3$. So, if you don't switch, then only 1/3 of the time will you win the car. (This makes sense if you consider that originally, you had 1/3 chance of guessing the correct door. When the host opens another door it may change the probabilities of the other doors concealing the car, but it won't change the probability that you guessed right in the first place, which is still 1/3.)

(b) If the car is actually behind C, then the host had to open door B, so $P(\text{host opens B} \mid \text{car behind C}) = 1$. Then $P(\text{win if switch} \mid \text{host opens B}) = P(\text{car behind C} \mid \text{host opens B}) = [P(\text{car behind C}) / P(\text{host opens B})]\, P(\text{host opens B} \mid \text{car behind C}) = [(1/3) / (1/2)]\,(1) = 2/3$. (This makes sense since we must have $P(\text{win if don't switch} \mid \text{host opens B}) + P(\text{win if switch} \mid \text{host opens B}) = 1$.)

(c) Many people find this very surprising. To do an experiment, hide a pebble under one of three cups (say), let a volunteer guess one cup, then reveal an unselected non-pebbled cup, and give a volunteer the option to switch to the other cup or stick with the original cup. Do this repeatedly, and compute what fraction of the time they win if they do or do not switch.

(d) In this case, we would instead have $P(\text{host opens B} \mid \text{car behind C}) = 1$. Also, we would have $P(\text{host opens B}) = P(\text{host opens B, car behind A}) + P(\text{host opens B, car behind C}) = 1/3 + 1/3 = 2/3$. So, in this case, $P(\text{win if don't switch} \mid \text{host opens B}) = P(\text{car behind A} \mid \text{host opens B}) = [P(\text{car behind A}) / P(\text{host opens

B)] P(host opens B | car behind A) = [(1/3) / (2/3)] (1) = 1/2. Also, P(win if switch | host opens B) = P(car behind C | host opens B) = [P(car behind C) / P(host opens B)] P(host opens B | car behind C) = [(1/3) / (2/3)] (1) = 1/2. So in *this* case, it doesn't matter if you switch or not.
(e) This is a standard conditional probability calculation. We have P(win if don't switch | car not behind B) = P(car behind A | car not behind B) = P(car behind A, car not behind B) / P(car not behind B) = (1/3) / (2/3) = 1/2. Similarly, P(win if switch | car not behind B) = P(car behind C | car not behind B) = P(car behind C, car not behind B) / P(car not behind B) = (1/3) / (2/3) = 1/2. So in *this* case also, it doesn't matter if you switch or not. (When the original Monty Hall problem was first proposed, many people incorrectly interpreted it as this case, leading to confusion over whether the correct answer was 1/3 or 1/2.)

1.6 Continuity of P

Exercises

1.6.1 For the first way, let $A_n = \{2, 4, 6, \ldots, 2n\}$. Then by finite additivity, $P(A_n) = P(2) + P(4) + \cdots + P(2n) = 2^{-2} + 2^{-4} + \ldots + 2^{-2n} = (1/4)[1 - (1/4)^n] / [1 - (1/4)] = (1/3)[1 - (1/4)^n]$. But also $\{A_n\} \nearrow A$. Hence, $P(A) = \lim_{n\to\infty} P(A_n) = \lim_{n\to\infty}(1/3)[1 - (1/4)^n] = 1/3$. For the second way, by countable additivity, $P(A) = P(2) + P(4) + P(6) + \cdots + = 2^{-2} + 2^{-4} + 2^{-6} + \cdots + 2^{-2n} = (1/4) / [1 - (1/4)] = 1/3$.

1.6.2 Let $A_n = [1/4,\ 1 - e^{-n}]$. Then $\{A_n\} \nearrow A$, where $A = [1/4,\ 1)$. Hence, $\lim_{n\to\infty} P([1/4,\ 1 - e^{-n}]) = \lim_{n\to\infty} P(A_n) = P(A) = P([1/4, 1)) = 1 - 1/4 = 3/4$.

1.6.3 Let $A_n = \{1, 2, \ldots, n\}$. Then $\{A_n\} \nearrow A$, where $A = \{1, 2, 3, \ldots\} = S$. Hence, $\lim_{n\to\infty} P(A_n) = P(A) = P(S) = 1$.

1.6.4
(a) Let $A_n = (0,\ 1/n)$. Then $\{A_n\} \searrow A$, where $A = \emptyset$ is the empty set. Hence, $\lim_{n\to\infty} P(A_n) = P(A) = P(\emptyset) = 0$.
(b) Suppose $P(\{0\}) = 1$, so that P puts all of the probability at the single value 0. Then $P([0, 1/n)) = 1$ for all n, so $\lim_{n\to\infty} P([0, 1/n)) = 1 > 0$. (However, here $P((0, 1/n)) = 0$ for all n.)

Challenge

1.6.5 Let $A_1, A_2, A_3, \ldots$ be disjoint, and let $B = \bigcup_{n=1}^{\infty} A_n$. We must prove that $\sum_{n=1}^{\infty} P(A_n) = P(B)$. Well, let $B_n = A_1 \cup A_2 \cup \cdots \cup A_n$. Then $P(B_n) = P(A_1) + P(A_2) + \cdots + P(A_n)$ by finite additivity. Also, $B_n \subseteq B_{n+1}$, and $\bigcup_n B_n = B$, so that $\{B_n\} \nearrow B$. It follows that $\lim_{N\to\infty} P(B_N) = P(B)$. But $\lim_{N\to\infty} P(B_N) = \lim_{N\to\infty}[P(A_1) + P(A_2 + \cdots + P(A_N)] =$
$\lim_{N\to\infty} \sum_{n=1}^{N} P(A_n) = \sum_{n=1}^{\infty} P(A_n)$. So, $\sum_{n=1}^{\infty} P(A_n) = P(B)$.

Chapter 2

Random Variables and Distributions

2.1 Random Variables

Exercises

2.1.1
(a) $\min_{s \in S} X(s) = X(1) = 1$ since $X(s) > 1$ for all other $s \in S$.
(b) $\max_{s \in S} X(s)$ does not exist since $\lim_{s \to \infty} X(s) = \infty$ but $X(s) \neq \infty$ for all $s \in S$.
(c) $\min_{s \in S} Y(s)$ does not exist since $\lim_{s \to \infty} Y(s) = 0$ but $Y(s) \neq 0$ for all $s \in S$.
(d) $\max_{s \in S} Y(s) = Y(1) = 1$ since $Y(s) < 1$ for all other $s \in S$.

2.1.2
(a) No, since $X(\text{low}) > Y(\text{low})$.
(b) No, sincc $X(\text{low}) > Y(\text{low})$.
(c) No, since $Y(\text{middle}) = Z(\text{middle})$.
(d) Yes, since $Y(s) \leq Z(s)$ for all $s \in S$.
(e) No, since $X(\text{middle})\, Y(\text{middle}) = Z(\text{middle})$.
(f) Yes, since $X(\text{middle})\, Y(\text{middle}) \leq Z(\text{middle})$ for all $s \in S$.

2.1.3
(a) For example, let $X(s) = s$ and $Y(s) = s^2$ for all $s \in S$.
(b) For the above example, $Z(1) = X(1) + Y(1)^2 = 1 + 1^2 = 2$, $Z(2) = X(2) + Y(2)^2 = 2 + 4^2 = 18$, $Z(3) = X(3) + Y(3)^2 = 3 + 9^2 = 84$, $Z(4) = X(4) + Y(4)^2 = 4 + 16^2 = 260$, and $Z(5) = X(5) + Y(5)^2 = 5 + 25^2 = 630$.

2.1.4 Here $Z(1) = X(1)\, Y(1) = (1)(1^3 + 2) = 3$, $Z(2) = X(2)\, Y(2) = (2)(2^3 + 2) = 20$, $Z(3) = X(3)\, Y(3) = (3)(3^3{+}2) = 87$, $Z(4) = X(4)\, Y(4) = (4)(4^4{+}2) = 1032$, $Z(5) = X(5)\, Y(5) = (5)(5^5 + 2) = 15{,}635$, and $Z(6) = X(6)\, Y(6) = (6)(6^6 + 2) = 279{,}948$.

2.1.5 Yes, X is an indicator function of the event $A \cap B$, i.e., $X = I_{A\cap B}$.

Problems

2.1.6
(a) No, we could have $X(s) < 0$ for some $s \in S$.
(b) No, if S is infinite, then it could be that for all c there is some $s \in S$ with $X(s) < c$, so that $X(s) + c < 0$.
(c) Yes, if S is finite, then we can take $c = -\min_{s\in S} X(s) < \infty$ and then $X(s) + c \geq 0$ for all $s \in S$.

2.1.7 No, if S is finite, then $\max_{s\in S} |X(s)|$ must be finite, so X must be bounded.

2.1.8 Yes, then $X = I_A$, where $A = \{s \in S;\ X(s) = 1\}$.

2.1.9 If $|S| = m$, then the number of subsets of S is 2^m (since each $s \in S$ can be either included or not). Since subsets are in one-to-one correspondence with indicator functions, this means there are 2^m indicator functions as well.

2.1.10 No, if $X(s) < 0$ for some $s \in S$, then $Y(s) = \sqrt{X(s)}$ is undefined (or, at least, not a real number), so Y is not a random variable.

2.2 Distributions of Random Variables

Exercises

2.2.1 Clearly, X must equal 0, 1, or 2. Also, $X = 0$ if and only if the coins are both tails, which has probability $(1/2)^2 = 1/4$. Similarly, $X = 2$ if and only if the coins are both heads, which also has probability $(1/2)^2 = 1/4$. Hence, $P(X = 0) = P(X = 2) = 1/4$ and $P(X = 1) = 1 - P(X = 0) - P(X = 2) = 1/2$, with $P(X = x) = 0$ for $x \neq 0, 1, 2$.

2.2.2
(a) Clearly, $P(X = x) = 0$ for $x \neq 0, 1, 2, 3$. For $x \in \{0, 1, 2, 3\}$, there are $\binom{3}{x}$ ways we could end up with x heads, and each has probability $(1/2)^3 = 1/8$. Hence, $P(X = x) = \binom{3}{x}/8$ for $x = 0, 1, 2, 3$, so that $P(X = 0) = P(X = 3) = 1/8$, and $P(X = 1) = P(X = 2) = 3/8$.
(b) Here $P(X \in B) = (1/8)I_B(0) + (3/8)I_B(1) + (3/8)I_B(2) + (1/8)I_B(3)$.

2.2.3
(a) Here $P(Y = y) = 0$ for $y \neq 2, 3, 4, 5, 6, 7, 8, 9, 10, 11, 12$. For

$$y \in \{2, 3, 4, 5, 6, 7, 8, 9, 10, 11, 12\}$$

$P(Y = y)$ equals the number of ways the two dice can add up to y, divided by 36. Thus, $P(Y = 2) = 1/36$, $P(Y = 3) = 2/36$, $P(Y = 4) = 3/36$, $P(Y = 5) = 4/36$, $P(Y = 6) = 5/36$, $P(Y = 7) = 6/36$, $P(Y = 8) = 5/36$, $P(Y = 9) = 4/36$, $P(Y = 10) = 3/36$, $P(Y = 11) = 2/36$, and $P(Y = 12) = 1/36$.

(b) Here

$$P(Y \in B) = (1/36)I_B(2) + (2/36)I_B(3) + (3/36)I_B(4) + (4/36)I_B(5) + (5/36)I_B(6) + (6/36)I_B(7) + (5/36)I_B(8) + (4/36)I_B(9) + (3/36)I_B(10) + (2/36)I_B(11) + (1/36)I_B(12).$$

2.2.4 Here $P(Z = z) = 1/6$ for $z = 1, 2, 3, 4, 5, 6$. Hence:
(a) $P(W = w) = 1/6$ for $w = 5, 12, 31, 68, 129, 220$, with $P(W = w) = 0$ otherwise.
(b) $P(V = v) = 1/6$ for $v = 1, \sqrt{2}, \sqrt{3}, 2, \sqrt{5}, \sqrt{6}$, with $P(V = v) = 0$ otherwise.
(c) $P(ZW = x) = 1/6$ for $w = 5, 24, 93, 272, 645, 1320$, with $P(ZW = x) = 0$ otherwise.
(d) $P(VW = y) = 1/6$ for $y = 5, 12\sqrt{2}, 31\sqrt{3}, 136, 129\sqrt{5}, 220\sqrt{6}$, with $P(VW = y) = 0$ otherwise.
(e) $P(V + W = r) = 1/6$ for $r = 6, 12 + \sqrt{2}, 31 + \sqrt{3}, 72, 129 + \sqrt{5}, 220 + \sqrt{6}$, with $P(V + W = r) = 0$ otherwise.

2.2.5
(a) $P(X = 1) = .3, P(X = 2) = .2, P(X = 3) = .5$, and $P(X = x) = 0$ for all $x \notin \{1, 2, 3\}$.
(b) $P(Y = 1) = .3, P(Y = 2) = .2, P(Y = 3) = .5$, and $P(Y = y) = 0$ for all $y \notin \{1, 2, 3\}$.
(c) $P(W = 2) = (.3)^2 = 0.09$, $P(W = 3) = 2\,(.3)\,(.2) = 0.12$, $P(W = 4) = (.2)^2 + 2\,(.3)\,(.5) = 0.34$, $P(W = 5) = 2\,(.2)\,(.5) = 0.2$, $P(W = 6) = (.5)^2 = 0.25$ and $P(W = w) = 0$ for all other choices of w.

2.2.6
(a) $P(X = x) = 4/52 = 1/13$ for $x \in \{1, 2, \ldots, 13\}$ and $P(X = x) = 0$ otherwise.
(b) $P(Y = y) = 13/52 = 1/4$ for $y \in \{1, 2, 3, 4\}$ and $P(Y = y) = 0$ otherwise.
(c) $P(W = 2) = P(X = 1, Y = 1) = 1/52$, $P(W = 3) = P(X = 1, Y = 2) + P(X = 2, Y = 1) = 2/52$, $P(W = 4) = 3/52$, $P(W = 5) = 4/52$, $P(W = 6) = 4/52$, $P(W = 7) = 4/52$, $P(W = 8) = 4/52$, $P(W = 9) = 4/52$, $P(W = 10) = 4/52$, $P(W = 11) = 4/52$, $P(W = 12) = 4/52$, $P(W = 13) = 4/52$, and $P(W = 14) = 4/52$, $P(W = 15) = 3/52$, $P(W = 16) = 2/52$, $P(W = 17) = 1/52$.

2.2.7 $P(X = 25) = .45, P(X = 30) = .55$, and $P(X = x) = 0$ otherwise.

2.2.8 Note that each number $w \in \{0, 1, \ldots, 99\}$ can occur and

$$P(W = w) = P(X_2 = \lfloor w/10 \rfloor, X_1 = w - 10 \lfloor w/10 \rfloor) = (1/10)^2 = 1/100.$$

Problem

2.2.9 Note that each number $w \in \{0, 1, \ldots .99\} \cap \{0, 11, 22, \ldots, 99\}$ can occur and so

$$P(W = w) = P(X_2 = \lfloor w/10 \rfloor, X_1 = w - 10 \lfloor w/10 \rfloor) = (1/10)(1/9) = 1/90.$$

Challenge

2.2.10 Clearly, $P(Z = z) = 0$ unless $z \in \{-5, -4, \ldots, 2, 3\}$, in which case

$$\begin{aligned} P(Z = z) &= \sum_{0 \le x \le 3,\ 0 \le y \le 5,\ x - y = z} P(X = x,\ Y = y) \\ &= \sum_{0 \le x \le 3,\ 0 \le y \le 5,\ x - y = z} \binom{3}{x} (1/2)^3 \binom{5}{y} (1/2)^5 \\ &= (1/2)^8 \sum_{0 \le x \le 3,\ 0 \le y \le 5,\ x - y = z} \binom{3}{x} \binom{5}{y}. \end{aligned}$$

Hence,

$$P(Z = 3) = (1/2)^8 \binom{3}{3}\binom{5}{0} = 1/2^8$$

$$P(Z = 2) = (1/2)^8 \left[\binom{3}{3}\binom{5}{1} + \binom{3}{2}\binom{5}{0}\right] = 8/2^8$$

$$P(Z = 1) = (1/2)^8 \left[\binom{3}{3}\binom{5}{2} + \binom{3}{2}\binom{5}{1} + \binom{3}{1}\binom{5}{0}\right] = 28/2^8$$

$$P(Z = 0) = (1/2)^8 \left[\binom{3}{3}\binom{5}{3} + \binom{3}{2}\binom{5}{2} + \binom{3}{1}\binom{5}{1} + \binom{3}{0}\binom{5}{0}\right] = 56/2^8$$

$$P(Z = -1) = (1/2)^8 \left[\binom{3}{3}\binom{5}{4} + \binom{3}{2}\binom{5}{3} + \binom{3}{1}\binom{5}{2} + \binom{3}{0}\binom{5}{1}\right] = 70/2^8$$

$$P(Z = -2) = (1/2)^8 \left[\binom{3}{3}\binom{5}{5} + \binom{3}{2}\binom{5}{4} + \binom{3}{1}\binom{5}{3} + \binom{3}{0}\binom{5}{2}\right] = 56/2^8$$

$$P(Z = -3) = (1/2)^8 \left[\binom{3}{2}\binom{5}{5} + \binom{3}{1}\binom{5}{4} + \binom{3}{0}\binom{5}{3}\right] = 28/2^8$$

$$P(Z = -4) = (1/2)^8 \left[\binom{3}{1}\binom{5}{5} + \binom{3}{0}\binom{5}{4}\right] = 8/2^8$$

$$P(Z = -5) = (1/2)^8 \binom{3}{0}\binom{5}{5} = 1/2^8.$$

2.3 Discrete Distributions

Exercises

2.3.1 Here $p_Y(2) = 1/36$, $p_Y(3) = 2/36$, $p_Y(4) = 3/36$, $p_Y(5) = 4/36$, $p_Y(6) = 5/36$, $p_Y(7) = 6/36$, $p_Y(8) = 5/36$, $p_Y(9) = 4/36$, $p_Y(10) = 3/36$, $p_Y(11) = 2/36$, and $p_Y(12) = 1/36$, with $p_Y(y) = 0$ otherwise.

2.3.2
(a) $p_Z(1) = p_Z(3) = 1/2$, with $p_Z(z) = 0$ otherwise.
(b) $p_W(2) = p_W(12) = 1/2$, with $p_W(w) = 0$ otherwise.

2.3.3 Here $p_Z(1) = p_Z(5) = 1/4$, with $p_Z(0) = 1/4 + 1/4 = 1/2$ and $p_Z(z) = 0$ otherwise.

2.3.4
(a) $p_Z(0) = p_Z(2) = 1/4$, with $p_Z(1) = 1/4 + 1/4 = 1/2$ and $p_Z(z) = 0$ otherwise.
(b) $p_W(1) = 1/4$, with $p_W(0) = 1/4+1/4+1/4 = 3/4$ and $p_W(w) = 0$ otherwise.

2.3.5 Here $p_W(1) = 1/36$, $p_W(2) = 2/36$, $p_W(3) = 2/36$, $p_W(4) = 2/36 + 1/36 = 3/36$, $p_W(5) = 2/36$, $p_W(6) = 2/36 + 2/36 = 4/36$, $p_W(8) = 2/36$, $p_W(9) = 1/36$, $p_W(10) = 2/36$, $p_W(12) = 2/36 + 2/36 = 4/36$, $p_W(15) = 2/36$, $p_W(16) = 1/36$, $p_W(18) = 2/36$, $p_W(20) = 2/36$, $p_W(24) = 2/36$, $p_W(25) = 1/36$, $p_W(30) = 2/36$, and $p_W(36) = 1/36$, with $p_W(w) = 0$ otherwise.

2.3.6 $P(5 \le Z \le 9) = \sum_{z=5}^{9}(1-p)^z p = p\frac{(1-p)^5-(1-p)^{10}}{1-(1-p)} = (1-p)^5 - (1-p)^{10}$.

2.3.7 Here $P(X = 11) = \binom{12}{11}p^{11}(1-p)^1 = 12p^{11}(1-p)$. This has derivative $12 \cdot 11p^{10}(1-p) - 12p^{11}$, which equals 0 if either $p = 0$, or $11(1-p) = p$ whence $p = 11/12$. We see that $p = 11/12$ maximizes $P(X = 11)$.

2.3.8 Here $P(W = 11) = e^{-\lambda}\lambda^{11} / 11!$. This is maximized when $e^{-\lambda}\lambda^{11}$ is maximized. This has derivative $-e^{-\lambda}\lambda^{11} + e^{-\lambda}11\,\lambda^{10}$, which equals 0 if either $\lambda = 0$, or $\lambda = 11$. We see that $\lambda = 11$ maximizes $P(W = 11)$.

2.3.9

$$
\begin{aligned}
P(Z \le 2) &= P(Z = 0) + P(Z = 1) + P(Z = 2) \\
&= \binom{2}{0}(1/4)^3(1-1/4)^0 + \binom{3}{1}(1/4)^3(1-1/4)^1 \\
&\quad + \binom{4}{2}(1/4)^3(1-1/4)^2 \\
&= 1/4^3 + 9/4^4 + 54/4^5 = 53/512
\end{aligned}
$$

2.3.10

$$
\begin{aligned}
P(X^2 \le 15) &= P(X \le \sqrt{15}) = P(X \le 3) = \sum_{k=0}^{3} P(X = k) \\
&= \sum_{k=0}^{3}(1-1/5)^k(1/5) = \frac{1-(4/5)^4}{1-(4/5)}(1/5) = 369/625
\end{aligned}
$$

2.3.11 $P(Y = 10) = \binom{10}{10}p^{10}(1-p)^{10-10} = p^{10}$.

2.3.12 $p_Y(y) = P(Y = y) = P(X - 7 = y) = P(X = y + 7) = e^{-\lambda}\lambda^{y+7} / (y+7)!$ for $y = -7, -6, -5, \ldots$, with $p_Y(y) = 0$ otherwise.

2.3.13 $p_X(3) = \binom{7}{3}\binom{13}{5}/\binom{20}{8} = 0.35759$ and $P(X = 8) = 0$ since there are only seven elements in the population with the label in question.

2.3.14
(a) Binomial$(20, 2/3)$.
(b) $P(X = 5) = \binom{20}{5} (2/3)^5 (1/3)^{15} = 1.4229 \times 10^{-4}$.

2.3.15
(a) $\binom{10}{3} (.35)^3 (.65)^7 = 0.25222$.
(b) $(.35) (.65)^9 = 7.2492 \times 10^{-3}$.
(c) $\binom{9}{1} (.35)^2 (.65)^8 = 3.5131 \times 10^{-2}$.

2.3.16
(a) $\binom{15}{5} (4/9)^5 (5/9)^{10} = 0.14585$.
(b) $(4/9) (5/9)^{14} = 1.1858 \times 10^{-4}$.
(c) $\binom{15}{4} (4/9)^5 (5/9)^{10} = 6.6297 \times 10^{-2}$.

2.3.17
(a) Hypergeometric$(9, 4, 2)$.
(b) Hypergeometric$(9, 5, 2)$.

2.3.18
(a) $P(X = 5) = \left((2 \cdot 2)^5 /5!\right) \exp\{-2 \cdot 2\} = 0.15629$.
(b) $P(X = 5, Y = 5) = P(X = 5)P(Y = 5) = (0.15629)^2 = 2.4427 \times 10^{-2}$.
(c) $P(X = 0) = (2 \cdot 10)^0 \exp\{-2 \cdot 10\} = 2.0612 \times 10^{-9}$.

2.3.19 The number of black balls observed is distributed Binomial$(10, 1/1000)$. Then

$$P(X = 5) \approx \frac{(100/1000)^5}{5!} \exp\{-100/1000\} = 7.5403 \times 10^{-8}.$$

2.3.20 This is the probability that the test fails 4 times and passes on the fifth test, so this probability is $(1/3) (2/3)^4 = 6.5844 \times 10^{-2}$.

Computer Exercises

2.3.21 The tabulation is given by
0 0.000357
1 0.009526
2 0.075018
3 0.240057
4 0.350083
5 0.240057
6 0.075018
7 0.009526
8 0.000357
and the plot is as below.

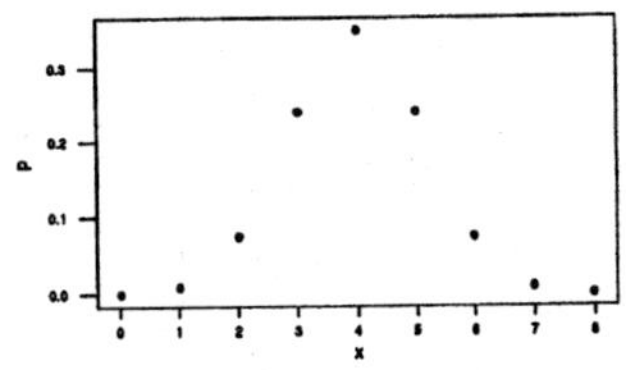

2.3.22 The Binomial(30, .3) probability function is plotted below.

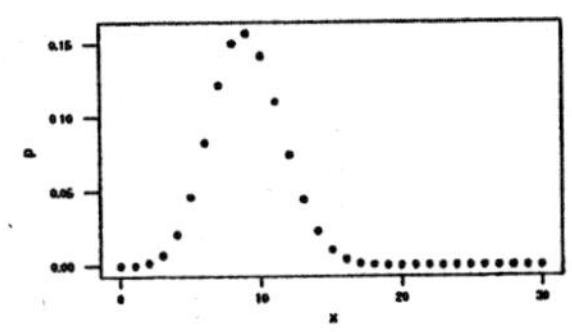

The Binomial(30, .7) probability function is plotted below.

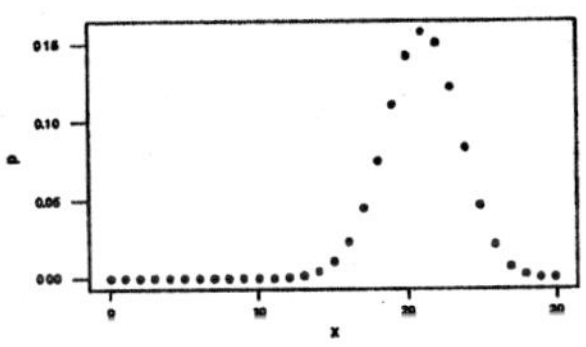

We have that

$$\binom{n}{x} p^x (1-p)^{n-x} = \binom{n}{n-x} (1-p)^{n-x} (1-(1-p))^x .$$

Problems

2.3.23
(a) $p_Y(y) = 2^{-\sqrt{y}}$ for $y = 1, 4, 9, 16, 25, \ldots$ (i.e., for y a positive perfect square), with $p_Y(y) = 0$ otherwise.
(b) $p_Z(z) = 2^{-z-1}$ for $z = 0, 1, 2, \ldots$, with $p_Z(z) = 0$ otherwise. Hence, $Z \sim$ Geometric(1/2).

2.3.24 Here $Z \sim$ Binomial$(n_1 + n_2, p)$. This is because X corresponds to the number of heads on the first n_1 coins (where each coin has probability p of being heads), and Y corresponds to the number of heads on the next n_2 coins, so Z corresponds to the number of heads on the first $n_1 + n_2$ coins.

2.3.25 $Z \sim$ Negative Binomial$(2, \theta)$ since $X + Y$ is equal to the number of tails until the second head is observed in independent tosses with heads occurring with probability θ. With r coins the sum will be distributed Negative Binomial(r, θ).

2.3.26

$$P(X \le Y) = \sum_{y=0}^{\infty} P(X \le y)\theta(1-\theta)^y = \sum_{y=0}^{\infty}\left(\sum_{x=0}^{y}\theta_1(1-\theta_1)^x\right)\theta_2(1-\theta_2)^y$$
$$= \sum_{y=0}^{\infty}\left(1-(1-\theta_1)^{y+1}\right)\theta_2(1-\theta_2)^y = 1-\theta_2(1-\theta_1)\sum_{y=0}^{\infty}(1-\theta_1)^y(1-\theta_2)^y$$
$$= 1-\frac{\theta_2(1-\theta_1)}{1-(1-\theta_1)(1-\theta_2)}$$

This is the probability that in tossing two coins that have probability θ_1 and θ_2 of yielding a head respectively, the first head occurs on the first coin.

2.3.27 $\lim_{n\to\infty} P(X \le n) = \lim_{n\to\infty} 1-(1-\lambda/n)^{n+1} = 1-e^{-\lambda}$.

2.3.28 $Z \sim$ Negative Binomial$(r+s,\theta)$ since $X+Y$ is equal to the number of tails until the $(r+s)$th head is observed in independent tosses with heads occurring with probability θ.

2.3.29 The probability is $\binom{M_1}{f_1}\binom{M_2}{f_2}\binom{N-M_1-M_2}{f_3}/\binom{N}{n}$, provided
$\max\{0, n-(N-M_1)\} \le f_1 \le \min\{M_1, n\}$,
$\max\{0, n-(N-M_2)\} \le f_2 \le \min\{M_2, n\}$,
$\max\{0, n-(M_1+M_2)\} \le f_3 \le \min\{N-M_1-M_2, n\}$,
and $f_1+f_2+f_3=n$.

2.3.30 $P(T>t) = P(\text{no units arrive in } (0,t]) = \left((\lambda t)^0/0!\right)\exp\{-\lambda t\}$.

2.4 Continuous Distributions

Exercises

2.4.1
(a) $P(U \le 0) = 0$.
(b) $P(U = 1/2) = 0$.
(c) $P(U < -1/3) = 0$.
(d) $P(U \le 2/3) = 2/3$.
(f) $P(U < 1) = 1$.
(g) $P(U \le 17) = 1$.

2.4.2
(a) $P(W \ge 5) = 0$.
(b) $P(W \ge 2) = 2/3$.
(c) $P(W^2 \le 9) = P(W \le 3) = 2/3$.
(d) $P(W^2 \le 2) = P(W \le \sqrt{2}) = (\sqrt{2}-1)/3$.

2.4.3
(a) $P(Z \ge 5) = e^{-20}$.
(b) $P(Z \ge -5) = 1$.

(c) $P(Z^2 \geq 9) = P(Z \geq 3) = e^{-12}$.
(d) $P(Z^2 - 17 \geq 9) = P(Z^2 \geq 25) = P(Z \geq 5) = e^{-25}$.

2.4.4
(a) $1 = \int_0^1 cx\, dx = c/2$ so $c = 2$.
(b) $1 = \int_0^1 cx^n\, dx = c/(n+1)$ so $c = n+1$.
(c) $1 = \int_0^2 cx^{1/2}\, dx = c\,(2/3)\, x^{3/2}\big|_0^2 = c\,(2/3)\, 2^{3/2}$ so $c = 3/\left(\sqrt{24}\right)$.
(d) $1 = \int_0^{\pi/2} c \sin x\, dx = -c \cos x|_0^{\pi/2} = c$.

2.4.5 This is not a density because it takes negative values.

Problems

2.4.6 $P(Y - h \geq y \,|\, Y \geq h) = P(Y \geq y+h) \,/\, P(Y \geq y) = e^{-\lambda(y+h)} \,/\, e^{-\lambda h} = e^{-\lambda y} = P(Y \geq y)$.

2.4.7
(a) Using integration by parts, $\Gamma(\alpha+1) = \int_0^\infty t^\alpha e^{-t}\, dt = 0 - \int_0^\infty (\alpha t^{\alpha-1})(-e^{-t})\, dt = \alpha \int_0^\infty t^{\alpha-1} e^{-t}\, dt = \alpha\, \Gamma(\alpha)$.
(b) $\Gamma(1) = \int_0^\infty t^0 e^{-t}\, dt = (-e^{-t})\Big|_{t=0}^{t=\infty} = 1$.
(c) We use induction on n. By part (b), the statement is true for $n = 1$. By part (a), if the statement is true for n, then it is also true for $n+1$.

2.4.8 Using the substitution $t = x^2/2$,

$$\begin{aligned}\int_{-\infty}^{\infty} \phi(x)\, dx &= 2\int_0^\infty \phi(x)\, dx = 2\int_0^\infty \frac{1}{\sqrt{2\pi}} e^{-x^2/2}\, dx \\ &= \frac{2}{\sqrt{2\pi}} \int_0^\infty e^{-t} \frac{1}{2} (2t)^{-1/2}\, dt = \frac{2}{\sqrt{2\pi}} \frac{1}{2} 2^{-1/2}\, \Gamma(1/2) = 1\end{aligned}$$

since $\Gamma(1/2) = \sqrt{\pi}$.

2.4.9 Using the substitution $t = \lambda x$, $\int_0^\infty f(x)\, dx = \int_0^\infty \frac{\lambda^\alpha x^{\alpha-1}}{\Gamma(\alpha)} e^{-\lambda x}\, dx = \frac{\lambda^\alpha}{\Gamma(\alpha)} \int_0^\infty x^{\alpha-1} e^{-\lambda x}\, dx = \frac{\lambda^\alpha}{\Gamma(\alpha)} \int_0^\infty (t/\lambda)^{\alpha-1} e^{-t}\, (1/\lambda)\, dt = \frac{1}{\Gamma(\alpha)} \int_0^\infty t^{\alpha-1} e^{-t}\, dt = \frac{1}{\Gamma(\alpha)} \Gamma(\alpha) = 1$.

2.4.10 We have that $f(x) \geq 0$ for every x and putting $u = e^{-x}, du = -e^{-x}\, dx$ we have $\int_{-\infty}^\infty e^{-x} \left(1 + e^{-x}\right)^{-2} dx = \int_0^\infty (1+u)^{-2}\, du = -\left(1+u\right)^{-1}\Big|_0^\infty = 1$.

2.4.11 We have that $f(x) \geq 0$ for every x and putting $u = x^\alpha, du = \alpha x^{\alpha-1}\, dx$ we have $\int_0^\infty \alpha x^{\alpha-1} e^{-x^\alpha}\, dx = \int_0^\infty e^{-u}\, du = -e^{-u}|_0^\infty = 1$.

2.4.12 We have that $f(x) \geq 0$ for every x, and we have $\int_0^\infty \alpha\, (1+x)^{-\alpha-1}\, dx = -\left(1+x\right)^{-\alpha}\Big|_0^\infty = 1$.

2.4.13 We have that $f(x) \geq 0$ for every x, and we have $\int_{-\infty}^\infty \frac{1}{\pi(1+x^2)}\, dx = \frac{1}{\pi} \arctan x|_{-\infty}^\infty = \frac{1}{\pi}\left(\frac{\pi}{2} - \left(\frac{-\pi}{2}\right)\right) = 1$.

2.4.14 We have that $f(x) \geq 0$ for every x, and we have

$$\int_{-\infty}^{\infty} \frac{1}{2} e^{-|x|} \, dx = \int_{0}^{\infty} \frac{1}{2} e^{-x} \, dx + \int_{-\infty}^{0} \frac{1}{2} e^{x} \, dx$$
$$= \frac{1}{2} + \frac{1}{2} \left(e^{x} |_{-\infty}^{0} \right) = \frac{1}{2} + \frac{1}{2} = 1.$$

2.4.15 We have that $f(x) \geq 0$ for every x, and we have $\int_{-\infty}^{\infty} e^{-x} \exp\{-e^{-x}\} \, dx = \exp\{-e^{-x}\}|_{-\infty}^{\infty} = 1 - 0 = 1$.

2.4.16

(a) We have that $f(x) \geq 0$ for every x, and $\int_0^1 B^{-1}(a,b) x^{a-1} (1-x)^{b-1} \, dx = B^{-1}(a,b) B(a,b) = 1$.

(b) $f(x) = 1$ for $0 < x < 1$ and this is the Uniform(0,1) distribution.

(c) $f(x) = B^{-1}(2,1) x = (\Gamma(2)\Gamma(1)/\Gamma(3))^{-1} x = 2x$ for $0 < x < 1$.

(d) $f(x) = B^{-1}(1,2)(1-x) = (\Gamma(1)\Gamma(2)/\Gamma(3))^{-1} x = 2(1-x)$ for $0 < x < 1$.

(e) $f(x) = B^{-1}(2,2) x (1-x) = (\Gamma(2)\Gamma(2)/\Gamma(4))^{-1} x = 6x(1-x)$ for $0 < x < 1$.

Challenge

2.4.17 The transformation $u = x + y, v = x/u$ has inverse $x = uv, y = u(1-v)$ and therefore Jacobian

$$\left| \det \begin{pmatrix} v & u \\ 1-v & -u \end{pmatrix} \right| = uv + u(1-v) = 1$$

so we have that

$$\Gamma(a)\Gamma(b)$$
$$= \int_0^{\infty} \int_0^{\infty} x^{a-1} y^{b-1} e^{-x-y} \, dx \, dy = \int_0^1 \int_0^{\infty} (uv)^{a-1} u^{b-1} (1-v)^{b-1} e^{-u} \, du \, dv$$
$$= \int_0^1 v^{a-1} (1-v)^{b-1} \left(\int_0^{\infty} u^{a+b-1} e^{-u} \, du \right) dv$$
$$= \Gamma(a+b) \int_0^1 v^{a-1} (1-v)^{b-1} \, dv.$$

2.5 Cumulative Distribution Functions (cdfs)

Exercises

2.5.1 Properties (a) and (b) follow by inspection. Properties (c) and (d) follow since $F_X(x) = 0$ for $x < 1$, and $F_X(x) = 1$ for $x > 6$.

2.5.2

$$F_X(x) = \begin{cases} 0 & x < 1 \\ 1/6 & 1 \le x < 4 \\ 2/6 & 4 \le x < 9 \\ 3/6 & 9 \le x < 16 \\ 4/6 & 16 \le x < 25 \\ 5/6 & 25 \le x < 36 \\ 1 & 36 \le x \end{cases}$$

Properties (a) and (b) follow by inspection. Properties (c) and (d) follow since $F_Y(y) = 0$ for $y < 1$, and $F_Y(y) = 1$ for $y > 36$.

2.5.3
(a) No, since $F(x) > 1$ for $x > 1$.
(b) Yes.
(c) Yes.
(d) No, since, e.g., $F(2) = 4 > 1$.
(e) Yes.
(f) Yes. (This distribution has mass 1/9 at the point 1.)
(g) No, since $F(-1) > F(0)$ so F is not non-decreasing.

2.5.4
(a) $P(X \le -5) = \Phi(-5) \doteq 2.87 \times 10^{-7}$.
(b) $P(-2 \le X \le 7) = \Phi(7) - \Phi(-2) \doteq 0.977$.
(c) $P(X \ge 3) = 1 - P(X < 3) = 1 - \Phi(3) \doteq 0.00135$.

2.5.5 Here $(Y+8)/2 \sim \text{Normal}(0,1)$. Hence:
(a) $P(Y \le -5) = P((Y+8)/2 \le (-5+8)/2) = \Phi((-5+8)/2) = \Phi(3/2) \doteq 0.933$.
(b) $P(-2 \le Y \le 7) = P((-2+8)/2 \le (Y+8)/2 \le (7+8)/2) = \Phi((7+8)/2) - \Phi((-2+8)/2) = \Phi(15/2) - \Phi(3) \doteq 0.00135$.
(c) $P(Y \ge 3) = P((Y+8)/2 \ge (3+8)/2) = 1 - \Phi((3+8)/2) = 1 - \Phi(11/2) \doteq 1.90 \times 10^{-8}$.

2.5.6 Here $G(x) \ge 0$ since $F_i(x) \ge 0$ and $\alpha_i \ge 0$. Also, $G(x) \le 1$ since $F_i(x) \le 1$ and $\sum_i \alpha_i(1) \le 1$. Also $\lim_{x\to\infty} G(x) = \sum_i \alpha_i \lim_{x\to\infty} F_i(x) = \sum_i \alpha_i(1) = 1$. Also, $\lim_{x\to-\infty} G(x) = \sum_i \alpha_i \lim_{x\to-\infty} F_i(x) = \sum_i \alpha_i(0) = 0$.

Problems

2.5.7 Since F is non-decreasing, $\lim_{n\to\infty} |F(2n) - F(n)| = \lim_{n\to\infty}[F(2n) - F(n)] = \lim_{n\to\infty} F(2n) - \lim_{n\to\infty} F(n) = 1 - 1 = 0$. (Hence, $\lim_{n\to\infty} P(n < X \le 2n) = 0$ for any X.)

2.5.8 Let X have cdf F, let A be the event $\{X \le x\}$, and let A_n be the event $\{X \le x + \frac{1}{n}\}$. Then $A_{n+1} \subseteq A_n$ and $\bigcap_n A_n = A$. Hence, $\{A_n\} \searrow A$, so by continuity of probabilities, $\lim_{n\to\infty} P(A_n) = P(A)$, i.e., $\lim_{n\to\infty} P(X \le x + \frac{1}{n}) = P(X \le x)$, i.e., $\lim_{n\to\infty} F(x + \frac{1}{n}) = F(x)$.

2.5.9 Since F is non-decreasing, then F is continuous at a if and only if $F(a^+) = F(a^-)$. But the previous exercise shows $F(a^+) = F(a)$. Hence, F is continuous at a if and only if $F(a) = F(a^-)$, i.e., $F(a) - F(a^-) = 0$. The result follows since $P(X = a) = F(a) - F(a^-)$.

2.5.10 Note that $\phi(-x) = \phi(x)$. Hence, using the substitution $s = -t$, we have $\Phi(-x) = \int_{-\infty}^{-x} \phi(t)\,dt = -\int_{x}^{\infty} \phi(s)\,(-ds) = \int_x^\infty \phi(s)\,ds = \Phi(\infty) - \Phi(x) = 1 - \Phi(x)$.

2.5.11 $F(x) = \int_{-\infty}^{x} e^{-z}\left(1+e^{-z}\right)^{-2} dz = \left(1+e^{-x}\right)^{-1}.$

2.5.12 $F(x) = \int_0^x \alpha z^{\alpha-1} \exp\left\{-z^\alpha\right\} dz = 1 - \exp\left\{-x^\alpha\right\}.$

2.5.13 $F(x) = \alpha \int_0^x (1+z)^{-\alpha-1} dz = 1 - (1+x)^{-\alpha}.$

2.5.14 $F(x) = \pi^{-1} \int_{-\infty}^{x} \left(1+z^2\right)^{-1} dz = \left(\arctan(x) + \pi/2\right)/\pi.$

2.5.15

$$F(x) = \begin{cases} \frac{1}{2}\int_{-\infty}^{x} e^z\,dz = \frac{1}{2}e^x & x \le 0 \\ \frac{1}{2} + \frac{1}{2}\int_0^x e^{-z}\,dz = \frac{1}{2} + \frac{1}{2}\left(1 - e^{-x}\right) & x > 0 \end{cases}$$

2.5.16 $F(x) = \int_{-\infty}^{x} e^{-z} \exp\left\{-e^{-z}\right\} dz = \exp\left\{-e^{-z}\right\}\big|_{-\infty}^{x} = \exp\left\{-e^{-x}\right\}.$

2.5.17
(b) $F(x) = \int_0^x dz = x$ for $0 < x < 1$.
(c) $F(x) = \int_0^x 2z\,dz = x^2$ for $0 < x < 1$.
(d) $F(x) = \int_0^x 2(1-z)\,dz = 1 - (1-x)^2$ for $0 < x < 1$.
(e) $F(x) = \int_0^x 6z(1-z)\,dz = \int_0^x 6\left(z - z^2\right) dz = 3x^2 - 2x^3$ for $0 < x < 1$.

2.6 One-dimensional Change of Variable

Exercises

2.6.1 Let $h(x) = cx + d$. Then $Y = h(X)$ and h is strictly increasing, so $f_Y(y) = f_X(h^{-1}(y))\,/\,|h'(h^{-1}(y))| = f_X((y-d)/c)\,/\,c$, which equals $1/(R-L)c$ for $L \le (y-d)/c \le R$, i.e., $cL + d \le y \le cR + d$, otherwise equals 0. Hence, $Y \sim \text{Uniform}[cL + d,\ cR + d]$.

2.6.2 Let $h(x) = cx + d$. Then $Y = h(X)$ and h is strictly decreasing, so $f_Y(y) = f_X(h^{-1}(y))\,/\,|h'(h^{-1}(y))| = f_X((y-d)/c)\,/\,|c|$, which equals $1/(R - L)|c| = 1/(cL - cR)$ for $L \le (y-d)/c \le R$, i.e., $cR + d \le y \le cL + d$, otherwise equals 0. Hence, $Y \sim \text{Uniform}[cR + d,\ cL + d]$.

2.6.3 Let $h(x) = cx + d$. Then $Y = h(X)$ and h is strictly increasing, so $f_Y(y) = f_X(h^{-1}(y))\,/\,|h'(h^{-1}(y))| = e^{-[((y-d)/c)-\mu]^2/2\sigma^2}\,/\,\sigma\sqrt{2\pi}c$
$= e^{-[y-d-c\mu]^2/2c^2\sigma^2}/c\sigma\sqrt{2\pi}$. Hence, $Y \sim \text{Normal}(c\mu + d,\ c^2\sigma^2)$.

2.6.4 Let $h(x) = cx$. Then $Y = h(X)$ and h is strictly increasing, so $f_Y(y) = f_X(h^{-1}(y))\,/\,|h'(h^{-1}(y))| = f_X(y/c)\,/\,c$, which equals $\lambda e^{-\lambda y/c}/c = (\lambda/c)e^{-(\lambda/c)y}$ for $y > 0$, otherwise equals 0. Hence, $Y \sim \text{Exponential}(\lambda/c)$.

2.6.5 Let $h(x) = x^3$. Then $Y = h(X)$ and h is strictly increasing, and $h^{-1}(y) = y^{1/3}$. Hence, $f_Y(y) = f_X(h^{-1}(y)) \,/\, |h'(h^{-1}(y))| = f_X(y^{1/3}) \,/\, 3(y^{1/3})^2$, which equals $\lambda e^{-\lambda y^{1/3}}/3y^{2/3} = (\lambda/3)y^{-2/3}e^{-\lambda y^{1/3}}$ for $y > 0$, otherwise equals 0.

2.6.6 Let $h(x) = x^{1/4}$. Then $Y = h(X)$, and h is strictly increasing over the region $\{x > 0\}$, where $f_X(x) > 0$. Also, $h^{-1}(y) = y^4$ on this region. Hence, for $y > 0$, $f_Y(y) = f_X(h^{-1}(y)) \,/\, |h'(h^{-1}(y))| = f_X(y^4) \,/\, (1/4)(y^4)^{-3/4} = \lambda e^{-\lambda y^4}/(1/4)y^{-3} = 4\lambda y^3 e^{-\lambda y^4}$, with $f_Y(y) = 0$ for $y \le 0$.

2.6.7 Let $h(x) = x^2$. Then $Y = h(X)$, and h is strictly increasing over the region $\{0 < x < 3\}$, where $f_X(x) > 0$. Also, $h^{-1}(y) = y^{1/2}$ on this region. Hence, $f_Y(y) = 0$ unless $y > 0$ and $0 < y^{1/2} < 3$, i.e., $0 < y < 9$, in which case $f_Y(y) = f_X(h^{-1}(y)) \,/\, |h'(h^{-1}(y))| = f_X(y^{1/2}) \,/\, 2y^{1/2} = 1/3\,(2y^{1/2}) = 1/6y^{1/2}$ for $0 < y < 9$.

2.6.8 The transformation is $y = h(x) = 2\mu - x$ and so $h^{-1}(y) = 2\mu - y$ and $h'(x) = -1$ and so the density of Y is given by $f_Y(y) = f_X\left(h^{-1}(y)\right) = f_X(2\mu - y) = f_X(\mu + (\mu - y)) = f_X(\mu - (\mu - y)) = f_X(y)$ so X and Y have the same distribution. Since the $N\left(\mu, \sigma^2\right)$ density is symmetric about μ, this proves that $Y \sim N\left(\mu, \sigma^2\right)$.

Problems

2.6.9
(a) First, let $h(x) = x^3$. Then $Y = h(X)$ and h is strictly increasing and $h^{-1}(y) = y^{1/3}$. Hence, $f_Y(y) = f_X(h^{-1}(y)) \,/\, |h'(h^{-1}(y))| = f_X(y^{1/3}) \,/\, 3(y^{1/3})^2$, which equals $(1/5)/3y^{2/3} = y^{-2/3}/15$ for $2 < y^{1/3} < 7$, i.e., $8 < y < 343$, otherwise equals 0. Second, let $h(y) = y^{1/2}$. Then $Z = h(Y)$ and h is strictly increasing over the region $\{8 < y < 343\}$, where $f_Y(y) > 0$. Also, $h^{-1}(z) = z^2$ on this region. Hence, for $\sqrt{8} < z < \sqrt{343}$, $f_Z(z) = f_Y(h^{-1}(z)) \,/\, |h'(h^{-1}(z))| = f_Y(z^2) \,/\, (1/2)(z^2)^{-1/2} = [(z^2)^{-2/3}/15] \,/\, (1/2)(z^2)^{-1/2} = 2z^{-1/3}/15$, with $f_Z(z) = 0$ otherwise.
(b) Let $h(x) = x^{3/2}$. Then $Z = h(X)$ and h is strictly increasing over the region $\{2 < x < 7\}$, where $f_X(x) > 0$. Hence, $f_Z(z) = f_X(h^{-1}(y)) \,/\, |h'(h^{-1}(y))| = f_X(z^{2/3}) \,/\, (3/2)(z^{2/3})^{1/2}$, which equals $(1/5)/(3/2)z^{1/3} = 2z^{-1/3}/15$ for $2 < z^{2/3} < 7$, i.e., $2^{3/2} < z < 7^{3/2}$, otherwise equals 0.

2.6.10 Here h is strictly decreasing on $x \le c$, and is strictly increasing on $x \ge c$. Hence, we can apply Theorem 2.6.2 if $c \le L < R$ and Theorem 2.6.3 if $L < R \le c$.

2.6.11 Let $h(x) = cx + d$. Then $Y = h(X)$ and h is strictly decreasing, so $f_Y(y) = f_X(h^{-1}(y)) \,/\, |h'(h^{-1}(y))| = \frac{1}{\sigma\sqrt{2\pi}} e^{-[((y-d)/c)-\mu]^2/2\sigma^2} \,/\, |c|$
$= \frac{1}{|c|\sigma\sqrt{2\pi}} e^{-[y-d-c\mu]^2/2c^2\sigma^2}$. Hence, $Y \sim \text{Normal}(c\mu + d,\ c^2\sigma^2)$.

2.6.12 The transformation $y = h(x) = e^x$ has $h'(x) = e^x$ and $h^{-1}(y) = \ln y$. Therefore, Y has density

$$\frac{1}{\sqrt{2\pi}\tau} \exp\left(-\frac{(\ln y)^2}{2\tau^2}\right) e^{-\ln y} = \frac{1}{\sqrt{2\pi}\tau} \exp\left(-\frac{(\ln y)^2}{2\tau^2}\right) \frac{1}{y}$$

for $y > 0$.

2.6.13 The transformation $y = h(x) = x^{\beta}$ has $h'(x) = \beta x^{\beta-1}$ and $h^{-1}(y) = y^{1/\beta}$. Therefore, Y has density $(\alpha/\beta)\left(y^{1/\beta}\right)^{\alpha-1} e^{-\left(y^{1/\beta}\right)^{\alpha}} / \left(y^{1/\beta}\right)^{\beta-1}$ $= (\alpha/\beta) y^{(\alpha/\beta)-1} e^{-y^{\alpha/\beta}}$ for $y > 0$, so $Y \sim$ Weibull(α/β).

2.6.14 The transformation $y = h(x) = (1+x)^{\beta} - 1$ has $h'(x) = \beta(1+x)^{\beta-1}$ and $h^{-1}(y) = (1+y)^{1/\beta} - 1$. Therefore, Y has density

$$\frac{\alpha}{\beta}\left(1 + (1+y)^{1/\beta} - 1\right)^{-\alpha-1} / \left(1 + (1+y)^{1/\beta} - 1\right)^{\beta-1} = \frac{\alpha}{\beta}(1+y)^{-(\alpha/\beta)-1}$$

for $y > 0$, so $Y \sim$ Pareto(α/β).

2.6.15 The transformation $y = h(x) = e^{-x}$ has $h'(x) = -e^{-x}$ and $h^{-1}(y) = -\ln y$. Therefore Y has density $e^{\ln y} \exp\left\{-e^{\ln y}\right\} / e^{\ln y} = e^{-y}$ for $y > 0$ and so $Y \sim$ Exponential(1).

Challenge

2.6.16 We have that, for $y > 0$,

$$\begin{aligned} f_Y(y) &= \frac{d}{dy}F_Y(y) = \frac{d}{dy}P(Y \le y) = \frac{d}{dy}P(Y \le y) = \frac{d}{dy}P(-\sqrt{y} \le X \le \sqrt{y}) \\ &= \frac{d}{dy}\left(\Phi(\sqrt{y}) - \Phi(-\sqrt{y})\right) = \frac{\varphi(\sqrt{y})}{2\sqrt{y}} + \frac{\varphi(-\sqrt{y})}{2\sqrt{y}} \\ &= \frac{\varphi(\sqrt{y})}{2\sqrt{y}} + \frac{\varphi(\sqrt{y})}{2\sqrt{y}} = \frac{\varphi(\sqrt{y})}{\sqrt{y}} = \frac{\exp\{-y/2\}}{\sqrt{2\pi}\sqrt{y}} \end{aligned}$$

for $y > 0$.

2.7 Joint Distributions

Exercises

2.7.1

$$F_{X,Y}(x,y) = \begin{cases} 0 & \min[x,\ (y+2)/4] < 0 \\ 1/3 & 0 \le \min[x,\ (y+2)/4] < 1 \\ 1 & \min[x,\ (y+2)/4] \ge 1 \end{cases}$$

2.7.2

$$F_{X,Y}(x,y) = \begin{cases} 0 & x < 0 \text{ and } y < 0 \\ 0 & x < 1 \text{ and } y < -7 \\ 1/4 & 0 \le x < 1 \text{ and } y \ge 0 \\ 3/4 & x \ge 1 \text{ and } -7 \le y < 0 \\ 1 & x \ge 1 \text{ and } y \ge 0 \end{cases}$$

2.7.3
(a) $p_X(2) = p_X(3) = p_X(-3) = p_X(-2) = p_X(17) = 1/5$, with $p_X(x) = 0$ otherwise.
(b) $p_Y(3) = p_Y(2) = p_Y(-2) = p_Y(-3) = p_Y(19) = 1/5$, with $p_Y(y) = 0$ otherwise.
(c) $P(Y > X) = p_{X,Y}(2,3) + p_{X,Y}(-3,-2) + p_{X,Y}(17,19) = 3/5$.
(d) $P(Y = X) = 0$ since this never occurs.
(e) $P(XY < 0) = 0$ since this never occurs.

2.7.4
(a) $C = 4$, and $P(X \le 0.8,\ Y \le 0.6) \doteq 0.0863$.
(b) $C = 18/5$ and $P(X \le 0.8,\ Y \le 0.6) \doteq 0.209$.
(c) $C = 9/1024003600$ and $P(X \le 0.8,\ Y \le 0.6) \doteq 5.09 \times 10^{-10}$.
(d) $C = 9/1024000000$ and $P(X \le 0.8,\ Y \le 0.6) \doteq 2.99 \times 10^{-12}$.

2.7.5 Since $\{X \le x,\ Y \le y\} \subseteq \{X \le x\}$ and $\{X \le x,\ Y \le y\} \subseteq \{Y \le y\}$, then $P(X \le x,\ Y \le y) \le P(X \le x)$ and $P(X \le x,\ Y \le y) \le P(Y \le y)$, i.e., $F_{X,Y}(x,y) \le F_X(x)$ and $F_{X,Y}(x,y) \le F_Y(y)$, so $F_{X,Y}(x,y) \le \min\left(F_X(x),\ F_Y(y)\right)$.

Problems

2.7.6 $F_{X,Y}(x,y) = P(X \le x,\ Y \le y) = P(X \le x,\ X^3 \le y) = P(X \le x,\ X \le y^{1/3}) = P(X \le \min(x,\ y^{1/3}))$, which equals $1 - e^{-\lambda \min(x,\ y^{1/3})}$ for $x, y > 0$, otherwise equals 0.

2.7.7 We know $F_{X,Y}(x,y) \le F_X(x)$ and that $\lim_{x\to-\infty} F_X(x) = 0$. Hence, $\lim_{x\to-\infty} F_{X,Y}(x,y) \le \lim_{x\to-\infty} F_X(x) = 0$.

2.7.8 Let $z = (x - \mu_1)/\sigma_1$ and $w = [(y - \mu_2)/\sigma_2] - [\rho(x - \mu_1)/\sigma_1]$. Then $f_{X,Y}(x,y) = (2\pi\sigma_1\sigma_2\sqrt{1-\rho^2})^{-1} \exp\{-[2(1-\rho^2)]^{-1}[(1-\rho^2)z^2 + w^2]\}$. Also, $dy = \sigma_2\, dw$. Hence,

$$\begin{aligned}
&\int_{-\infty}^{\infty} f_{X,Y}(x,y)\, dy \\
&= (2\pi\sigma_1\sigma_2\sqrt{1-\rho^2})^{-1} \int_{-\infty}^{\infty} \exp\{-[2(1-\rho^2)]^{-1}[(1-\rho^2)z^2 + w^2]\}\, \sigma_2\, dw \\
&= (2\pi\sigma_1\sigma_2\sqrt{1-\rho^2})^{-1} \exp\{-[2(1-\rho^2)]^{-1}(1-\rho^2)z^2\}[\sqrt{2\pi(1-\rho^2)}\sigma_2] \\
&= \frac{1}{\sigma_1\sqrt{2\pi}} e^{-z^2/2} = \frac{1}{\sigma_1\sqrt{2\pi}} e^{-(x-\mu_1)^2/2\sigma_1}.
\end{aligned}$$

2.7.9
(a) $\int_0^1 \int_0^1 Cye^{-xy}\, dx\, dy = \int_0^1 -Ce^{-xy}\big|_0^1\, dy = C\int_0^1 (1 - e^{-y})\, dy$
$= C\left(1 + e^{-y}\big|_0^1\right) = Ce^{-1}$ and so $C = e$
(b) $e\int_{1/2}^1 \int_{1/2}^1 ye^{-xy}\, dx\, dy = e\int_{1/2}^1 e^{-xy}\big|_{1/2}^1\, dy = e\int_{1/2}^1 \left(e^{-y/2} - e^{-y}\right) dy$
$= e\left(-2e^{-y/2}\big|_{1/2}^1 + e^{-y}\big|_{1/2}^1\right) = e\left(2e^{-1/4} - 2e^{-1/2} + e^{-1} - e^{-1/2}\right) = 0.28784$
(c) Using integration by parts with $u = y, du = 1, dv = e^{-xy+1}$, and $v = -e^{-xy+1}/x$, we have that $f_X(x) = \int_0^1 ye^{-xy+1}\, dy = -\frac{y}{x}e^{-xy+1}\big|_0^1 +$

$\frac{1}{x}\int_0^1 e^{-xy+1}\,dy = -\frac{e^{-x+1}}{x} - \frac{1}{x^2}\,e^{-xy+1}\big|_0^1 = e\left(\frac{1}{x^2} - \frac{e^{-x}}{x} - \frac{e^{-x}}{x^2}\right)$ for $0 < x < 1$. Also, we have that $f_Y(y) = \int_0^1 ye^{-xy+1}\,dx = -e^{-xy+1}\big|_0^1 = e\,(1 - e^{-y})$ for $0 < y < 1$.

2.7.10

(a) $\int_0^1\int_0^y Cye^{-xy}\,dx\,dy = \int_0^1 -Ce^{-xy}\big|_0^y\,dy = C\int_0^1\left(1 - e^{-y^2}\right)dy$
$= C\left(1 + \sqrt{\pi}\left(\Phi(0) - \Phi(\sqrt{2})\right)\right)$ and so $C = \left(1 + \sqrt{\pi}\left(\Phi(0) - \Phi(\sqrt{2})\right)\right)^{-1}$
(b)

$$
\begin{aligned}
C\int_{1/2}^1\int_{1/2}^y ye^{-xy}\,dx\,dy &= C\int_{1/2}^1 e^{-xy}\big|_{1/2}^y\,dy = C\int_{1/2}^1\left(e^{-y/2} - e^{-y^2}\right)dy\\
&= C\left(-2e^{-y/2}\Big|_{1/2}^1 + \sqrt{\pi}\left(\Phi\left(\sqrt{2}/2\right) - \Phi\left(\sqrt{2}\right)\right)\right)\\
&= C\left(2e^{-1/4} - 2e^{-1/2} + \sqrt{\pi}\left(\Phi\left(\sqrt{2}/2\right) - \Phi\left(\sqrt{2}\right)\right)\right)
\end{aligned}
$$

(c) Using integration by parts with $u = y, du = 1, dv = e^{-xy+1}, v = -e^{-xy+1}/x$ we have that

$$
\begin{aligned}
f_X(x) &= \int_x^1 ye^{-xy+1}\,dy = -\frac{y}{x}e^{-xy+1}\Big|_x^1 + \frac{1}{x}\int_x^1 e^{-xy+1}\,dy\\
&= e^{-x^2+1} - \frac{e^{-x+1}}{x} - \frac{1}{x^2}\,e^{-xy+1}\Big|_x^1 = e\left(e^{-x^2} - \frac{e^{-x}}{x} + \frac{e^{-x^2}}{x^2} - \frac{e^{-x}}{x^2}\right)
\end{aligned}
$$

for $0 < x < 1$. Also, we have that $f_Y(y) = \int_0^y ye^{-xy+1}\,dx = -e^{-xy+1}\big|_0^y = e\left(1 - e^{-y^2}\right)$ for $0 < y < 1$.

2.7.11

(a) $\int_0^\infty\int_0^y Ce^{-(x+y)}\,dx\,dy = C\int_0^\infty -e^{-(x+y)}\big|_0^y\,dx\,dy = C\int_0^\infty e^{-y}\,(1 - e^{-y})\,dy = C\,(1 - \int_0^\infty e^{-2y}\,dy) = C\,(1 - 1/2) = C/2$, so $C = 2$.
(b) We have that $f_X(x) = 2\int_x^\infty e^{-(x+y)}\,dy = 2e^{-x}\int_x^\infty e^{-y}\,dy = 2e^{-2x}$so $X \sim$ Exponential(2) and $f_Y(y) = 2\int_0^y e^{-(x+y)}\,dx = 2e^{-y}\int_0^y e^{-x}\,dx = 2e^{-y}\,(1 - e^{-y})$ for $y > 0$.

2.7.12

(a) We need to calculate $\int_0^1\int_0^{1-x_2} x_1^{\alpha_1-1}x_2^{\alpha_2-1}\,(1 - x_1 - x_2)^{\alpha_3-1}\,dx_1\,dx_2$
$= \int_0^1 x_2^{\alpha_2-1}\,(1 - x_2)^{\alpha_1+\alpha_3-2}\left(\int_0^{1-x_2}\left(\frac{x_1}{1-x_2}\right)^{\alpha_1-1}\left(1 - \frac{x_1}{1-x_2}\right)^{\alpha_3-1}dx_1\right)dx_2$ and, making the transformation $u = x_1/(1 - x_2), du = (1 - x_2)^{-1}\,dx_1$, we have that

this integral equals

$$\int_0^1 x_2^{\alpha_2-1}(1-x_2)^{\alpha_1+\alpha_3-1}\left(\int_0^1 u^{\alpha_1-1}(1-u)^{\alpha_3-1}\,du\right)dx_2$$
$$=\frac{\Gamma(\alpha_1)\Gamma(\alpha_3)}{\Gamma(\alpha_1+\alpha_3)}\int_0^1 x_2^{\alpha_2-1}(1-x_2)^{\alpha_1+\alpha_3-1}\,dx_2$$
$$=\frac{\Gamma(\alpha_1)\Gamma(\alpha_3)}{\Gamma(\alpha_1+\alpha_3)}\frac{\Gamma(\alpha_2)\Gamma(\alpha_1+\alpha_3)}{\Gamma(\alpha_1+\alpha_2+\alpha_3)}=\frac{\Gamma(\alpha_1)\Gamma(\alpha_2)\Gamma(\alpha_3)}{\Gamma(\alpha_1+\alpha_2+\alpha_3)}$$

by two applications of (2.4.10). This establishes that f_{X_1,X_2} is a density.
(b) We have that

$$f_{X_1}(x_1)$$
$$=\frac{\Gamma(\alpha_1+\alpha_2+\alpha_3)}{\Gamma(\alpha_1)\Gamma(\alpha_2)\Gamma(\alpha_3)}\int_0^{1-x_1} x_1^{\alpha_1-1}x_2^{\alpha_2-1}(1-x_1-x_2)^{\alpha_3-1}\,dx_2$$
$$=\frac{\Gamma(\alpha_1+\alpha_2+\alpha_3)}{\Gamma(\alpha_1)\Gamma(\alpha_2)\Gamma(\alpha_3)}x_1^{\alpha_1-1}(1-x_1)^{\alpha_2+\alpha_3-2}$$
$$\times\int_0^{1-x_1}\left(\frac{x_2}{1-x_1}\right)^{\alpha_2-1}\left(1-\frac{x_2}{1-x_1}\right)^{\alpha_3-1}dx_2$$
$$=\frac{\Gamma(\alpha_1+\alpha_2+\alpha_3)}{\Gamma(\alpha_1)\Gamma(\alpha_2)\Gamma(\alpha_3)}x_1^{\alpha_1-1}(1-x_1)^{\alpha_2+\alpha_3-1}\times\int_{02}^{1-x_1} u^{\alpha_2-1}(1-u)^{\alpha_3-1}\,du$$
$$=\frac{\Gamma(\alpha_1+\alpha_2+\alpha_3)}{\Gamma(\alpha_1)\Gamma(\alpha_2)\Gamma(\alpha_3)}x_1^{\alpha_1-1}(1-x_1)^{\alpha_2+\alpha_3-1}\frac{\Gamma(\alpha_2)\Gamma(\alpha_3)}{\Gamma(\alpha_2+\alpha_3)}$$
$$=\frac{\Gamma(\alpha_1+\alpha_2+\alpha_3)}{\Gamma(\alpha_1)\Gamma(\alpha_2+\alpha_3)}x_1^{\alpha_1-1}(1-x_1)^{\alpha_2+\alpha_3-1},$$

so $X_1\sim\text{Beta}(\alpha_1,\alpha_2+\alpha_3)$. Similarly, $X_2\sim\text{Beta}(\alpha_2,\alpha_1+\alpha_3)$.

2.7.13 We have that

$$\int_0^1\cdots\int_0^{1-x_3-\cdots-x_k}\int_0^{1-x_2-\cdots-x_k} x_1^{\alpha_1-1}x_2^{\alpha_2-1}\cdots x_k^{\alpha_k-1}$$
$$\times(1-x_1-x_2-\cdots-x_k)^{\alpha_{k+1}-1}\,dx_1\,dx_2\cdots dx_k$$
$$=\int_0^1\cdots\int_0^{1-x_3-\cdots-x_k} x_2^{\alpha_2-1}\cdots x_k^{\alpha_k-1}(1-x_2-\cdots-x_k)^{\alpha_1+\alpha_{k+1}-2}\times$$
$$\left(\int_0^{1-x_2-\cdots-x_k}\left(\frac{x_1}{1-x_2-\cdots-x_k}\right)^{\alpha_1-1}\left(1-\frac{x_1}{1-x_2-\cdots-x_k}\right)^{\alpha_{k+1}-1}dx_1\right)$$
$$\times\,dx_2\cdots dx_k$$
$$=\int_0^1\cdots\int_0^{1-x_3-\cdots-x_k} x_2^{\alpha_2-1}\cdots x_k^{\alpha_k-1}(1-x_2-\cdots-x_k)^{\alpha_1+\alpha_{k+1}-1}$$
$$\times\left(\int_0^1 u^{\alpha_1-1}(1-u)^{\alpha_{k+1}-1}\,du\right)dx_2\cdots dx_k$$

and this in turn equals

$$\frac{\Gamma(\alpha_1)\Gamma(\alpha_{k+1})}{\Gamma(\alpha_1+\alpha_{k+1})}\int_0^1\cdots\int_0^{1-x_3-\cdots-x_k} x_2^{\alpha_2-1}\cdots x_k^{\alpha_k-1}$$
$$\times(1-x_2-\cdots-x_k)^{\alpha_1+\alpha_{k+1}-1}\,dx_2\cdots dx_k$$
$$=\ldots.=\frac{\Gamma(\alpha_1)\Gamma(\alpha_{k+1})}{\Gamma(\alpha_1+\alpha_{k+1})}\frac{\Gamma(\alpha_2)\Gamma(\alpha_1+\alpha_{k+1})}{\Gamma(\alpha_1+\alpha_2+\alpha_{k+1})}\cdots\frac{\Gamma(\alpha_k)\Gamma(\alpha_1+\cdots+\alpha_k)}{\Gamma(\alpha_1+\cdots+\alpha_{k+1})}$$
$$=\frac{\Gamma(\alpha_1)\Gamma(\alpha_2)\cdots\Gamma(\alpha_{k+1})}{\Gamma(\alpha_1+\cdots+\alpha_{k+1})}$$

and this establishes that $f_{X_1,\ldots,X_k}$ is a density.

Challenge

2.7.14 For example, take X and Y to be i.i.d. $\sim$ Normal(0, 1), with $h(x)=-x$. Then $F_{X,Y}(x,h(x)) = P(X\le x,\ Y\le -x)\le P(Y\le -x)=\Phi(-x)\to 0$ as $x\to\infty$.

2.8 Conditioning and Independence

Exercises

2.8.1
(a) $p_X(-2) = p_{X,Y}(-2,3)+p_{X,Y}(-2,5) = 1/6+1/12 = 1/4$. $p_X(9) = p_{X,Y}(9,3)+p_{X,Y}(9,5) = 1/6+1/12 = 1/4$. $p_X(13) = p_{X,Y}(13,3)+p_{X,Y}(13,5) = 1/3+1/6 = 1/2$. Otherwise, $p_X(x)=0$.
(b) $p_Y(3) = p_{X,Y}(-2,3)+p_{X,Y}(9,3)+p_{X,Y}(13,3) = 1/6+1/6+1/3 = 2/3$. $p_Y(5) = p_{X,Y}(-2,5)+p_{X,Y}(9,5)+p_{X,Y}(13,5) = 1/12+1/12+1/6 = 1/3$. Otherwise, $p_Y(y)=0$.
(c) Yes, since $p_X(x)\,p_Y(y) = p_{X,Y}(x,y)$ for all x and y.

2.8.2
(a) $p_X(-2) = p_{X,Y}(-2,3)+p_{X,Y}(-2,5) = 1/16+1/4 = 5/16$. $p_X(9) = p_{X,Y}(9,3)+p_{X,Y}(9,5) = 1/2+1/16 = 9/16$. $p_X(13) = p_{X,Y}(13,3)+p_{X,Y}(13,5) = 1/16+1/16 = 1/8$. Otherwise, $p_X(x)=0$.
(b) $p_Y(3) = p_{X,Y}(-2,3)+p_{X,Y}(9,3)+p_{X,Y}(13,3) = 1/16+1/2+1/16 = 5/8$. $p_Y(5) = p_{X,Y}(-2,5)+p_{X,Y}(9,5)+p_{X,Y}(13,5) = 1/4+1/16+1/16 = 3/8$. Otherwise, $p_Y(y)=0$.
(c) No, since, e.g., $p_X(-2)\,p_Y(3)\ne p_{X,Y}(-2,3)$.

2.8.3

(a) For $0\le x\le 1$, $f_X(x)=\int_0^1 (12/49)(2+x+xy+4y^2)\,dy = (18x/49)+(40/49)$, otherwise $f_X(x)=0$.
(b) For $0\le y\le 1$, $f_Y(y)=\int_0^1 (12/49)(2+x+xy+4y^2)\,dx = (48y^2+6y+30)/49$, otherwise $f_Y(y)=0$.
(c) No, since $f_X(x)\,f_Y(y)\ne f_{X,Y}(x,y)$.

2.8.4
(a) For $0 \le x \le 1$, $f_X(x) = \int_0^1 (2/5(2+e))(3+e^x+3y+3ye^y+ye^x+ye^{x+y})\,dy = (3+e^x)/(2+e)$, otherwise $f_X(x) = 0$.
(b) For $0 \le y \le 1$, $f_Y(y) = \int_0^1 (2/5(2+e))(3+e^x+3y+3ye^y+ye^x+ye^{x+y})\,dx = 2(1+y+ye^y)/5$, otherwise $f_Y(y) = 0$.
(c) Yes, since $f_X(x)\,f_Y(y) = f_{X,Y}(x,y)$ for all x and y.

2.8.5
(a) $P(Y=4\,|\,X=9) = P(X=9,\,Y=4)\,/\,P(X=9) = (1/9)\,/\,(3/9+2/9+1/9) = 1/6$.
(b) $P(Y=-2\,|\,X=9) = P(X=9,\,Y=-2)\,/\,P(X=9) = (3/9)\,/\,(3/9+2/9+1/9) = 1/2$.
(c) $P(Y=0\,|\,X=-4) = P(X=-4,\,Y=0)\,/\,P(X=-4) = 0\,/\,(1/9) = 0$.
(d) $P(Y=-2\,|\,X=5) = P(X=5,\,Y=-2)\,/\,P(X=5) = (2/9)\,/\,(2/9) = 1$.
(e) $P(X=5\,|\,Y=-2) = P(X=5,\,Y=-2)\,/\,P(Y=-2) = (2/9)\,/\,(1/9+2/9+3/9) = 1/3$.

2.8.6 $P(Z=0) = P(X=0,\,Y=0) = (1-p)p$. For z, a positive integer, $P(Z=z) = P(X=0,\,Y=z) + P(X=1,\,Y=z-1) = (1-p)(1-p)^z p + p(1-p)^{z-1}p$ $= (1-p)^{z-1}[(1-p)^2 p + p^2] = p(1-p)^{z-1}[1-p+p^2]$.

2.8.7
(a) Recall $C=4$. Hence, $f_X(x) = \int_0^1 (2x^2y+4y^5)\,dy = x^2+2/3$ and $f_Y(y) = \int_0^1 (2x^2y+4y^5)\,dx = 4y^5+2y/3$. Then for $0 \le x \le 1$ and $0 \le y \le 1$, $f_{Y|X}(y\,|\,x) = f_{X,Y}(x,y)\,/\,f_X(x) = (2x^2y+4y^5)\,/\,(x^2+2/3)$ (otherwise $f_{Y|X}(y\,|\,x) = 0$). Thus, X and Y are not independent since $f_{Y|X}(y\,|\,x) \ne f_Y(y)$.
(b) Here $f_X(x) = \int_0^1 C(xy+x^5y^5)\,dy = C(x^5/6+x/2)$ and $f_Y(y) = \int_0^1 C(xy+x^5y^5)\,dx = C(y^5/6+y/2)$. Then for $0 \le x \le 1$ and $0 \le y \le 1$, $f_{Y|X}(y\,|\,x) = f_{X,Y}(x,y)\,/\,f_X(x) = C(xy+x^5y^5)\,/\,C(x^5/6+x/2) = (xy+x^5y^5)\,/\,(x^5/6+x/2)$ (otherwise $f_{Y|X}(y\,|\,x) = 0$). Thus, X and Y are not independent since $f_{Y|X}(y\,|\,x) \ne f_Y(y)$.
(c) Here $f_X(x) = \int_0^{10} C(xy+x^5y^5)\,dy = C(50000x^5/3+50x)$ and $f_Y(y) = \int_0^4 C(xy+x^5y^5)\,dx = C(2048y^5/3+8y)$. Then for $0 \le x \le 4$ and $0 \le y \le 10$, $f_{Y|X}(y\,|\,x) = f_{X,Y}(x,y)\,/\,f_X(x) = C(xy+x^5y^5)\,/\,C(50000x^5/3+50x) = (xy+x^5y^5)\,/\,(50000x^5/3+50x)$ (otherwise $f_{Y|X}(y\,|\,x) = 0$). Thus, X and Y are not independent since $f_{Y|X}(y\,|\,x) \ne f_Y(y)$.
(d) Here $f_X(x) = \int_0^{10} C(x^5y^5)\,dy = C(50000x^5/3)$ and $f_Y(y) = \int_0^4 C(xy+x^5y^5)\,dx = C(2048y^5/3)$. Then for $0 \le x \le 4$ and $0 \le y \le 10$, $f_{Y|X}(y\,|\,x) = f_{X,Y}(x,y)\,/\,f_X(x) = C(x^5y^5)\,/\,C(50000x^5/3) = 3y^5\,/\,50000$ (otherwise $f_{Y|X}(y\,|\,x) = 0$). Here X and Y are independent since $f_{Y|X}(y\,|\,x) = f_Y(y)$ for all x and y.

2.8.8 We have that $e^{-3x} = P(Y > 5\,|\,X = x) = \int_5^\infty f_{Y|X}(y\,|\,x)\,dy$. Hence, $P(Y > 5) = P(Y > 5,\ X > 0) = \int_0^\infty \int_5^\infty f_X(x)\,f_{Y|X}(y\,|\,x)\,dy\,dx$ $= \int_0^\infty 2e^{-2x}\,e^{-3x}\,dx = -(2/5)e^{-5x}\,\Big|_{x=0}^{x=\infty} = 2/5$.

2.8.9 For example, suppose $P(X=1,\,Y=1) = P(X=1,\,Y=2) = P(X=2,\,Y=1) = P(X=3,\,Y=3) = 1/4$. Then $P(X=1) = P(Y=1) = 1/2$, so $P(X = 1)\,P(Y = 1) = 1/4 = P(X = 1, Y = 1)$. On the other hand, $P(X = 3)\,P(Y = 3) = (1/4)(1/4) \neq 1/4 = P(X = 3,\, Y = 3)$, so X and Y are not independent.

2.8.10 Here $P(X = 1, Y = 0) = P(X = 1) - P(X = 1, Y = 1) = P(X = 1) - P(X = 1)\,P(Y = 1) = P(X = 1)(1 - P(Y = 1)) = P(X = 1)P(Y = 0)$. Similarly, $P(X = 0,\, Y = 1) = P(Y = 1) - P(X = 1,\, Y = 1) = P(Y = 1)P(X = 0)$. Finally, $P(X = 0,\, Y = 0) = P(X = 0) - P(X = 0,\, Y = 1) = P(X = 0) - P(X = 0)\,P(Y = 1) = P(X = 0)(1 - P(Y = 1)) = P(X = 0)P(Y = 0)$. Hence, $P(X = x,\; Y = y) = P(X = x)\,P(Y = y)$ for all x and y, so X and Y are independent.

2.8.11 If $X = C$ is constant, then $P(X \in B_1) = I_{B_1}(C)$ and $P(X \in B_1,\; Y \in B_2) = I_{B_1}(C)\,P(Y \in B_2)$. Hence, $P(X \in B_1,\; Y \in B_2) = P(X \in B_1)\,P(Y \in B_2) = I_{B_1}(C)\,P(Y \in B_2)$ for any subsets B_1 and B_2, so X and Y are independent.

2.8.12 Since X and Y are independent, $P(X = 1 \,|\, Y = 5) = P(X = 1) = 1/3$.

Problems

2.8.13 We compute that $f_X(x) = C_1(x^2 + C_2/6)$ and $f_Y(y) = C_1(C_2 y^5 + 2y/3)$, with $\int_0^1 \int_0^1 f_{X,Y}(x,y)\,dx\,dy = C_1(C_2/6 + 1/3)$. So, we require that $C_1(C_2/6 + 1/3) = 1$ and that $C_1(x^2 + C_2/6)\,C_1(C_2 y^5 + 2y/3) = C_1(2x^2 y + C_2 y^5)$ for $0 \le x \le 1$ and $0 \le y \le 1$. The second condition requires that $C_2 = 0$, while the first requires that $C_1 = 3$, which gives the solution.

2.8.14 Let $C_1 = \sum_x g(x)$, and $C_2 = \sum_y h(y)$. then $\sum_{x,y} p_{X,Y}(x,y) = C_1 C_2 = 1$. Also, $p_X(x) = \sum_y p_{X,Y}(x,y) = g(x) \sum_y h(y) = g(x) C_2$ and $p_Y(y) = \sum_x p_{X,Y}(x,y) = h(y) \sum_x g(x) = h(y) C_1$. Hence, $p_X(x) p_Y(y) = g(x) C_2 h(y) C_1 = (C_1 C_2) g(x) h(y) = g(x) h(y) = p_{X,Y}(x,y)$, so X and Y are independent.

2.8.15 Let $C_1 = \int_{-\infty}^{\infty} g(x)$, and $C_2 = \int_{-\infty}^{\infty} h(y)$. then $\int_{-\infty}^{\infty}\int_{-\infty}^{\infty} f_{X,Y}(x,y) = C_1 C_2 = 1$. Also, $f_X(x) = \int_{-\infty}^{\infty} f_{X,Y}(x,y)\,dy = g(x)\int_{-\infty}^{\infty} h(y)\,dy = g(x) C_2$ and $f_Y(y) = \int_{-\infty}^{\infty} f_{X,Y}(x,y)\,dx = h(y)\int_{-\infty}^{\infty} g(x)\,dx = h(y) C_1$. Hence, $f_X(x) f_Y(y) = g(x) C_2 h(y) C_1 = (C_1 C_2) g(x) h(y) = g(x) h(y) = f_{X,Y}(x,y)$, so X and Y are independent.

2.8.16 If X and Y were independent, then we would have $P(Y = 1) = P(Y = 1 \,|\, X = 1) = 3/4$, and $P(Y = 2) = P(Y = 2 \,|\, X = 2) = 3/4$. This is impossible since we must always have $P(Y = 1) + P(Y = 2) \le 1$.

2.8.17 We have from Problem 2.7.8 that $f_X(x) = \left(\sigma_1 \sqrt{2\pi}\right)^{-1} e^{-(x-\mu_1)^2/2\sigma_1}$ and, similarly, $f_Y(y) = \left(\sigma_2 \sqrt{2\pi}\right)^{-1} e^{-(y-\mu_2)^2/2\sigma_2}$. Multiplying these together, we see that they are equal to the expression for $f_{X,Y}(x,y)$, except with $\rho = 0$. Hence, $f_X(x) f_Y(y) = f_{X,Y}(x,y)$ if and only if $\rho = 0$.

2.8.18 We have that $P(X_1 = f_1) = \sum_{f_2=0}^{n-f_1} \binom{n}{f_1\; f_2\; n-f_1-f_2} \theta_1^{f_1} \theta_2^{f_2} \theta_3^{n-f_1-f_2} = \sum_{f_2=0}^{n-f_1} \binom{n}{f_1}\binom{n-f_1}{f_2} \theta_1^{f_1} \theta_2^{f_2} (1 - \theta_1 - \theta_2)^{n-f_1-f_2} = \binom{n}{f_1} \theta_1^{f_1} (1 - \theta_1)^{n-f_1} \times$

$\sum_{f_2=0}^{n-f_1}\binom{n-f_1}{f_2}\left(\frac{\theta}{1-\theta_1}\right)^{f_2}\left(1-\frac{\theta_2}{1-\theta_1}\right)^{n-f_1-f_2} = \binom{n}{f_1}\theta_1^{f_1}(1-\theta_1)^{n-f_1}\times$ $\left(\frac{\theta}{1-\theta_1}+1-\frac{\theta_2}{1-\theta_1}\right)^{n-f_1} = \binom{n}{f_1}\theta_1^{f_1}(1-\theta_1)^{n-f_1}$, so $X_1 \sim \text{Binomial}(n,\theta_1)$.

2.8.19 We have that

$$\begin{aligned} &P(X_2 = f_2 \mid X_1 = f_1) \\ &= \binom{n}{f_1\, f_2\, n-f_1-f_2}\theta_1^{f_1}\theta_2^{f_2}\theta_3^{n-f_1-f_2}\Big/\binom{n}{f_1}\theta_1^{f_1}(1-\theta_1)^{n-f_1} \\ &= \binom{n-f_1}{f_2}\left(\frac{\theta_2}{1-\theta_1}\right)^{f_2}\left(1-\frac{\theta_2}{1-\theta_1}\right)^{n-f_1-f_2}, \end{aligned}$$

so $X_2 \mid X_1 = f_1 \sim \text{Binomial}(n-f_1, \theta_2/(1-\theta_1))$.

2.8.20 The cdf of the Exponential(λ) is given by $F(x) = 1-e^{-\lambda x}$ for $x > 0$ and is 0 otherwise. Therefore for $X > 0$, $P(X_{(n)} \le x) = (1-e^{-\lambda x})^n$, so $f_{X_{(n)}}(x) = \frac{d}{dx}(1-e^{-\lambda x})^n = n\lambda(1-e^{-\lambda x})^{n-1}$. Also, $P(X_{(1)} \le x) = 1 - e^{-n\lambda x}$, so $f_{X_{(1)}}(x) = \frac{d}{dx}(1-e^{-n\lambda x}) = n\lambda(1-e^{-\lambda x})^{n-1}$.

2.8.21 We have that $P(X_{(i)} \le x) = P(\text{at least } i \text{ sample values are } \le x) = \sum_{j=i}^{n} P(\text{exactly } j \text{ sample values are } \le x) = \sum_{j=i}^{n}\binom{n}{j}F^j(x)(1-F(x))^{n-j}$.

2.8.22 From Problem 2.8.21 the distribution function of $X_{(3)}$, for $0 < x < 1$, is given by $P(X_{(3)} \le x) = \sum_{j=3}^{5}\binom{5}{j}x^j(1-x)^{5-j} = 10x^3(1-x)^2 + 5x^4(1-x) + x^5 = 10x^3 - 15x^4 + 6x^5$, so $f(x) = 30x^2 - 60x^3 + 30x^4 = 30x^2(x-1)^2$. This is the Beta(3,3) density.

2.8.23 From (2.7.1) we have that $X = \mu_1 + \sigma_1 Z_1$, $Y = \mu_2 + \sigma_2(\rho Z_1 + \sqrt{1-\rho^2}Z_2)$, so specifying $X = x$ implies that $Z_1 = (x-\mu_1)/\sigma_1$, so $Y = \mu_2 + \rho\sigma_2(x-\mu_1) + \sigma_2\sqrt{1-\rho^2}Z_2$, and this immediately implies the result.

By symmetry we can also write that the distribution of (X,Y) is obtained from $Y = \mu_2 + \sigma_2 Z_1$, $X = \mu_1 + \sigma_1(\rho Z_1 + \sqrt{1-\rho^2}Z_2)$, so the conditional distribution of X given $Y = y$ is $N(\mu_1 + \rho\sigma_1(y-\mu_2), (1-\rho^2)\sigma_1^2)$.

Challenge

2.8.24

(a) The "only if" part follows from Theorem 2.8.4(a). For the "if" part, the condition says that $P(X = x, Y = y) = P(X = x)\,P(Y = y)$ whenever $P(X = x) > 0$. But if $P(X = x) = 0$, then $P(X = x, Y = y) \le P(X = x) = 0$, so $P(X = x, Y = y) = P(X = x)\,P(Y = y) = 0$. We conclude that $P(X = x, Y = y) = P(X = x)\,P(Y = y)$ for all x and y. Hence, X and Y are independent.

(b) Very similar to (a).

2.9 Multi-dimensional Change of Variable

Exercises

2.9.1 We compute that

$$\frac{\partial h_1}{\partial u_1} = -\cos(2\pi u_2)\,/\,u_1\sqrt{2\log(1/u_1)}, \quad \frac{\partial h_1}{\partial u_2} = -2\sqrt{2\pi}\sin(2\pi u_2)\sqrt{2\log(1/u_1)}$$
$$\frac{\partial h_2}{\partial u_1} = -\sin(2\pi u_2)\,/\,u_1\sqrt{2\log(1/u_1)}, \quad \frac{\partial h_2}{\partial u_2} = -2\sqrt{2\pi}\cos(2\pi u_2)\sqrt{2\log(1/u_1)}.$$

Then $J(u_1, u_2) = \frac{\partial h_1}{\partial u_1}\frac{\partial h_2}{\partial u_2} - \frac{\partial h_2}{\partial u_1}\frac{\partial h_1}{\partial u_2} = -2\pi/u_1$.

2.9.2
(a) $f_{X,Y}(x,y) = e^{-x}$ for $x \ge 0$ and $1 \le y \le 4$, otherwise $f_{X,Y}(x,y) = 0$.
(b) $h(x,y) = (x+y,\ x-y)$.
(c) $h^{-1}(z,w) = ((z+w)/2,\ (z-w)/2)$.
(d) Here $J(x,y) = \frac{\partial h_1}{\partial x}\frac{\partial h_2}{\partial y} - \frac{\partial h_2}{\partial x}\frac{\partial h_1}{\partial y} = |(1)(-1) - (1)(1)| = 2$, so $f_{Z,W}(z,w) = f_{X,Y}(h^{-1}(z,w))\,/\,|J(h^{-1}(z,w))| = f_{X,Y}((z+w)/2,\ (z-w)/2)\,/\,2$, which equals $e^{-(z+w)/2}$ for $(z+w)/2 \ge 0$ and $1 \le (z-w)/2 \le 4$, i.e., for $z \ge 1$ and $\max(-z,\ z-8) \le w \le z-2$, otherwise $f_{Z,W}(z,w) = 0$.

2.9.3
(b) $h(x,y) = (x^2+y^2,\ x^2-y^2)$
(c) $h^{-1}(z,w) = (\sqrt{(z+w)/2},\ \sqrt{(z-w)/2})$, at least for $z+w \ge 0$ and $z-w \ge 0$
(d) Here $J(x,y) = \frac{\partial h_1}{\partial x}\frac{\partial h_2}{\partial y} - \frac{\partial h_2}{\partial x}\frac{\partial h_1}{\partial y} = |(2x)(-2y) - (2y)(2x)| = 4xy$ for $x, y \ge 0$, so

$$\begin{aligned} f_{Z,W}(z,w) &= f_{X,Y}(h^{-1}(z,w))\,/\,|J(h^{-1}(z,w))| \\ &= f_{X,Y}(\sqrt{(z+w)/2},\ \sqrt{(z-w)/2})\,/\,4\sqrt{(z+w)/2}\sqrt{(z-w)/2} \\ &= f_{X,Y}(\sqrt{(z+w)/2},\ \sqrt{(z-w)/2})\,/\,2\sqrt{z^2-w^2} \end{aligned}$$

which equals $e^{-\sqrt{(z+w)/2}}\,/\,2\sqrt{z^2-w^2}$ for $\sqrt{(z+w)/2} \ge 0$ and $1 \le \sqrt{(z-w)/2}$ ≤ 4, i.e., for $z \ge 4$ and $\max(-z,\ z-64) \le w \le z-4$, otherwise $f_{Z,W}(z,w) = 0$.

2.9.4
(b) $h(x,y) = (x+4,\ y-3)$
(c) $h^{-1}(z,w) = (z-4,\ w+3)$
(d) Here $J(x,y) = \frac{\partial h_1}{\partial x}\frac{\partial h_2}{\partial y} - \frac{\partial h_2}{\partial x}\frac{\partial h_1}{\partial y} = |(1)(1) - (0)(0)| = 1$, so $f_{Z,W}(z,w) = f_{X,Y}(h^{-1}(z,w))\,/\,|J(h^{-1}(z,w))| = f_{X,Y}(z-4,\ w+3)\,/\,1$, which equals $e^{-(z-4)}$ for $z-4 \ge 0$ and $1 \le w+3 \le 4$, i.e., for $z \ge 4$ and $-2 \le w \le 1$, otherwise $f_{Z,W}(z,w) = 0$.

2.9.5
(b) $h(x,y) = (y^4,\ x^4)$
(c) $h^{-1}(z,w) = (w^{1/4},\ z^{1/4})$
(d) Here $J(x,y) = \frac{\partial h_1}{\partial x}\frac{\partial h_2}{\partial y} - \frac{\partial h_2}{\partial x}\frac{\partial h_1}{\partial y} = |(0)(0) - (4y^3)(4x^3)| = 4x^3y^3$, at least for $x, y \ge 0$, so $f_{Z,W}(z,w) = f_{X,Y}(h^{-1}(z,w))\,/\,|J(h^{-1}(z,w))| =$

$f_{X,Y}(w^{1/4},\ z^{1/4})\,/\,4w^{3/4}z^{3/4}$, which equals $e^{-w^{1/4}}$ for $w^{1/4} \geq 0$ and $1 \leq z^{1/4} \leq 4$, i.e., for $w \geq 0$ and $1 \leq z \leq 256$, otherwise $f_{Z,W}(z, w) = 0$.

2.9.6

(a) $p_{Z,W}(5,5) = 1/7$; $p_{Z,W}(8,2) = 1/7$; $p_{Z,W}(9,1) = 1/7$; $p_{Z,W}(8,0) = 3/7$; $p_{Z,W}(12,4) = 1/7$; $p_{Z,W}(z,w) = 0$ otherwise.

(b) $p_{A,B}(25,10) = 1/7$; $p_{A,B}(34,-17) = 1/7$; $p_{A,B}(41,-38) = 1/7$; $p_{A,B}(64,16) = 3/7$; $p_{A,B}(80,-32) = 1/7$; $p_{A,B}(a,b) = 0$ otherwise.

(c) $p_{Z,A}(5,25) = 1/7$; $p_{Z,A}(8,34) = 1/7$; $p_{Z,A}(9,41) = 1/7$; $p_{Z,A}(8,64) = 3/7$; $p_{Z,A}(12,80) = 1/7$; $p_{Z,A}(z,a) = 0$ otherwise.

(d) $p_{Z,B}(5,10) = 1/7$; $p_{Z,B}(8,-17) = 1/7$; $p_{Z,B}(9,-38) = 1/7$; $p_{Z,B}(8,16) = 3/7$; $p_{Z,B}(12,-32) = 1/7$; $p_{Z,B}(z,b) = 0$ otherwise.

2.9.7 $p_Z(2) = (1/3)(1/6) = 1/18$; $p_Z(4) = (1/2)(1/6) = 1/12$; $p_Z(5) = (1/3)(1/12) + (1/6)(1/6) = 1/18$; $p_Z(7) = (1/2)(1/12) = 1/24$; $p_Z(8) = (1/6)(1/12) = 1/72$; $p_Z(9) = (1/3)(3/4) = 1/4$; $p_Z(11) = (1/2)(3/4) = 3/8$; $p_Z(12) = (1/6)(3/4) = 1/8$; $p_Z(z) = 0$ otherwise.

2.9.8 If w is an integer between 2 and 4, then $p_W(w) = P(Y = 2,\ X = w-2) = (1/6)(3/4)^{w-2}(1/4) = (3/4)^{w-2}/24$. If w is an integer between 5 and 8, then $p_W(w) = P(Y = 2,\ X = w-2) + P(Y = 5,\ X = w-5) = (1/6)(3/4)^{w-2}(1/4) + (1/12)(3/4)^{w-5}(1/4)$. If w is an integer ≥ 9, then $p_W(w) = P(Y = 2,\ X = w-2) + P(Y = 5,\ X = w-5) + P(Y = 9,\ X = w-9) = (1/6)(3/4)^{w-2}(1/4) + (1/12)(3/4)^{w-5}(1/4) + (3/4)(3/4)^{w-9}(1/4)$. Otherwise, $p_W(w) = 0$.

Problems

2.9.9 For z, an integer between 0 and $n_1 + n_2$,

$$\begin{aligned} P(Z = z) &= \sum_x p_X(x) p_Y(z-x) \\ &= \sum_{x=\max(0,\,z-n_2)}^{\min(z,\,n_1)} \binom{n_1}{x} p^x (1-p)^{n_1-x} \binom{n_2}{z-x} p^{z-x} (1-p)^{n_2-(z-x)} \\ &= p^z (1-p)^{n_1+n_2-z} \sum_{x=\max(0,\,z-n_2)}^{\min(z,\,n_1)} \binom{n_1}{x}\binom{n_2}{z-x}. \end{aligned}$$

Now this sum represents the number of ways of choosing z positions out of n_1+n_2 positions, so it equals $\binom{n_1+n_2}{z}$. (Indeed, of the z positions chosen, some number x of them must be among the first n_1 positions, with the remaining $z-x$ choices among the final n_2 positions.) Thus, $P(Z = z) = \binom{n_1+n_2}{z} p^z (1-p)^{n_1+n_2-z}$ for z, an integer between 0 and $n_1 + n_2$. Hence, $Z \sim \text{Binomial}(n_1 + n_2, p)$.

2.9.10 For z a non-negative integer,

$$\begin{aligned} P(Z=z) &= \sum_x p_X(x) p_Y(z-x) \\ &= \sum_{x=0}^{z} \binom{r_1-1+x}{x} p^{r_1}(1-p)^x \binom{r_2-1+z-x}{z-x} p^{r_2}(1-p)^{z-x} \\ &= p^{r_1+r_2}(1-p)^z \sum_{x=0}^{z} \binom{r_1-1+x}{x}\binom{r_2-1+z-x}{z-x}. \end{aligned}$$

Now this sum represents the number of ways of lining up z red balls and r_1+r_2 black balls, such that a black ball comes last. (Indeed, all balls up to and including the r_1th black ball are responsible for the first factor, with the remaining balls responsible for the second factor.) Thus,

$$\sum_{x=0}^{z} \binom{r_1-1+x}{x}\binom{r_2-1+z-x}{z-x} = \binom{r_1+r_2-1+z}{z}.$$

Hence, $P(Z=z) = p^{r_1+r_2}(1-p)^z \binom{r_1+r_2-1+z}{z}$, for z, a non-negative integer. Hence, $Z \sim$ Negative-Binomial(r_1+r_2, p).

2.9.11 We have that $f_Z(z) = \int_{-\infty}^{\infty} f_X(x)\, f_Y(z-x)\, dx = \int_{-\infty}^{\infty} \frac{1}{\sigma_1\sqrt{2\pi}} e^{-(x-\mu_1)^2/2\sigma_1}$ $\times \frac{1}{\sigma_2\sqrt{2\pi}} e^{-(z-x-\mu_2)^2/2\sigma_2}\, dx$. Squaring out the exponents, and remembering that $\int_{-\infty}^{\infty} e^{-t^2/2}\, dt = \sqrt{2\pi}$, we compute that

$$f_Z(z) = \left(2\pi(\sigma_1^2+\sigma_2^2)\right)^{-1/2} \exp\left(-(z-\mu_1-\mu_2)^2/2\sqrt{\sigma_1^2+\sigma_2^2}\right)$$

so that $Z \sim$ Normal$(\mu_1+\mu_2,\, \sigma_1^2+\sigma_2^2)$.

2.9.12 We have that

$$\begin{aligned} f_Z(z) &= \int_{-\infty}^{\infty} f_X(x)\, f_Y(z-x)\, dx \\ &= \int_0^z \Gamma(\alpha_1)^{-1}\lambda^{\alpha_1} x^{\alpha_1-1} e^{-\lambda x} \Gamma(\alpha_2)^{-1}\lambda^{\alpha_2}(z-x)^{\alpha_2-1} e^{-\lambda(z-x)}\, dx \\ &= \frac{1}{\Gamma(\alpha_1)\Gamma(\alpha_2)} \lambda^{\alpha_1+\alpha_2} e^{-\lambda z} \int_0^z x^{\alpha_1-1}(z-x)^{\alpha_2-1}\, dx. \end{aligned}$$

We recognize this integral as a Beta integral, with

$$\int_0^z x^{\alpha_1-1}(z-x)^{\alpha_2-1}\, dx = z^{\alpha_1+\alpha_2-1}\Gamma(\alpha_1)\Gamma(\alpha_2)/\Gamma(\alpha_1+\alpha_2).$$

Hence, $f_Z(z) = \Gamma(\alpha_1+\alpha_2)^{-1}\lambda^{\alpha_1+\alpha_2} z^{\alpha_1+\alpha_2-1} e^{-\lambda z}$, so that $Z \sim$ Gamma$(\alpha_1+\alpha_2, \lambda)$.

2.9.13 The joint density of (Z_1, Z_2) is $(2\pi)^{-1} \exp\{-(z_1^2 + z_2^2)/2\}$. The inverse of the transformation given by (2.7.1) is $Z_1 = (X - \mu_1)/\sigma_1$, $Z_2 = ((Y - \mu_2)/\sigma_2 - \rho(X - \mu_1)/\sigma_1)/\sqrt{1-\rho^2}$, and this has Jacobian

$$\left|\det\begin{pmatrix} 1/\sigma_1 & 0 \\ -\rho/\left(\sigma_1\sqrt{1-\rho^2}\right) & 1/\left(\sigma_2\sqrt{1-\rho^2}\right)\end{pmatrix}\right| = \left(\sigma_1\sigma_2\sqrt{1-\rho^2}\right)^{-1}.$$

So the joint density of (X, Y) is given by

$$\frac{1}{2\pi\sigma_1\sigma_2\sqrt{1-\rho^2}} \exp\left\{-\frac{1}{2(1-\rho^2)}\begin{pmatrix} (1-\rho^2)\left(\frac{X-\mu_1}{\sigma_1}\right)^2 + \rho^2\left(\frac{X-\mu_1}{\sigma_1}\right)^2 - \\ 2\rho\left(\frac{X-\mu_1}{\sigma_1}\right)\left(\frac{Y-\mu_2}{\sigma_2}\right) + \left(\frac{Y-\mu_2}{\sigma_2}\right)^2\end{pmatrix}\right\}$$

$$= \frac{1}{2\pi\sigma_1\sigma_2\sqrt{1-\rho^2}} \exp\left\{-\frac{1}{2(1-\rho^2)}\begin{pmatrix} \left(\frac{X-\mu_1}{\sigma_1}\right)^2 - \\ 2\left(\frac{X-\mu_1}{\sigma_1}\right)\left(\frac{Y-\mu_2}{\sigma_2}\right) + \left(\frac{Y-\mu_2}{\sigma_2}\right)^2\end{pmatrix}\right\}$$

and this proves the result.

2.10 Simulating Probability Distributions

Exercises

2.10.1 We can let $Z = -7$ if $U \le 1/2$, $Z = -2$ if $1/2 < U \le 5/6$, and $Z = 5$ if $U > 5/6$.

2.10.2
(a) Here $F^{-1}(t) = t$, so we can let $X = U$.
(b) Here $F^{-1}(t) = \sqrt{t}$, so we can let $X = \sqrt{U}$.
(c) Here $F^{-1}(t) = 3\sqrt{t}$, so we can let $X = 3\sqrt{U}$.
(d) Here $F^{-1}(t) = 3\sqrt{t}$ for $t \ge 1/9$, with $F^{-1}(t) = 1$ for $t \le 1/9$. Hence, we can let $X = 1$ for $U \le 1/9$, with $X = 3\sqrt{U}$ for $U > 1/9$.
(e) Here $F^{-1}(t) = 5t^{1/5}$, so we can let $X = 5U^{1/5}$.
(f) Here $F^{-1}(t)$ equals 0 for $t \le 1/3$, and equals 7 for $1/3 < t \le 3/4$, and equals 11 for $t > 3/4$. Hence, we can let $X = 0$ for $U \le 1/3$, $X = 7$ for $1/3 < U \le 3/4$, and $X = 11$ for $U > 3/4$.

Problems

2.10.4 First choose a random variable I, independent of all the X_i, such that $I\{1, 2, \ldots, k\}$, with $P(I = i) = \alpha_i$. Then set $Y = X_I$. [That is, Y is equal to X_i for the choice $i = I$.] Then $P(Y \le y) = \sum_i P(I = i)\, P(Y \le y \,|\, I = i) = \sum_i \alpha_i\, F_i(y) = G(y)$, as desired.

2.10.5 Here $F_X(x) = 0$ for $x < 1$, while for $x \ge 1$, $F_X(x) = \int_{-\infty}^{x} f_X(t)\,dt = \int_1^x t^{-2}\,dt = -t^{-1}\big|_{t=1}^{t=x} = 1 - (1/x)$. Hence, $F^{-1}(t) = 1/(1-t)$. Thus, we can let $Z = 1/(1-U)$.

2.10.6 From Problem 2.5.11 we have that $F(x) = (1 + e^{-x})^{-1} = u$, so inverting this we have that $x = F^{-1}(u) = \ln(u/(1-u))$ for $0 \le u \le 1$.

2.10.7 From Problem 2.5.12 we have that $F(x) = 1 - \exp\{-x^\alpha\} = u$ for $x > 0$, so inverting this we have that $x = F^{-1}(u) = (-\ln(1-u))^{1/\alpha}$ for $0 \le u \le 1$.

2.10.8 From Problem 2.5.13 we have that $F(x) = 1 - (1+x)^{-\alpha} = u$ for $x > 0$, so inverting this we have that $x = F^{-1}(u) = (1-u)^{-1/\alpha} - 1$ for $0 \le u \le 1$.

2.10.9 From Problem 2.5.14 we have that $F(x) = (\arctan(x) + \pi/2)/\pi = u$, so inverting this we have that $x = F^{-1}(u) = \tan(\pi u - \pi/2)$ for $0 \le u \le 1$.

2.10.10 From Problem 2.5.15 we have that $F(x) = \frac{1}{2}\int_{-\infty}^{x} e^{z}\,dz = \frac{1}{2}e^{x} = u$ for $x \le 0$, and $F(x) = \frac{1}{2} + \frac{1}{2}\int_0^x e^{-z}\,dz = \frac{1}{2} + \frac{1}{2}(1 - e^{-x}) = u$ for $x > 0$. So, for $0 \le u \le 1/2$, inverting this we have that $x = F^{-1}(u) = \ln(2u)$ and, for $1/2 \le u \le 1$, $x = F^{-1}(u) = -\ln 2(1-u)$.

2.10.11 From Problem 2.5.16 we have that $F(x) = \exp\{-e^{-x}\} = u$, so inverting this we have that $x = F^{-1}(u) = -\ln(-\ln u)$ for $0 \le u \le 1$.

2.10.12 From Problem 2.5.17 we have that
(b) $u = F(x) = x$ for $0 < x < 1$, so $x = u$ for $0 \le u \le 1$.
(c) $u = F(x) = x^2$ for $0 < x < 1$, so $x = \sqrt{u}$ for $0 \le u \le 1$.
(d) $u = F(x) = 1 - (1-x)^2$ for $0 < x < 1$, so $x = 1 - \sqrt{1-u}$ for $0 \le u \le 1$.

2.10.13 We have that $P(Y \le y) = \int_{-\infty}^{\infty}\left(\int_{-\infty}^{y} f_{Y|X}(z\,|\,x)\,dz\right) f_X(x)\,dx = \int_{-\infty}^{y}\int_{-\infty}^{\infty} f(x,z)\,dx\,dz = \int_{-\infty}^{y} f_Y(z)\,dz = F_Y(y)$.

Challenge

2.10.14
(a)

$$P(a \le Y \le b \,|\, f(Y) \ge Ucg(Y)) = \frac{P(a \le Y \le b, f(Y) \ge Ucg(Y))}{P(f(Y) \ge Ucg(Y))}$$
$$= \frac{E(P(a \le y \le b, f(Y) \ge Ucg(u) \,|\, Y = y))}{E(f(y) \ge Ucg(y) \,|\, Y = y))} = \frac{E(I_{(a,b)}(Y) f(Y)/cg(Y))}{E(f(Y)/cg(Y))}$$
$$= \frac{E(P(a \le y \le b, f(Y) \ge Ucg(u) \,|\, Y = y))}{E(f(y) \ge Ucg(y) \,|\, Y = y))} = \frac{E(I_{(a,b)}(Y) f(Y)/cg(Y))}{E(f(Y)/cg(Y))}$$
$$= \int_a^b \frac{f(y)}{cg(y)} g(y)\,dy \Big/ \int_{-\infty}^{\infty} \frac{f(y)}{cg(y)} g(y)\,dy = \int_a^b f(y)\,dy$$

(b) Let $p = P(f(Y) \ge Ucg(Y))$. Then, using (a) and the independence of the U_i and Y_i, we have that

$$P(X_{i_1} \le x) = \sum_{j=1}^{\infty} P(Y_j \le x, i_1 = j)$$
$$= \sum_{j=1}^{\infty} P\left(\begin{array}{c} Y_j \le x, f(Y_1) < U_1 cg(Y_1), \ldots, f(Y_{j-1}) < U_{j-1} cg(Y_{j-1}), \\ f(Y_j) \ge U_{j-1} cg(Y_j) \end{array}\right)$$

$$= \sum_{j=1}^{\infty} P\left(Y_j \le x, f(Y_j) \ge U_j c g(Y_j)\right) P\left(\begin{array}{c} f(Y_1) < U_1 c g(Y_1), \ldots, \\ f(Y_{j-1}) < U_{j-1} c g(Y_{j-1}) \end{array}\right)$$

$$= \sum_{j=1}^{\infty} P\left(Y_j \le x \,|\, f(Y_j) \ge U_j c g(Y_j)\right) p(1-p)^{j-1} = \int_{-\infty}^{x} f(y)\, dy,$$

so $X_{i_1} \sim f$.

Further, we have that

$$P(X_{i_1} \le x_1, , X_{i_2} \le x_2) = \sum_{j_1=1}^{\infty} \sum_{j_2=j_1+1}^{\infty} P\left(Y_{j_1} \le x_1, i_1 = j_1, Y_{j_2} \le x_2, i_2 = j_2\right)$$

$$= \sum_{j_1=1}^{\infty} \sum_{j_2=j_1+1}^{\infty} P\left(\begin{array}{c} Y_{j_1} \le x_1, f(Y_1) < U_1 c g(Y_1), \ldots, \\ f(Y_{j_1-1}) < U_{j_1-1} c g(Y_{j_1-1}), f(Y_{j_1}) \ge U_{j_1} c g(Y_{j_1}), \\ Y_{j_2} \le x_2, f(Y_{j_1+1}) < U_{j_1+1} c g(Y_{j_1+1}), \ldots, \\ f(Y_{j_2-1}) < U_{j_2-1} c g(Y_{j_2-1}), f(Y_{j_2}) \ge U_{j_2} c g(Y_{j_2}) \end{array}\right)$$

$$= \sum_{j_1=1}^{\infty} \sum_{j_2=j_1+1}^{\infty} \{P\left(Y_{j_1} \le x_1, f(Y_{j_1}) \ge U_{j_1} c g(Y_{j_1})\right) \times$$
$$P\left(f(Y_1) < U_1 c g(Y_1), \ldots, f(Y_{j_1-1}) < U_{j_1-1} c g(Y_{j_1-1})\right) \times$$
$$P\left(Y_{j_2} \le x_2, f(Y_{j_2}) \le U_{j_2-1} c g(Y_{j_2})\right)) \times$$
$$P\left(f(Y_{j_1+1}) < U_{j_1+1} c g(Y_{j_1+1}), \ldots, f(Y_{j_2-1}) < U_{j_2-1} c g(Y_{j_2-1})\right)\}$$

$$= \sum_{j_1=1}^{\infty} \sum_{j_2=j_1+1}^{\infty} \{P\left(Y_{j_1} \le x_1 \,|\, f(Y_{j_1}) \ge U_{j_1} c g(Y_{j_1})\right) p(1-p)^{j_1-1} \times$$
$$P\left(Y_{j_2} \le x_2 \,|\, f(Y_{j_2}) \le U_{j_2-1} c g(Y_{j_2})\right) p(1-p)^{j_2-j_1-1}\}$$
$$= \left(\int_{-\infty}^{x_1} f(y)\, dy\right) \left(\int_{-\infty}^{x_2} f(y)\, dy\right).$$

so $X_{i_1} \sim f$ independently of $X_{i_2} \sim f$. Continuing in this fashion proves that $X_{i_1}, X_{i_2}, \ldots$ is an i.i.d. sequence from the distribution, with density given by f.

Chapter 3

Expectation

3.1 The Discrete Case

Exercises

3.1.1
(a) $E(X) = (-4)(1/7) + (0)(2/7) + (3)(4/7) = 8/7$.
(b) We recognize that $X \sim$ Geometric(1/2). Hence, $E(X) = (1-(1/2))\,/\,(1/2) = 1$.
(c) Using the substitution $y = x - 7$, we have $E(X) = \sum_{x=7}^{\infty} x\,2^{-x+6} = \sum_{y=0}^{\infty}(y+7)\,2^{-y-1} = 7 + \sum_{y=0}^{\infty} y\,2^{-y-1} = 7 + 1 = 8$ since $\sum_{y=0}^{\infty} y\,2^{-y-1} = 1$ is the mean of a Geometric(1/2) distribution.

3.1.2
(a) $E(X) = (5)(1/7) + (5)(1/7) + (5)(1/7) + (8)(3/7) + (8)(1/7) = 47/7$.
(b) $E(Y) = (0)(1/7) + (3)(1/7) + (4)(1/7) + (0)(3/7) + (4)(1/7) = 11/7$.
(c) By linearity, $E(3X + 7Y) = 3E(X) + 7E(Y) = 3(47/7) + 7(11/7) = 218/7$.
(d) $E(X^2) = (5)^2(1/7) + (5)^2(1/7) + (5)^2(1/7) + (8)^2(3/7) + (8)^2(1/7) = 331/7$.
(e) $E(Y^2) = (0)^2(1/7) + (3)^2(1/7) + (4)^2(1/7) + (0)^2(3/7) + (4)^2(1/7) = 41/7$.
(f) $E(XY) = (5)(0)(1/7) + (5)(3)(1/7) + (5)(4)(1/7) + (8)(0)(3/7) + (8)(4)(1/7) = 67/7$.
(g) By linearity, $E(XY + 14) = E(XY) + 14 = 67/7 + 14 = 165/7$.

3.1.3
(a) $E(X) = (2)(1/2) + (-7)(1/6) + (2)(1/12) + (-7)(1/12) + (2)(1/12) + (-7)(1/12) = -173/12 \doteq -14.4$.
(b) $E(Y) = (10)(1/2) + (10)(1/6) + (12)(1/12) + (12)(1/12) + (14)(1/12) + (14)(1/12) = 11$.
(c) $E(X^2) = (2)^2(1/2) + (-7)^2(1/6) + (2)^2(1/12) + (-7)^2(1/12) + (2)^2(1/12) + (-7)^2(1/12) = 19$.
(d) $E(Y^2) = (10)^2(1/2) + (10)^2(1/6) + (12)^2(1/12) + (12)^2(1/12) + (14)^2(1/12) + (14)^2(1/12) = 370/3 \doteq 123.3$.
(e) $E(X^2 + Y^2) = E(X^2) + E(Y^2) = 19 + 370/3 = 427/3 \doteq 142.3$.

(f) $E(XY-4Y) = (2\cdot 10 - 4\cdot 10)(1/2) + ((-7)\cdot 10 - 4\cdot 10)(1/6) + (2\cdot 12 - 4\cdot 12)(1/12) + ((-7)\cdot 12 - 4\cdot 12)(1/12) + (2\cdot 14 - 4\cdot 14)(1/12) + ((-7)\cdot 14 - 4\cdot 14)(1/12) = -113/2 = -56.5$.

3.1.4 $E(4X-3Y) = 4E(X) - 3E(Y) = 4(p_1) - 3(np_2)$.

3.1.5 $E(8X - Y + 12) = 8E(X) - E(Y) + 12 = 8((1-p)/p) - \lambda + 12$.

3.1.6 $E(Y+Z) = E(Y) + E(Z) = (100)(0.3) + (7) = 37$.

3.1.7 Since X and Y are independent, $E(XY) = E(X)E(Y) = ((80)(1/4))(3/2) = 30$.

3.1.8 Let Z be the number showing on the die. Then $X = 1 + 3Z$, so $E(X) = 1 + 3E(Z) = 1 + 3(3.5) = 11.5$.

3.1.9 Let $Y = 1$ if the coin comes up tails, otherwise $Y = 0$ if the coin comes up heads. Then $X = 8 - 4Y$ and $E(Y) = 1(1/2) + 0(1/2) = 1/2$. Hence, $E(X) = 8 - 4E(Y) = 8 - 4(1/2) = 6$.

Problems

3.1.10 Let Z be the number of heads before the first tail. Then we know that $P(Z=k) = 1/2^{k+1}$ for $k = 0, 1, 2, \ldots$, and $E(Z) = (1-(1/2))/(1/2) = 1$. Now, $X = 1 + 2Z$, so $E(X) = 1 + 2E(Z) = 1 + 2(1) = 3$.

3.1.11 Again, let Z be the number of heads before the first tail, so $P(Z = k) = 1/2^{k+1}$ for $k = 0, 1, 2, \ldots$. Then $X = 2^Z$, so $E(X) = \sum_{k=0}^{\infty} 2^k (1/2^{k+1}) = \sum_{k=0}^{\infty}(1/2) = \infty$. Hence, $E(X)$ is infinite in this case.

3.1.12
(a) $E(Y) = \sum_{x=0}^{\infty} \min(x, 100)\,(1-\theta)^x\theta = \theta\sum_{x=0}^{100} x\,(1-\theta)^x + \theta(100)\sum_{x=101}^{\infty}(1-\theta)^x = \theta S + 100(1-\theta)^{101}$, where $S = \sum_{x=0}^{100} x\,(1-\theta)^x$. Then $(1-\theta)S = \sum_{x=0}^{100} x\,(1-\theta)^{x+1} = \sum_{y=1}^{101}(y-1)\,(1-\theta)^y$. Hence, $\theta S = S - (1-\theta)S = \sum_{x=1}^{100}(1-\theta)^x - 100(1-\theta)^{101} = \theta^{-1}(1-\theta-(1-\theta)^{101}) - 100(1-\theta)^{101}$.
(b) $E(Y-X) = E(Y) - E(X) = -(1-\theta)^{101}(1/\theta + 100)$.

3.1.13 Any X with $X \le 100$ will do since then $\min(X, 100) = X$. For example, $X = 29$, or $X \sim$ Bernoulli$(80, 1/3)$.

3.1.14 For one example, let $P(X = 200) = 1$. For another, let $P(X = 300) = P(X = 100) = 1/2$.

3.1.15 Let $p_{X,Y}(1,1) = P_{X,Y}(1,0) = 1/4$, with $p_{X,Y}(0,0) = 1/2$. Then $P(X=1) = 1/2$ and $P(Y=1) = 1/4$, but $E(XY) = 1/4$.

3.1.16 We have that

$$E(X) = \sum_{x=\max(0,n+M-N)}^{\min(n,M)} x\frac{\binom{M}{x}\binom{N-M}{n-x}}{\binom{N}{n}} = \sum_{x=\max(1,n+M-N)}^{\min(n,M)} x\frac{\binom{M}{x}\binom{N-M}{n-x}}{\binom{N}{n}}$$

$$= n\frac{M}{N}\sum_{x=\max(1,n+M-N)}^{\min(n,M)} \frac{\binom{M-1}{x-1}\binom{N-1-(M-1)}{n-1-(x-1)}}{\binom{N-1}{n-1}}$$

$$= n\frac{M}{N} \sum_{x=\max(0,\, n-1+(M-1)-(N-1))}^{\min(n-1,M-1)} \frac{\binom{M-1}{x}\binom{N-1-(M-1)}{n-1-x}}{\binom{N-1}{n-1}} = n\frac{M}{N}$$

since the final sum is the sum of all Hypergeometric$(N-1, M-1, n)$ probabilities.

3.1.17 We have that if $X_1, \ldots, X_r$ are i.i.d. Geometric(θ), then $X = X_1 + \cdots + X_r \sim$ Negative Binomial(r, θ) so $E(X) = E(X_1 + \cdots + X_r) = r(1-\theta)/\theta$.

3.1.18 This follows immediately since $X_i \sim$ Binomial(n, θ_i).

Challenges

3.1.19 Here $E(X^2) = \sum_k k^2 P(X = k) = \sum_{k=0}^{\infty} k^2 (1-p)^k p$. Hence, $(1-p)E(X^2) = \sum_{k=0}^{\infty} k^2 (1-p)^{k+1} p = \sum_{j=1}^{\infty} (j-1)^2 (1-p)^j p$. Then $pE(X^2) = E(X^2) - (1-p)E(X^2) = \sum_{k=1}^{\infty} [k^2 - (k-1)^2](1-p)^k p = \sum_{k=1}^{\infty} [2k-1](1-p)^k p = 2E(X) - (1-p) = 2(1-p)/p - (1-p) = 2(1-p)/p - (1-p)$. Hence, $E(X^2) = 2(1-p)/p^2 - (1-p)/p$.

3.1.20 Let $Y = X - \min(X, M)$. Then Y is also discrete. Also, since $\min(X, M) \le X$, we have $Y \ge 0$. Now, if $E(\min(X, M)) = E(X)$, then $E(Y) = 0$, so that $0 = \sum_y y P(Y = y) = \sum_{y \ge 0} y P(Y = y)$. But the only way a sum of non-negative terms can be 0 is if each term is 0, i.e., $y P(Y = y) = 0$ for all $y \in R^1$. This means that $P(Y = y) = 0$ for $y \ne 0$, so that $P(Y = 0) = 1$. But $\{Y = 0\} = \{\min(X, M) = X\} = \{X \le M\}$, so $P(X \le M) = 1$, i.e., $P(X > M) = 0$.

3.2 The Absolutely Continuous Case

Exercises

3.2.1

(a) $1 = \int_{-\infty}^{\infty} f_X(x)\, dx = \int_5^9 C\, dx = 4C$, where $C = 1/4$. Then $E(X) = \int_{-\infty}^{\infty} x f_X(x)\, dx = \int_5^9 x (1/4)\, dx = (9^2 - 5^2)/8 = 7$.

(b) $1 = \int_{-\infty}^{\infty} f_X(x)\, dx = \int_6^8 C(x+1)\, dx = C(9^2 - 7^2)/2 = 16C$, where $C = 1/16$. Then $E(X) = \int_{-\infty}^{\infty} x f_X(x)\, dx = \int_6^8 x (1/16)(x+1)\, dx = (8^3 - 6^3)/48 + (8^2 - 6^2)/32 = 169/24 \doteq 7.04$.

(c) $1 = \int_{-\infty}^{\infty} f_X(x)\, dx = \int_{-5}^{-2} Cx^4\, dx = C((-2)^5 - (-5)^5)/5 = C\,3093/5$, where $C = 5/3093$. Then $E(X) = \int_{-\infty}^{\infty} x f_X(x)\, dx = \int_{-5}^{-2} x (5/3093)(x^4)\, dx = (5/3093)((-2)^6 - (-5)^6)/6 = -8645/2062 \doteq -4.19$.

3.2.2

(a) $E(X) = \int_0^1 \int_0^1 x (4x^2 y + 2y^5)\, dx\, dy = 2/3$.

(b) $E(Y) = \int_0^1 \int_0^1 y (4x^2 y + 2y^5)\, dx\, dy = 46/63$.

(c) $E(3X+7Y)=3E(X)+7E(Y)=3(2/3)+7(46/63)=64/9$.
(d) $E(X^2)=\int_0^1\int_0^1 x^2\,(4x^2y+2y^5)\,dx\,dy=23/45$.
(e) $E(Y^2)=\int_0^1\int_0^1 y^2\,(4x^2y+2y^5)\,dx\,dy=7/12$.
(f) $E(XY)=\int_0^1\int_0^1 xy\,(4x^2y+2y^5)\,dx\,dy=10/21$.
(g) $E(XY+14)=E(XY)+14=(10/21)+14=304/21$.

3.2.3
(a) $E(X)=\int_{-\infty}^{\infty}\int_{-\infty}^{\infty} x\,f_{X,Y}(x,y)\,dx\,dy=\int_0^3\int_0^1 x((4xy+3x^2y^2)/18)\,dx\,dy=17/24$.
(b) $E(Y)=\int_{-\infty}^{\infty}\int_{-\infty}^{\infty} y\,f_{X,Y}(x,y)\,dx\,dy=\int_0^3\int_0^1 y((4xy+3x^2y^2)/18)\,dx\,dy=17/8$.
(c) $E(X^2)=\int_{-\infty}^{\infty}\int_{-\infty}^{\infty} x^2\,f_{X,Y}(x,y)\,dx\,dy=\int_0^3\int_0^1 x^2((4xy+3x^2y^2)/18)\,dx\,dy=11/20$.
(d) $E(Y^2)=\int_{-\infty}^{\infty}\int_{-\infty}^{\infty} y^2\,f_{X,Y}(x,y)\,dx\,dy=\int_0^3\int_0^1 y^2((4xy+3x^2y^2)/18)\,dx\,dy=99/20$.
(e) $E(Y^4)=\int_{-\infty}^{\infty}\int_{-\infty}^{\infty} y^4\,f_{X,Y}(x,y)\,dx\,dy=\int_0^3\int_0^1 y^4((4xy+3x^2y^2)/18)\,dx\,dy=216/7$.
(f) $E(X^2Y^3)=\int_{-\infty}^{\infty}\int_{-\infty}^{\infty} x^2y^3\,f_{X,Y}(x,y)\,dx\,dy=\int_0^3\int_0^1 x^2y^3((4xy+3x^2y^2)/18)\times dx\,dy=27/4$.

3.2.4
(a) $E(X)=\int_{-\infty}^{\infty}\int_{-\infty}^{\infty} x\,f_{X,Y}(x,y)\,dx\,dy=\int_0^1\int_y^1 x(6xy+(9/2)x^2y^2)\,dx\,dy=57/70$.
(b) $E(Y)=\int_{-\infty}^{\infty}\int_{-\infty}^{\infty} y\,f_{X,Y}(x,y)\,dx\,dy=\int_0^1\int_y^1 y(6xy+(9/2)x^2y^2)\,dx\,dy=157/280$.
(c) $E(X^2)=\int_{-\infty}^{\infty}\int_{-\infty}^{\infty} x^2\,f_{X,Y}(x,y)\,dx\,dy=\int_0^1\int_y^1 x^2(6xy+(9/2)x^2y^2)\,dx\,dy=11/16$.
(d) $E(Y^2)=\int_{-\infty}^{\infty}\int_{-\infty}^{\infty} y^2\,f_{X,Y}(x,y)\,dx\,dy=\int_0^1\int_y^1 y^2(6xy+(9/2)x^2y^2)\,dx\,dy=29/80$.
(e) $E(Y^4)=\int_{-\infty}^{\infty}\int_{-\infty}^{\infty} y^4\,f_{X,Y}(x,y)\,dx\,dy=\int_0^1\int_y^1 y^4(6xy+(9/2)x^2y^2)\,dx\,dy=53/280$.
(f) $E(X^2Y^3)=\int_{-\infty}^{\infty}\int_{-\infty}^{\infty} x^2y^3\,f_{X,Y}(x,y)\,dx\,dy=\int_0^1\int_y^1 x^2y^3(6xy+(9/2)x^2y^2)\times dx\,dy=133/660$.

3.2.5 $E(-5X-6Y)=-5E(X)-6E(Y)=-5((3+7)/2)-3(1/9)=-76/3$.

3.2.6 $E(11X+14Y+3)=11E(X)+14E(Y)+3=11(((-12)+(-9))/2)+14(-8)+3=-449/2$.

3.2.7 $E(Y+Z)=E(Y)+E(Z)=(1/9)+(1/8)=17/72$.

3.2.8 $E(Y+Z)=E(Y)+E(Z)=(1/9)+(5/4)=49/36$.

Problems

3.2.9 Letting $t=\lambda x$, we have that $E(X)=\int_0^\infty x\,\frac{\lambda^\alpha x^{\alpha-1}}{\Gamma(\alpha)}e^{-\lambda x}\,dx$
$=\int_0^\infty \frac{\lambda^\alpha x^\alpha}{\Gamma(\alpha)}e^{-\lambda x}\,dx=\int_0^\infty \frac{\lambda^\alpha t^\alpha}{\lambda^\alpha\Gamma(\alpha)}e^{-t}\,(1/\lambda)\,dt=\frac{1}{\lambda\Gamma(\alpha)}\int_0^\infty t^\alpha e^{-t}\,dx$
$=\frac{1}{\lambda\Gamma(\alpha)}\Gamma(\alpha+1)=\frac{1}{\lambda\Gamma(\alpha)}\alpha\Gamma(\alpha)=\alpha/\lambda$.

3.2.10 We have $E(X) = \int_{-\infty}^{0} xe^{-x}(1+e^{-x})^{-2}\,dx + \int_0^\infty xe^{-x}(1+e^{-x})^{-2}\,dx$ $= -\int_0^\infty xe^{-x}(1+e^{-x})^{-2}\,dx + \int_0^\infty xe^{-x}(1+e^{-x})^{-2}\,dx$, so $E(X) = 0$, provided $\int_0^\infty xe^{-x}(1+e^{-x})^{-2}\,dx < \infty$. This is the case because $\int_0^\infty xe^{-x}(1+e^{-x})^{-2}\,dx \le \int_0^\infty xe^{-x}\,dx = 1$.

3.2.11 We have that $E(X) = \int_0^\infty x\alpha x^{\alpha-1}e^{-x^\alpha}\,dx = \int_0^\infty \alpha x^\alpha e^{-x^\alpha}\,dx$ and putting $u = x^\alpha, x = u^{1/\alpha}, du = \alpha x^{\alpha-1}dx$ we have that $E(X) = \int_0^\infty u^{1/\alpha}e^{-u}\,du = \Gamma(1/\alpha+1)$.

3.2.12 We have that

$$\begin{aligned}
E(X) &= \int_0^\infty x\alpha(1+x)^{-\alpha-1}\,dx = \int_0^\infty \alpha(1+x)^{-\alpha}\,dx - 1 \\
&= \begin{cases} \frac{\alpha}{-\alpha+1}(1+x)^{-\alpha+1}\Big|_0^\infty - 1 & \alpha \ne 1 \\ \alpha\ln(1+x)|_0^\infty & \alpha = 1 \end{cases} \\
&= \begin{cases} \infty & 0 < \alpha \le 1 \\ 1/(\alpha-1) & \text{if } \alpha > 1. \end{cases}
\end{aligned}$$

3.2.13 We have that $\int_0^\infty x\pi^{-1}(1+x^2)^{-1}\,dx = (\ln(1+x^2))/2|_0^\infty = \infty$ and $\int_{-\infty}^0 x\pi^{-1}(1+x^2)^{-1}\,dx = -\infty$, so $E(X)$ doesn't exist.

3.2.14 We have that

$$E(X) = \int_{-\infty}^0 xe^{x}\,dx + \int_0^\infty xe^{-x}\,dx = -\int_0^\infty xe^{-x}\,dx + \int_0^\infty xe^{-x}\,dx = -1+1 = 0.$$

3.2.15 We have that

$$\begin{aligned}
E(X) &= \int_0^1 x\frac{\Gamma(a+b)}{\Gamma(a)\Gamma(b)}x^{a-1}(1-x)^{b-1}\,dx = \frac{\Gamma(a+b)}{\Gamma(a)\Gamma(b)}\int_0^1 x^a(1-x)^{b-1}\,dx \\
&= \frac{\Gamma(a+b)}{\Gamma(a)\Gamma(b)}\frac{\Gamma(a+1)\Gamma(b)}{\Gamma(a+b+1)} = \frac{\Gamma(a+b)}{\Gamma(a)\Gamma(b)}\frac{a\Gamma(a)\Gamma(b)}{(a+b)\Gamma(a+b)} = \frac{a}{a+b}.
\end{aligned}$$

3.2.16 We have that

$$\begin{aligned}
&E(X_1) \\
&= \int_0^1\int_0^{1-x_2} x_1\frac{\Gamma(\alpha_1+\alpha_2+\alpha_3)}{\Gamma(\alpha_1)\Gamma(\alpha_2)\Gamma(\alpha_3)}x_1^{\alpha_1-1}x_2^{\alpha_2-1}(1-x_1-x_2)^{\alpha_3-1}\,dx_1\,dx_2 \\
&= \frac{\Gamma(\alpha_1+\alpha_2+\alpha_3)}{\Gamma(\alpha_1)\Gamma(\alpha_2)\Gamma(\alpha_3)}\int_0^1\int_0^{1-x_2} x_1^{\alpha_1}x_2^{\alpha_2-1}(1-x_1-x_2)^{\alpha_3-1}\,dx_1\,dx_2 \\
&= \frac{\Gamma(\alpha_1+\alpha_2+\alpha_3)}{\Gamma(\alpha_1)\Gamma(\alpha_2)\Gamma(\alpha_3)}\frac{\Gamma(\alpha_1+1)\Gamma(\alpha_2)\Gamma(\alpha_3)}{\Gamma(\alpha_1+\alpha_2+\alpha_3+1)} = \frac{\alpha_1}{\alpha_1+\alpha_2+\alpha_3}.
\end{aligned}$$

3.3 Variance, Covariance, and Correlation

Exercises

3.3.1
(a) $\text{Cov}(X,Y) = E(XY) - E(X)\,E(Y) = 26 - (4)(19/3) = 2/3$.
(b) $E(X^2) = 3^2(1/2) + 3^2(1/6) + 6^2(1/6) + 6^2(1/6) = 18$, so $\text{Var}(X) = E(X^2) - E(X)^2 = 18 - 4^2 = 2$. Also $E(Y^2) = 5^2(1/2) + 9^2(1/6) + 5^2(1/6) + 9^2(1/6) = 131/3$, so $\text{Var}(Y) = E(Y^2) - E(Y)^2 = 131/3 - (19/3)^2 = 32/9$.
(c) $\text{Corr}(X,Y) = \text{Cov}(X,Y) \,/\, \sqrt{\text{Var}(X)\,\text{Var}(Y)} = (2/3) \,/\, \sqrt{(2)(32/9)} = 1/4$.

3.3.2
(a) $E(X) = (5)(1/7) + (5)(1/7) + (5)(1/7) + (8)(3/7) + (8)(1/7) = 47/7$. Also, $E(Y) = (0)(1/7) + (3)(1/7) + (4)(1/7) + (0)(3/7) + (4)(1/7) = 11/7$.
(b) $E(XY) = (5)(0)(1/7)+(5)(3)(1/7)+(5)(4)(1/7)+(8)(0)(3/7)+(8)(4)(1/7) = 67/7$. Then $\text{Cov}(X,Y) = E(XY) - E(X)\,E(Y) = 67/7 - (47/7)(11/7) = -48/49$.
(c) $E(X^2) = (5)^2(1/7) + (5)^2(1/7) + (5)^2(1/7) + (8)^2(3/7) + (8)^2(1/7) = 331/7$. Then $\text{Var}(X) = E(X^2) - E(X)^2 = 331/7 - (47/7)^2 = 108/49$. Also, $E(Y^2) = (0)^2(1/7)+(3)^2(1/7)+(4)^2(1/7)+(0)^2(3/7)+(4)^2(1/7) = 41/7$. Then $\text{Var}(Y) = E(Y^2) - E(Y)^2 = 41/7 - (11/7)^2 = 166/49$.
(d) $\text{Corr}(X,Y) = \text{Cov}(X,Y) \,/\, \sqrt{\text{Var}(X)\,\text{Var}(Y)} =$
$(-48/49) \,/\, \sqrt{(108/49)(166/49)} = -4\sqrt{2/249} \doteq -0.3585$.

3.3.3 We have that
$E(X) = \int_0^1 \int_0^1 x\,(4x^2y + 2y^5)\,dx\,dy = 2/3$,
$E(Y) = \int_0^1 \int_0^1 y\,(4x^2y + 2y^5)\,dx\,dy = 46/63$,
$E(X^2) = \int_0^1 \int_0^1 x^2(4x^2y + 2y^5)\,dx\,dy = \int_0^1 \left(\frac{4}{5}y + \frac{2}{3}y^5\right)\,dy = \frac{2}{5} + \frac{2}{9} = \frac{28}{45}$,
$E(Y^2) = \int_0^1 \int_0^1 y^2\,(4x^2y + 2y^5)\,dx\,dy = \int_0^1 \left(\frac{4}{3}y^3 + 2y^7\right)\,dy = \frac{7}{12}$,
$E(XY) = \int_0^1 \int_0^1 xy\,(4x^2y + 2y^5)\,dx\,dy = \int_0^1 \left(y^2 + y^6\right)\,dy = \frac{10}{21}$,

$$\text{Corr}\,(X,Y) = \frac{\frac{10}{21} - \left(\frac{2}{3}\right)\left(\frac{46}{63}\right)}{\sqrt{\frac{23}{45} - \left(\frac{2}{3}\right)^2}\sqrt{\frac{7}{12} - \left(\frac{46}{63}\right)^2}} = -0.18292.$$

3.3.4 Here
$E(X) = \int_0^1 \int_0^1 x\,(15x^3y^4 + 6x^2y^7)\,dx\,dy = 63/80$,
$E(Y) = \int_0^1 \int_0^1 y\,(15x^3y^4 + 6x^2y^7)\,dx\,dy = 61/72$. $E(X^2) = \int_0^1 \int_0^1 x^2\,(15x^3y^4 + 6x^2y^7)\,dx\,dy = 13/20$.
$E(Y^2) = \int_0^1 \int_0^1 y^2\,(15x^3y^4 + 6x^2y^7)\,dx\,dy = 103/140$. $\text{Var}(X) = E(X^2) - E(X)^2 = (13/20) - (63/80)^2 = 191/6400$.
$\text{Var}(Y) = E(Y^2) - E(Y)^2 = (103/140) - (61/72)^2 = 3253/181440$. $E(XY) = \int_0^1 \int_0^1 xy\,(15x^3y^4 + 6x^2y^7)\,dx\,dy = 2/3$.
$\text{Cov}(X,Y) = E(XY) - E(X)E(Y) = (2/3) - (63/80)(61/72) = -1/1920$.

$$\text{Corr}(X,Y) = \text{Cov}(X,Y) \,/\, \sqrt{\text{Var}(X)\,\text{Var}(Y)}$$
$$= (-1/1920) \,/\, \sqrt{(191/6400)(3253/181440)} = -3\sqrt{35/621323} \doteq -0.0225.$$

3.3.5 If X and Y are independent, then $\text{Cov}(X,Y) = E(XY) - E(X)E(Y) = E(X)E(Y) - E(X)E(Y) = 0$, so $\text{Corr}(X,Y) = \text{Cov}(X,Y) / \sqrt{\text{Var}(X)\,\text{Var}(Y)} = 0$.

3.3.6 If X and Z are independent, then $\text{Cov}(X+Y, Z) = \text{Cov}(X,Z) + \text{Cov}(Y,Z) = 0 + \text{Cov}(Y,Z) = \text{Cov}(Y,Z)$.

3.3.7
(a) $\text{Cov}(X,Z) = \text{Cov}(X, X+Y) = \text{Cov}(X,X) + \text{Cov}(X,Y) = \text{Var}(X) + 0 = 1/3^2 = 1/9$.
(b) $\text{Corr}(X,Z) = \text{Cov}(X,Z) / \sqrt{\text{Var}(X)\,\text{Var}(Z)} = (1/9) / \sqrt{(1/9)((1/9)+5)} = 1/\sqrt{46} \doteq 0.147$.

3.3.8 We can write $X = L + (R-L)U$, where $U \sim \text{Uniform}[0,1]$. Then $E(X) = L + (R-L)E(U) = L + (R-L)/2 = (L+R)/2$ and $\text{Var}(X) = (R-L)^2\,\text{Var}(U)$. Now $E(U^2) = \int_0^1 u^2\,du = 1/3$, so $\text{Var}(U) = 1/3 - 1/4 = 1/12$.

3.3.9 $E(X(X-1)) = E(X^2) - E(X)$, so $E(X(X-1)) - E(X)(E(X)-1) = E(X^2) - (E(X))^2 = \text{Var}(X)$. Then, when $X \sim \text{Binomial}(n,\theta)$, we have that,

$$\begin{aligned}
E(X(X-1)) &= \sum_{x=0}^{n} x(x-1)\binom{n}{x}\theta^x(1-\theta)^{n-x} \\
&= n(n-1)\theta^2 \sum_{x=2}^{n} \binom{n-2}{x-2}\theta^{x-2}(1-\theta)^{n-2-(x-2)} \\
&= n(n-1)\theta^2 \sum_{x=0}^{n-2} \binom{n-2}{x}\theta^{x}(1-\theta)^{n-2-x} = n(n-1)\theta^2,
\end{aligned}$$

so $\text{Var}(X) = n(n-1)\theta^2 - n\theta(n\theta - 1) = n\theta(1-\theta)$.

Problems

3.3.10 Here $\text{Cov}(X,Y) = E(XY) - E(X)E(Y) = E(cX^2) - E(X)E(cX) = c(1) - (0)(0) = c$, and $\text{Corr}(X,Y) = \text{Cov}(X,Y) / \sqrt{\text{Var}(X)\,\text{Var}(Y)} = c / \sqrt{(1)(c^2)} = c/|c| = sgn(c)$, where $sgn(c) = 1$ for $c > 0$, $sgn(c) = 0$ for $c = 0$, and $sgn(c) = -1$ for $c < 0$. Hence,
(a) $\lim_{c\searrow 0} \text{Cov}(X,Y) = \lim_{c\searrow 0} c = 0$.
(b) $\lim_{c\nearrow 0} \text{Cov}(X,Y) = \lim_{c\nearrow 0} c = 0$.
(c) $\lim_{c\searrow 0} \text{Corr}(X,Y) = \lim_{c\searrow 0} sign(c) = 1$.
(d) $\lim_{c\nearrow 0} \text{Corr}(X,Y) = \lim_{c\nearrow 0} sign(c) = -1$.
(e) As c passes from positive to negative, $\text{Corr}(X,Y)$ is not continuous but rather "jumps" from $+1$ to -1.

3.3.11 We have that $E(X) = \mu_1$, $\text{Var}(X) = \sigma_1^2$, $E(Y) = \mu_2$, $\text{Var}(Y) = \sigma_2^2$, and using (2.7.1) we have that

$$E(XY) = E\left((\mu_1 + \sigma_1 Z_1)\left(\mu_2 + \sigma_2\left(\rho Z_1 + \sqrt{1-\rho^2}Z_2\right)\right)\right)$$

$$= E\left(\begin{array}{c} \mu_1\mu_2 + \sigma_1\mu_2 Z_1 + \rho\sigma_2\mu_1 Z_1 + \rho\sigma_1\sigma_2 Z_1^2 \\ +\sigma_2\mu_1\sqrt{1-\rho^2}Z_2 + \sigma_1\sigma_2\sqrt{1-\rho^2}Z_1 Z_2 \end{array}\right) = \mu_1\mu_2 + \rho\sigma_1\sigma_2$$

where we have used $E(Z_1) = E(Z_2) = E(Z_1 Z_2) = E(Z_1)E(Z_2) = 0$, $E(Z_1^2) = 1$. So $\text{Cov}(XY) = \rho\sigma_1\sigma_2$ and $\text{Corr}(X,Y) = \text{Cov}(X,Y)/\sqrt{\text{Var}(X)\text{Var}(Y)} = (\sigma_1\sigma_2\rho)/\sqrt{(\sigma_1^2)(\sigma_2^2)} = \rho$.

3.3.12 We have that

$$E(X(X-1)) = \theta\sum_{x=0}^{\infty} x(x-1)(1-\theta)^x = \theta(1-\theta)^2\sum_{x=2}^{\infty} x(x-1)(1-\theta)^{x-2}$$

$$= \theta(1-\theta)^2\sum_{x=2}^{\infty}\frac{d^2(1-\theta)^x}{d\theta^2} = \theta(1-\theta)^2\frac{d^2}{d\theta^2}\sum_{x=2}^{\infty}(1-\theta)^x$$

$$= \theta(1-\theta)^2\frac{d^2}{d\theta^2}\left(\frac{1}{\theta} - 1 - (1-\theta)\right) = \theta(1-\theta)^2\frac{2}{\theta^3} = \frac{(1-\theta)^2}{\theta}.$$

Therefore

$$\text{Var}(X) = \frac{2(1-\theta)^2}{\theta^2} - \frac{(1-\theta)}{\theta}\left(\frac{(1-\theta)}{\theta} - 1\right)$$

$$= \frac{(1-\theta)^2}{\theta^2} + \frac{(1-\theta)}{\theta} = \frac{(1-\theta)}{\theta}\left(\frac{(1-\theta)}{\theta} + 1\right) = \frac{(1-\theta)}{\theta^2}.$$

3.3.13 We have that when $X_1, \ldots, X_r$ are i.i.d. Geometric(θ) then $X = X_1 + \cdots + X_r \sim$ Negative Binomial(r, θ). Therefore, $\text{Var}(X) = r(1-\theta)/\theta^2$.

3.3.14 We have that

$$E(X^2) = \int_0^{\infty} x^2\frac{\lambda^\alpha x^{\alpha-1}}{\Gamma(\alpha)}e^{-\lambda x}\,dx = \int_0^{\infty}\frac{\lambda^\alpha x^{\alpha+1}}{\Gamma(\alpha)}e^{-\lambda x}\,dx$$

$$= \int_0^{\infty}\frac{\lambda^\alpha t^{\alpha+1}}{\lambda^{\alpha+1}\Gamma(\alpha)}e^{-t}(1/\lambda)\,dt = \frac{1}{\lambda^2\Gamma(\alpha)}\int_0^{\infty} t^{\alpha+1}e^{-t}\,dx = \frac{\Gamma(\alpha+2)}{\lambda^2\Gamma(\alpha)} = \frac{\alpha(\alpha+1)}{\lambda^2},$$

so $\text{Var}(X) = \alpha(\alpha+1)/\lambda^2 - \alpha^2/\lambda^2 = \alpha/\lambda^2$.

3.3.15 We have that $E(X^2) = \int_0^\infty x^2\alpha x^{\alpha-1}e^{-x^\alpha}\,dx = \int_0^\infty \alpha x^{\alpha+1}e^{-x^\alpha}\,dx$, and putting $u = x^\alpha, x = u^{1/\alpha}, du = \alpha x^{\alpha-1}dx$ we have that $E(X^2) = \int_0^\infty u^{2/\alpha}e^{-u}\,du = \Gamma(2/\alpha+1)$ and $\text{Var}(X) = E(X^2) - (E(X))^2 = \Gamma(2/\alpha+1) - \Gamma^2(1/\alpha+1)$.

3.3.16 We have that

$$E\left((X+1)^2\right) = \int_0^{\infty}(x+1)^2\alpha(1+x)^{-\alpha-1}\,dx = \int_0^{\infty}\alpha(1+x)^{-\alpha+1}\,dx$$

$$= \begin{cases} \frac{\alpha}{-\alpha+2}(1+x)^{-\alpha+2}\Big|_0^\infty & \alpha \neq 2 \\ \alpha\ln(1+x)|_0^\infty & \alpha = 2 \end{cases} = \begin{cases} \infty & 0 < \alpha \leq 2 \\ \alpha/(\alpha-2) & \text{if } \alpha > 2. \end{cases}$$

Therefore, when $\alpha > 2$,

$$\begin{aligned}\text{Var}(X) &= E\left((X+1)^2\right) - 2E(X) - 1 - (E(X))^2\\ &= \frac{\alpha}{\alpha-2} - \frac{2}{\alpha-1} - 1 - \frac{1}{(\alpha-1)^2}\\ &= \frac{\alpha(\alpha-1)^2 - 2(\alpha-1)(\alpha-2) - (\alpha-1)^2(\alpha-2) - (\alpha-2)}{(\alpha-1)^2(\alpha-2)}\\ &= \frac{\alpha}{(\alpha-1)^2(\alpha-2)}.\end{aligned}$$

3.2.17 We have that $E(X^2) = \int_0^\infty x^2 e^{-x}\,dx = \Gamma(3) = 2 = \text{Var}(X)$ since $E(X) = 0$.

3.3.18 We have that

$$\begin{aligned}E(X^2) &= \int_0^1 x^2 \frac{\Gamma(a+b)}{\Gamma(a)\Gamma(b)} x^{a-1}(1-x)^{b-1}\,dx = \frac{\Gamma(a+b)}{\Gamma(a)\Gamma(b)} \int_0^1 x^{a+1}(1-x)^{b-1}\,dx\\ &= \frac{\Gamma(a+b)}{\Gamma(a)\Gamma(b)} \frac{\Gamma(a+2)\Gamma(b)}{\Gamma(a+b+2)} = \frac{a(a+1)}{(a+b)(a+b+1)}.\end{aligned}$$

Therefore

$$\begin{aligned}\text{Var}(X) &= E(X^2) - (E(X))^2 = \frac{a(a+1)}{(a+b)(a+b+1)} - \left(\frac{a}{a+b}\right)^2\\ &= \frac{a(a+1)(a+b) - a(a+b+1)}{(a+b)^2(a+b+1)} = \frac{ab}{(a+b)^2(a+b+1)}.\end{aligned}$$

3.3.19 We have that $X_i \sim \text{Binomial}(n, \theta_i)$ so that $E(X_i) = n\theta_i$, $\text{Var}(X_i) = n\theta_i(1-\theta_i)$. Also,

$$\begin{aligned}E(X_1X_2) &= \sum_{x_1=0}^{n}\sum_{x_2=0}^{n-x_1} x_1x_2 \binom{n}{x_1\, x_2\, n-x_1-x_2} \theta_1^{x_1}\theta_2^{x_2}\theta_3^{n-x_1-x_2}\\ &= n(n-1)\theta_1\theta_2 \sum_{x_1=1}^{n}\sum_{x_2=1}^{n-x_1} \binom{n-2}{x_1-1\, x_2-1\, n-2-(x_1-1)-(x_2-1)}\\ &\times \theta_1^{x_1-1}\theta_2^{x_2-1}\theta_3^{n-2-(x_1-1)-(x_2-1)}\\ &= n\theta_1\theta_2 \sum_{x_1=0}^{n-2}\sum_{x_2=0}^{n-2-x_1} \binom{n-2}{x_1\, x_2\, n-2-x_1-x_2} \theta_1^{x_1}\theta_2^{x_2}\theta_3^{n-2-x_1-x_2} = n\theta_1\theta_2\end{aligned}$$

since the sum is the sum of all Multinomial$(n-2, \theta_1, \theta_2, \theta_3)$ probabilities. Therefore, $\text{Cov}(X_1, X_2) = n(n-1)\theta_1\theta_2 - n^2\theta_1\theta_2 = -n\theta_1\theta_2$.

3.3.20 We have that $X_1 \sim \text{Beta}(\alpha_1, \alpha_2 + \alpha_3)$, so $E(X_1) = \alpha_1/(\alpha_1 + \alpha_2 + \alpha_3)$ and $\text{Var}(X_1) = \alpha_1(\alpha_2 + \alpha_3)/(\alpha_1 + \alpha_2 + \alpha_3)^2(\alpha_1 + \alpha_2 + \alpha_3 + 1)$ by Problem

3.3.17. Also,

$$
\begin{aligned}
&E\left(X_1 X_2\right) \\
&= \int_0^1 \int_0^{1-x_2} x_1 x_2 \frac{\Gamma\left(\alpha_1+\alpha_2+\alpha_3\right)}{\Gamma\left(\alpha_1\right)\Gamma\left(\alpha_2\right)\Gamma\left(\alpha_3\right)} x_1^{\alpha_1-1} x_2^{\alpha_2-1}\left(1-x_1-x_2\right)^{\alpha_3-1} dx_1\, dx_2 \\
&= \frac{\Gamma\left(\alpha_1+\alpha_2+\alpha_3\right)}{\Gamma\left(\alpha_1\right)\Gamma\left(\alpha_2\right)\Gamma\left(\alpha_3\right)} \int_0^1 \int_0^{1-x_2} x_1^{\alpha_1} x_2^{\alpha_2}\left(1-x_1-x_2\right)^{\alpha_3-1} dx_1\, dx_2 \\
&= \frac{\Gamma\left(\alpha_1+\alpha_2+\alpha_3\right)}{\Gamma\left(\alpha_1\right)\Gamma\left(\alpha_2\right)\Gamma\left(\alpha_3\right)} \frac{\Gamma\left(\alpha_1+1\right)\Gamma\left(\alpha_2+1\right)\Gamma\left(\alpha_3\right)}{\Gamma\left(\alpha_1+\alpha_2+\alpha_3+2\right)} \\
&= \frac{\alpha_1\alpha_2}{\left(\alpha_1+\alpha_2+\alpha_3\right)\left(\alpha_1+\alpha_2+\alpha_3-1\right)}
\end{aligned}
$$

so

$$
\begin{aligned}
\operatorname{Cov}\left(X_1, X_2\right) &= \frac{\alpha_1\alpha_2}{\left(\alpha_1+\alpha_2+\alpha_3\right)\left(\alpha_1+\alpha_2+\alpha_3+1\right)} - \frac{\alpha_1\alpha_2}{\left(\alpha_1+\alpha_2+\alpha_3\right)^2} \\
&= \frac{-\alpha_1\alpha_2}{\left(\alpha_1+\alpha_2+\alpha_3\right)^2\left(\alpha_1+\alpha_2+\alpha_3+1\right)}.
\end{aligned}
$$

3.3.21 We have that

$$
\begin{aligned}
E(X(X-1)) &= \sum_{x=\max(0,\, n+M-N)}^{\min(n,M)} x(x-1) \frac{\binom{M}{x}\binom{N-M}{n-x}}{\binom{N}{n}} \\
&= \sum_{x=\max(2,\, n+M-N)}^{\min(n,M)} x(x-1) \frac{\binom{M}{x}\binom{N-M}{n-x}}{\binom{N}{n}} \\
&= n(n-1)\frac{M(M-1)}{N(N-1)} \sum_{x=\max(2,\, n+M-N)}^{\min(n,M)} \frac{\binom{M-2}{x-2}\binom{N-2-(M-2)}{n-2-(x-2)}}{\binom{N-2}{n-2}} \\
&= n(n-1)\frac{M(M-1)}{N(N-1)} \sum_{x=\max(0,\, n-2+(M-2)-(N-2))}^{\min(n-2,M-2)} \frac{\binom{M-2}{x}\binom{N-2-(M-2)}{n-2-x}}{\binom{N-2}{n-2}} \\
&= n(n-1)\frac{M(M-1)}{N(N-1)}
\end{aligned}
$$

as we are summing all Hypergeometric$(N-2, M-2, n-2)$ probabilities. Therefore,

$$
\begin{aligned}
\operatorname{Var}(X) &= n(n-1)\frac{M(M-1)}{N(N-1)} - n\frac{M}{N}\left(n\frac{M}{N}-1\right) \\
&= n\frac{M}{N}\frac{(n-1)(M-1)N-(N-1)(nM-N)}{N(N-1)} = n\frac{M}{N}\left(1-\frac{M}{N}\right)\frac{(N-n)}{(N-1)}.
\end{aligned}
$$

Challenges

3.3.22 Assume Y is discrete, with $Y \geq 0$ and $E(Y) = 0$. Then $0 = \sum_y y\,P(Y = y) = \sum_{y \geq 0} y\,P(Y = y)$. But the only way a sum of non-negative terms can be 0 is if each term is 0, i.e., $y\,P(Y = y) = 0$ for all $y \in R^1$. This means that $P(Y = y) = 0$ for $y \neq 0$, so that $P(Y = 0) = 1$.

3.3.23 Let $C = E(X)$, and let $Y = (X - C)^2$. Then $Y \geq 0$, and $E(Y) = \text{Var}(X) = 0$. Hence, from the previous challenge, $P(Y = 0) = 1$. But $Y = 0$ if and only if $X = C$. Hence, $P(X = C) = 1$.

3.3.24 Let $C = \sum_{k=1}^{\infty} 1/k^3$. (Then $C = \zeta(3) \doteq 1.202$, but C cannot be expressed precisely in elementary terms.) Let $P(Y = k) = 1/Ck^3$ for $k = 1, 2, 3, \ldots$. Let $X = Y + 5 - \pi^2/6$. Then $E(Y) = \sum_{k=1}^{\infty} k(1/k^3) = \sum_{k=1}^{\infty}(1/k^2) = \pi^2/6$, so $E(X) = E(Y) + 5 - \pi^2/6 = \pi^2/6 + 5 - \pi^2/6 = 5$. On the other hand, $E(Y^2) = \sum_{k=1}^{\infty} k^2(1/k^3) = \sum_{k=1}^{\infty}(1/k) = \infty$. It follows that $E(X^2) = \infty$ and that $\text{Var}(X) = \infty$.

3.4 Generating Functions

Exercises

3.4.1
(a) $r_Z(t) = E(t^Z) = \sum_{z=1}^{\infty} t^z(1/2^z) = (t/2)\,/\,[1 - (t/2)] = t/(2 - t)$. Hence, $r_Z'(t) = ((2 - t)(1) - (t)(-1))/(2 - t)^2 = 2/(2 - t)^2$, so $r_Z'(0) = 2/2^2 = 1/2 = P(Z = 1)$. Also, $r_Z''(t) = \frac{d}{dt}[2/(2 - t)^2] = -4/(t - 2)^3$, so so $r_Z''(0) = (-4)/(-2)^3 = 1/2 = 2(1/4) = 2P(Z = 2)$.
(b) Note that $Z = X + 1$, where $X \sim \text{Geometric}(1/2)$. Hence, $E(Z) = E(X) + 1 = 1 + 1 = 2$, and $\text{Var}(Z) = \text{Var}(X) = (1 - (1/2))/(1/2)^2 = 2$, where $E(Z^2) = \text{Var}(Z) + E(Z)^2 = 2 + 2^2 = 6$. On the other hand, $m_Z(t) = E(e^{tZ}) = r_Z(e^t) = e^t/(2 - e^t)$. Hence, $m_Z'(t) = [(2 - e^t)(e^t) - e^t(-e^t)]/(2 - e^t)^2 = 2e^t/(2 - e^t)^2$, so $m_Z'(0) = 2 = E(Z)$. Also, $m_Z''(t) = 2e^t(2 + e^t)/(2 - e^t)^3$, so $m_Z''(0) = 2(3)/1^3 = 6 = E(Z^2)$.

3.4.2 Here $m_X(s) = (e^s\theta + 1 - \theta)^n$. Hence, $m_X'(s) = e^s\theta n(e^s\theta + 1 - \theta)^{n-1}$, so $m_X'(0) = n\theta$. Then $m_X''(s) = e^{2s}\theta^2 n(n-1)(e^s\theta + 1 - \theta)^{n-2} + e^s\theta n(e^s\theta + 1 - \theta)^{n-1}$, so $m_X''(0) = \theta^2 n(n - 1) + \theta n$. Hence, $\text{Var}(X) = E(X^2) - E(X)^2 = m_X''(0) - (m_X'(0))^2 = \theta^2 n(n - 1) + \theta n - (n\theta)^2 = n\theta(1 - \theta)$.

3.4.3 Here $m_Y(s) = e^{\lambda(e^s - 1)}$. Hence, $m_Y'(s) = \lambda e^s e^{\lambda(e^s - 1)}$, so $m_Y'(s) = \lambda$. Also, $m_Y''(s) = (\lambda e^s + \lambda^2 e^{2s})e^{\lambda(e^s - 1)}$, so $m_Y''(s) = \lambda + \lambda^2$. Hence, $\text{Var}(Y) = E(Y^2) - E(Y)^2 = m_Y''(0) - (m_Y'(0))^2 = \lambda + \lambda^2 - (\lambda)^2 = \lambda$.

3.4.4 $r_Y(t) = E(t^Y) = E(t^{3X+4}) = t^4 E((t^3)^X) = t^4 r_X(t^3)$.

3.4.5 $m_Y(s) = E(e^{sY}) = E(e^{s(3X+4)}) = e^{4s}E(e^{3sX}) = e^{4s}m_X(3s)$.

3.4.6 We know $m_X''(s) = e^{2s}\theta^2 n(n-1)(e^s\theta + 1 - \theta)^{n-2} + e^s\theta n(e^s\theta + 1 - \theta)^{n-1}$, so $m_X'''(s) = e^s n\theta(1 - (e^s - 1)\theta)^{n-3}[1 - (e^s(3n - 1) + 2)\theta + (1 - e^s(3n - 1) + e^{2s}n^2)\theta^2]$, so $E(X^3) = m_X'''(0) = n\theta[1 - 3(n - 1)\theta + (n^2 - 3n + 2)\theta^2]$.

3.4.7 We know from previously that $m''_Y(s) = (\lambda e^s + \lambda^2 e^{2s})e^{\lambda(e^s-1)}$ so $m'''_Y(s) = e^{\lambda(e^s-1)}e^s\lambda(1+3e^s\lambda+e^{2s}\lambda^2)$, and $E(Y^3) = m'''_Y(0) = \lambda(1+3\lambda+\lambda^2)$.

3.4.8
(a) $r_X(t) = E(t^X) = (t^2)(1/2) + (t^5)(1/3) + (t^7)(1/6)$.
(b) $r'_X(t) = (2t)(1/2) + (5t^4)(1/3) + (7t^6)(1/6)$. Hence, $r'_X(0) = 0 = P(X=1)$. Also, $r''_X(t) = (2)(1/2) + (20t^3)(1/3) + (42t^5)(1/6)$. Hence, $r''_X(0) = (2)(1/2) = 1 = 2\,P(X=2)$.
(c) $m_X(s) = E(e^{sX}) = (e^{2s})(1/2) + (e^{5s})(1/3) + (e^{7s})(1/6)$.
$m'_X(s) = (2e^{2s})(1/2) + (5e^{5s})(1/3) + (7e^{7s})(1/6)$. Hence, $m'_X(0) = (2)(1/2) + (5)(1/3) + (7)(1/6) = E(X)$. Also, $m''_X(s) = (2^2e^{2s})(1/2) + (5^2e^{5s})(1/3) + (7^2e^{7s})(1/6)$. Hence, $m''_X(0) = (2^2)(1/2) + (5^2)(1/3) + (7^2)(1/6) = E(X^2)$.

Problems

3.4.9
(a) $m_X(s) = E(e^{sX}) = \int_0^{10} e^{sx}(1/10)dx = (1/10)(e^{10s}-1)/s$ for $s \neq 0$, with (of course) $m_X(0) = 1$.
(b) For $s \neq 0$, $m'_X(s) = (1/10)(s(10e^{10s}) - (e^{10s}-1))/s^2)$. We then compute using L'Hospital's Rule (twice) that $m'_X(0) = \lim_{s\to\infty} m'_X(s) = 5 = E(X)$.

3.4.10 We have that $r_X(t) = \sum_{x=0}^{\infty}(t(1-\theta))^x\theta = \theta(1-t(1-\theta))^{-1}$, provided $|t(1-\theta)| < 1$. Then $r'_X(t) = \theta(1-\theta)(1-t(1-\theta))^{-2}$, $r''_X(t) = 2\theta(1-\theta)^2(1-t(1-\theta))^{-3}$, so $r''_X(0)/2 = \theta(1-\theta)^2$.

3.4.11 We have that $r_X(t) = \sum_{x=0}^{\infty}\binom{r+x-1}{x}(t(1-\theta))^x\theta^r = \theta^r(1-t(1-\theta))^{-r}$, provided $|t(1-\theta)| < 1$. Then $r'_X(t) = r\theta^r(1-\theta)(1-t(1-\theta))^{-r-1}$, $r''_X(t) = r(r-1)\theta^r(1-\theta)^2(1-t(1-\theta))^{-r-2}$, so $r''_X(0)/2 = r(r+1)\theta^r(1-\theta)^2/2$
$= \binom{r+2-1}{2}\theta^r(1-\theta)^2$.

3.4.12
(a) $m_X(s) = E(e^{sX}) = \sum_{x=0}^{\infty} e^{sx}(1-\theta)^x\theta = \theta/(1-e^s(1-\theta))$, provided $|e^s(1-\theta)| < 1$, i.e., $s < -\log(1-\theta)$.
(b) $m'_X(s) = \frac{d}{ds}(\theta/[1-e^s(1-\theta)]) = \theta(1-\theta)e^s/[1-e^s(1-\theta)]^2$. Hence, $E(X) = m'_X(0) = \theta(1-\theta)/[1-(1-\theta)]^2 = (1-\theta)/\theta$.
(c) $m''_X(s) = (e^s\theta(1-\theta)(1+e^s(1-\theta)))/[1-e^s(1-\theta)]^3$. Hence, $E(X^2) = m''_X(0) = (\theta(1-\theta)(2-\theta))/[\theta]^3 = (1-\theta)(2-\theta)/\theta^2$, so $\text{Var}(X) = E(X^2) - E(X)^2 = [(1-\theta)(2-\theta)/\theta^2] - [(1-\theta)/\theta]^2 = (1-\theta)/\theta^2$.

3.4.13 We use the result of 3.4.12 and the fact that if $X_1, \ldots, X_r$ is a sample from the Geometric(θ) distribution, then $X = X_1 + \cdots + X_r \sim$ Negative-Binomial(r,θ).
(a) $m_X(s) = m_{X_1}(s)\cdots m_{X_r}(s) = \theta^r/(1-e^s(1-\theta))^r$.
(b) $m'_X(s) = re^s(1-\theta)\theta^r/(1-e^s(1-\theta))^{r+1}$, so $E(X) = m'_X(0) = r(1-\theta)\theta^r/\theta^{r+1} = r(1-\theta)/\theta$.
(c) $m''_X(s) = re^s(1-\theta)\theta^r/(1-e^s(1-\theta))^{r+1} +$
$r(r+1)e^{2s}(1-\theta)^2\theta^r/(1-e^s(1-\theta))^{r+2}$, so $\text{Var}(X) = m''_X(0) - (r(1-\theta)/\theta)^2$

$= r(1-\theta)\theta^r/\theta^{r+1} + r(r+1)(1-\theta)^2\theta^r/\theta^{r+2} - (r(1-\theta)/\theta)^2 = r(1-\theta)/\theta + r(1-\theta)^2/\theta^2 = r(1-\theta)/\theta^2$.

3.4.14 $r_Y(t) = E(t^Y) = E(t^{a+bX}) = E(t^a t^{bX}) = t^a E(t^{bX}) = t^a E\left((t^b)^X\right) = t^a r_X(t^b)$ and $m_Y(t) = E(e^{tY}) = E(e^{at+btX}) = E(e^{at}e^{btX}) = e^{at}E(e^{btX}) = e^{at}E(e^{(bt)X}) = e^{at}m_X(bt)$.

3.4.15 Write $Z = \mu + \sigma X$, where $X \sim \text{Normal}(0,1)$. Then $m_Z(s) = E(e^{sZ}) = E(e^{s(\mu+\sigma X)}) = e^{s\mu} + E(e^{s\sigma X}) = e^{s\mu} + m_X(\sigma s) = e^{s\mu} + e^{(\sigma s)^2/2} = e^{s\mu} + e^{\sigma^2 s^2/2}$.

3.4.16
(a) $m_Y(s) = E(e^{sY}) = \int_{-\infty}^{\infty} e^{sy}\, e^{|y|/2}\, dy = \int_0^\infty e^{sy}\, e^{y/2}\, dy + \int_{-\infty}^0 e^{sy}\, e^{-y/2}\, dy = \int_0^\infty e^{(s-1/2)y}\, dy + \int_{-\infty}^0 e^{(s+1/2)y}\, dy = 1/(1/2-s) + 1/(1/2+s)$, provided $|s| < 1/2$.
(b) $m'_Y(s) = (1/2-s)^{-2} - (1/2+s)^{-2}$, so $E(Y) = m'_Y(0) = 4 - 4 = 0$.
(c) $m''_Y(s) = 2(1/2-s)^{-3} + 2(1/2-s)^{-3}$, so $E(Y^2) = m''_Y(0) = 16 + 16 = 32$, where $\text{Var}(Y) = E(Y^2) - E(Y)^2 = 32$.

3.4.17 $E(X^k) = \int_0^\infty x^k \alpha x^{\alpha-1} e^{-x^\alpha}\, dx = \int_0^\infty \alpha x^{\alpha+k-1} e^{-x^\alpha}\, dx$ and putting $u = x^\alpha, x = u^{1/\alpha}, du = \alpha x^{\alpha-1}dx$ we have that $E(X) = \int_0^\infty u^{k/\alpha} e^{-u}\, du = \Gamma(k/\alpha + 1)$.

3.4.18 We put $u = 1/(1+x)$ so $x = (1/u) - 1, dx = -du/u^2$ so

$$
\begin{aligned}
E(X^k) &= \int_0^\infty x^k \alpha (1+x)^{-\alpha-1}\, dx = \alpha \int_0^\infty \left(\frac{1-u}{u}\right)^k u^{\alpha-1}\, du \\
&= \alpha \int_0^\infty u^{\alpha-k-1}(1-u)^k\, du \\
&= \begin{cases} \infty & 0 < \alpha \le k \\ \alpha \frac{\Gamma(\alpha-k)\Gamma(k+1)}{\Gamma(\alpha+1)} & \alpha > k. \end{cases} = \begin{cases} \infty & 0 < \alpha \le k \\ \frac{\Gamma(\alpha-k)\Gamma(k+1)}{\Gamma(\alpha)} & \alpha > k. \end{cases}
\end{aligned}
$$

3.4.19 Putting $z = \ln x$ so that $x = \exp(z), dx = \exp(z)dz$, we have that

$$
\begin{aligned}
&E(X^k) \\
&= \int_0^\infty x^k \frac{1}{\sqrt{2\pi}\tau} \exp\left\{-\frac{(\ln x)^2}{2\tau^2}\right\} \frac{1}{x}\, dx = \int_0^\infty x^{k-1} \frac{1}{\sqrt{2\pi}\tau} \exp\left\{-\frac{(\ln x)^2}{2\tau^2}\right\} dx \\
&= \int_0^\infty \exp(kz) \frac{1}{\sqrt{2\pi}\tau} \exp\left\{-\frac{z^2}{2\tau^2}\right\} dx = \exp\left\{\frac{\tau^2 k^2}{2}\right\}
\end{aligned}
$$

since this is the moment generating function of the $N(0, \tau^2)$ distribution at k.

3.4.20

$$
\begin{aligned}
m(s) &= \int_0^\infty e^{xt} \frac{(\lambda x)^{\alpha-1}}{\Gamma(\alpha)} e^{-\lambda x} \lambda\, dx = \lambda \int_0^\infty \frac{(\lambda x)^{\alpha-1}}{\Gamma(\alpha)} e^{-(\lambda-t)x}\, dx \\
&= \begin{cases} \infty & t \ge \lambda \\ \frac{\lambda^\alpha}{(\lambda-t)^\alpha} & t < k \end{cases}
\end{aligned}
$$

3.4.21 The mgf of the Poisson(λ_i) equals $m_i(s) = \exp\{\lambda_i(e^s - 1)\}$. Then the mgf of $Y = X_1 + \cdots + X_n$ is given by (Theorem 3.4.5)

$$m_Y(s) = \prod_{i=1}^n m_i(s) = \prod_{i=1}^n \exp\{\lambda_i(e^s - 1)\} = \exp\left\{\sum_{i=1}^n \lambda_i(e^s - 1)\right\}$$

and we recognize this as the mgf of the Poisson($\sum_{i=1}^n \lambda_i$). Therefore, the uniqueness theorem implies that this is the distribution of Y.

3.4.22 The mgf of the Geometric(θ) distribution is given by $\theta/(1 - e^s(1-\theta))$. Therefore, the Negative Binomial(r, θ) distribution has mgf given by $m(s) = \theta^r/(1 - e^s(1-\theta))^r$ since it can be obtained as the sum of r independent Geometric(θ) random variables and we use Theorem 3.4.5. Then X_i has mgf given by $m_i(s) = \theta^{r_i}/(1 - e^s(1-\theta))^{r_i}$ and, using Theorem 3.4.5 again, we have that Y has mgf

$$\begin{aligned} m_Y(s) &= \prod_{i=1}^n m_i(s) = \prod_{i=1}^n \theta^{r_i}/(1 - e^s(1-\theta))^{r_i} \\ &= \theta^{\sum_{i=1}^n r_i}/(1 - e^s(1-\theta))^{\sum_{i=1}^n r_i} \end{aligned}$$

and we recognize this as the mgf of the Negative Binomial(r_i, θ) distribution. Therefore, by the uniqueness theorem this is the distribution of Y.

3.4.23 The Gamma(α, λ) distribution has mgf $\lambda^\alpha/(\lambda - t)^\alpha$ for $t < \lambda$ by Problem 3.4.20. Therefore, by Theorem 3.4.5, Y has mgf

$$m_Y(s) = \prod_{i=1}^n m_i(s) = \prod_{i=1}^n \lambda^{\alpha_i}/(\lambda - t)^{\alpha_i} = \lambda^{\sum_{i=1}^n \alpha_i}/(\lambda - t)^{\sum_{i=1}^n \alpha_i}$$

and we recognize this as the mgf of the Gamma($\sum_{i=1}^n \alpha_i, \lambda$) distribution so, by the uniqueness theorem this must be the distribution of Y.

3.4.24 By Theorem 3.4.7 the mgf is given by (using $m_{X_i}(s) = \lambda/(\lambda - s)$ and $r_N(t) = \exp\{\lambda(t-1)\}$)

$$m_{S_N}(s) = r_N(m_{X_1}(s)) = \exp\left\{\lambda\left(\frac{\lambda}{\lambda - s} - 1\right)\right\} = \exp\left\{\frac{\lambda s}{\lambda - s}\right\}.$$

Then $m'_{S_N}(s) = \exp\left\{\frac{\lambda s}{\lambda - s}\right\}\left(\frac{\lambda s}{\lambda - s}\right)' = \exp\left\{\frac{\lambda s}{\lambda - s}\right\}\lambda^2(\lambda - s)^{-2}$ and $m'_{S_N}(0) = 1$ is the mean.

3.4.25 By Theorem 3.4.7 the mgf is given by (using $m_{X_i}(s) = \lambda/(\lambda - s)$ and $r_N(t) = \theta(1 - t(1-\theta))^{-1}$)

$$m_{S_N}(s) = r_N(m_{X_1}(s)) = \theta/\left(1 - \frac{\lambda}{\lambda - s}(1-\theta)\right).$$

Then

$$m'_{S_N}(s) = \frac{\theta}{\left(1 - \frac{\lambda}{\lambda - s}(1-\theta)\right)^2}(1-\theta)\frac{\lambda}{(\lambda - s)^2},$$

so $m'_{S_N}(0) = (1-\theta)\,/\,(\lambda\theta)$.

3.4.26 Here $c_X(s) = 1 - p + p\cos s + ip\sin s$, so $c'_X(s) = -p\sin s + ip\cos s$, $iE(X) = c'_X(0) = ip$, $E(X) = p$.

3.4.27
(a) We can write $Y = X_1 + X_2 + \ldots + X_n$, where the $\{X_i\}$ are i.i.d. $\sim$ Bernoulli(p). Hence, $c_Y(s) = c_{X_1}(s)c_{X_2}(s)\ldots c_{X_n}(s) = (1 - p + p\cos s + ip\sin s)^n$.
(b) $c'_Y(s) = n(1-p+p\cos s + ip\sin s)^{n-1}(-p\sin s + ip\cos s)$. Hence, $iE(Y) = c'_Y(s) = n1^{n-1}(ip) = inp$, so $E(Y) = np$.

3.4.28 The sample mean has characteristic function given by

$$\begin{aligned} c_{\bar{X}}(s) &= E\left(\exp\{is\bar{X}\}\right) = E\left(\exp\left\{i\frac{s}{n}\sum_{i=1}^{n} X_i\right\}\right) = \prod_{i=1}^{n} c_{X_1}\left(\frac{s}{n}\right) \\ &= \left(c_{X_1}\left(\frac{s}{n}\right)\right)^n = \left(\exp\left\{-\frac{|t|}{n}\right\}\right)^n = \exp\{-|t|\} \end{aligned}$$

and we recognize this as the cf of the Cauchy distribution. Then by the uniqueness theorem we must have that $\bar{X}$ is also distributed Cauchy. This implies that the sample mean in this case is just as variable as a single observation. So increasing the sample size does not make the distribution of $\bar{X}$ more concentrated. In fact, it doesn't change at all!

3.4.29 The cf of the $N(0,1)$ distribution is given by

$$\begin{aligned} c(t) &= \int_{-\infty}^{\infty} e^{itx}\frac{1}{\sqrt{2\pi}}e^{-x^2/2}\,dx = e^{-t^2/2}\int_{-\infty}^{\infty}\frac{1}{\sqrt{2\pi}}\exp\left\{-\frac{1}{2}(x - it)^2\right\}dx \\ &= e^{-t^2/2}. \end{aligned}$$

Therefore, if $X = \mu + \sigma Z$, where $Z \sim N(0,1)$, then $X \sim N(\mu, \sigma^2)$ and X has mgf

$$\begin{aligned} c_X(t) &= E\left(e^{itX}\right) = E\left(e^{it(\mu+\sigma Z)}\right) = e^{it\mu}E\left(e^{it\sigma Z}\right) = e^{it\mu}c(t\sigma) \\ &= \exp\left\{it\mu - \frac{\sigma^2 t^2}{2}\right\}. \end{aligned}$$

Then we have that $\ln c_X(t) = it\mu - \sigma^2 t^2/2$ so $(\ln c_X(t))' = i\mu - \sigma^2 t$, $(\ln c_X(0))'/i = \mu$ and the first cumulant is μ. Also, $(\ln c_X(t))'' = -\sigma^2$ and so $(\ln c_X(0))''/i^2 = \sigma^2$ and the second cumulant is σ^2. Also, all higher order derivatives of $\ln c_X(t)$ are 0, so all higher order cumulants are 0.

3.5 Conditional Expectation

Exercises

3.5.1

(a) $E(X \mid Y = 3) = \sum_x x\, P(X = x \mid Y = 3) = (2)((1/5)/(1/5 + 1/5 + 1/5)) + (3)((1/5)/(1/5 + 1/5 + 1/5)) = 5/2$.
(b) $E(Y \mid X = 3) = \sum_y y\, P(Y = y \mid X = 3) = (2)((1/5)/(1/5 + 1/5)) + (3)((1/5)/(1/5 + 1/5)) + (17)((1/5)/(1/5 + 1/5 + 1/5)) = 22/3$.
(c) $E(X \mid Y = 2) = \sum_x x\, P(X = x \mid Y = 2) = (2)((1/5)/(1/5 + 1/5 + 1/5)) + (3)((1/5)/(1/5 + 1/5 + 1/5)) = 5/2$. Also $E(X \mid Y = 17) = \sum_x x\, P(X = x \mid Y = 17) = (3)(1/1) = 3$. Hence,

$$E(X \mid Y) = \begin{cases} 5/2 & Y = 2 \\ 5/2 & Y = 3 \\ 3 & Y = 17 \end{cases}$$

(d) $E(Y \mid X = 2) = \sum_y y\, P(Y = y \mid X = 2) = (2)((1/5)/(1/5 + 1/5 + 1/5)) + (3)((1/5)/(1/5 + 1/5 + 1/5)) = 5/2$. Hence,

$$E(Y \mid X) = \begin{cases} 5/2 & X = 2 \\ 22/3 & X = 3 \end{cases}$$

3.5.2

(a) $f_X(x) = \int_{-\infty}^{\infty} f_{X,Y}(x,y)\, dy = \int_0^5 [9(xy + x^5y^5)/16000900]\, dy = (9x + 1875x^5)/1280072$.
(b) $f_Y(y) = \int_{-\infty}^{\infty} f_{X,Y}(x,y)\, dx = \int_0^4 [9(xy + x^5y^5)/16000900]\, dx = (18y + 1536y^5)/4000225$.
(c) For $0 \le y \le 5$,

$$E(X \mid Y = y) = \int_{-\infty}^{\infty} x\, f_{X|Y}(x|y)\, dx = \int_{-\infty}^{\infty} x\, (f_{X,Y}(x,y)/f_Y(y))\, dx$$
$$= \int_0^4 x\, (9(xy + x^5y^5)/16000900)/((18y + 1536y^5)/4000225)\, dx$$
$$= [8(7 + 768y^4)]\, /\, [7(3 + 256y^4)].$$

Hence, $E(X \mid Y) = [8(7 + 768Y^4)]\, /\, [7(3 + 256Y^4)]$ for $0 \le Y \le 5$.
(d) For $0 \le x \le 4$,

$$E(Y \mid X = x) = \int_{-\infty}^{\infty} y\, f_{Y|X}(y|x)\, dy = \int_{-\infty}^{\infty} y\, (f_{X,Y}(x,y)/f_X(x))\, dy$$
$$= \int_0^5 y\, (9(xy + x^5y^5)/16000900)/((9x + 1875x^5)/1280072)\, dx$$
$$= [70 + 18750x^4]\, /\, [21 + 4375x^4].$$

Hence, $E(Y \mid X) = [70 + 18750X^4]\, /\, [21 + 4375X^4]$.

(e)

$$
\begin{aligned}
E(E(X \mid Y)) &= E([8(7 + 768Y^4)] \,/\, [7(3 + 256Y^4)]) \\
&= \int_{-\infty}^{\infty} [8(7 + 768y^4)] \,/\, [7(3 + 256y^4)] \, f_Y(y)\, dy \\
&= \int_0^5 [8(7 + 768y^4)] \,/\, [7(3 + 256y^4)] \, [(18y + 1536y^5)/4000225] \, dy \\
&= 3840168/1120063 \doteq 3.42852857
\end{aligned}
$$

On the other hand, $E(X) = \int_{-\infty}^{\infty} x\, f_X(x)\, dx = \int_0^4 x\, [(9x{+}1875x^5)/1280072]\, dx = 3840168/1120063 = E(E(X \mid Y))$.

3.5.3
(a) $E(Y \mid X = 6) = \sum_y y\, P(Y = y \mid X = 6) = \sum_y y\, P(Y = y,\ X = 6)/P(X = 6) = (2)((1/11)/(4/11)) + (3)((1/11)/(4/11)) + (7)((1/11)/(4/11)) + (13)((1/11)/(4/11)) = 25/4 = 6.24$.
(b) $E(Y \mid X = -4) = \sum_y y\, P(Y = y,\ X = -4)/P(X = -4) = (2)((1/11)/(7/11)) + (3)((2/11)/(7/11)) + (7)((4/11)/(7/11)) = 36/7 \doteq 5.14$.
(c) $E(Y \mid X) = 25/4$ whenever $X = 6$ and $E(Y \mid X) = 36/7$ whenever $X = -4$.

3.5.4
(a) $E(X \mid Y = 2) = \sum_x x\, P(X = x \mid Y = 2) = \sum_x x\, P(X = x,\ Y = 2)/P(Y = 2) = (-4)((1/11)/(2/11)) + (6)((1/11)/(2/11)) = 1$.
(b) $E(X \mid Y = 3) = \sum_x x\, P(X = x,\ Y = 3)/P(Y = 3) = (-4)((2/11)/(3/11)) + (6)((1/11)/(3/11)) = -2/3$.
(c) $E(X \mid Y = 7) = \sum_x x\, P(X = x,\ Y = 7)/P(Y = 7) = (-4)((4/11)/(5/11)) + (6)((1/11)/(5/11)) = -2$.
(d) $E(X \mid Y = 13) = \sum_x x\, P(X = x,\ Y = 13)/P(Y = 13) = (6)((1/11)/(1/11)) = 6$.
(e) $E(X \mid Y) = 1$ whenever $Y = 2$; $E(X \mid Y) = -2/3$ whenever $Y = 3$, $E(X \mid Y) = -2$ whenever $Y = 7$, and $E(X \mid Y) = 6$ whenever $Y = 13$.

3.5.5 We have that $E(\text{earnings} \mid Y =$ "takes course"$) = \$(1000(.1) + 2000(.3) + 3000(.4) + 4000(.2)) = \$2700$, while $E(X \mid Y =$ "doesn't take course"$) = \$\,(1000(.3) + 2000(.4) + 3000(.2) + 4000(.1)) = \$2100$. Therefore, by TTE we have that $E(\text{earnings}) = \$\,(2700\,(.4) + 2100\,(.6)) = \$2340$.

Problems

3.5.6 We have that $E(Y \mid X)$ is given by $E(Y \mid X = 1) = 1(.3) + 2(.4) + 3(.3) = 2.0$, $E(Y \mid X = 0) = 1(.2) + 2(.5) + 3(.3) = 2.1$.

So $E(Y) = E(E(Y \mid X)) = 2\,(.75) + 2.1(.25) = 2.025$ and, of course, $E(X) = .75$.

The conditional distributions of X given Y are (using Baye's theorem):
$X \mid Y = 1 \sim \text{Bernoulli}(.3\,(.75)\,/\,(.3\,(.75) + .2(.25))) = \text{Bernoulli}\,(0.81818)$,
$X \mid Y = 2 \sim \text{Bernoulli}(.4\,(.75)\,/\,(.4\,(.75) + .5(.25))) = \text{Bernoulli}(0.70588)$, and
$X \mid Y = 3 \sim \text{Bernoulli}(.3\,(.75)\,/\,(.3\,(.75) + .3(.25))) = \text{Bernoulli}(0.75)$.

Therefore, $E(X \mid Y)$ is given by $E(X \mid Y = 1) = 0.81818, E(X \mid Y = 2) = 0.70588$ and $E(X \mid Y = 3) = .75$.

3.5.7 We have that $Y = X_1 + \cdots + X_5 \sim$ Negative Binomial$(5, \theta)$, and by symmetry, each of the conditional distributions, X_i given $Y = 10$, are the same. Then $E(X_1 \mid Y = 10) = E(Y - X_2 - \cdots - X_5 \mid Y = 10) = 10 - 4E(X_1 \mid Y = 10)$, so $5E(X_1 \mid Y = 10) = 10$ and $E(X_1 \mid Y = 10) = 2$. Note that this does not depend on θ.

3.5.8
(a) $E(Y \mid X)$ is given by $E(Y \mid X = 1) = .97$ and $E(Y \mid X = 2) = .98$.
(b) $E(Y \mid X, Z)$ is given by $E(Y \mid X = 1, Z = 0) = .99, E(Y \mid X = 2, Z = 0) = .987, E(Y \mid X = 1, Z = 1) = .962$, and $E(Y \mid X = 2, Z = 1) = .960$.
(c) The conditional expectations all correspond to the conditional probabilities of having a successful treatment, so the higher this probability is the better. The conditional expectations $E(Y \mid X)$ indicate that hospital 2 is better than hospital 1, while the conditional expectations $E(Y \mid X, Z)$ uniformly indicate that hospital 1 is better than hospital 2.
(d) We have that

$$\sum_z p_{Y|X,Z}(y \mid x, z)\, p_{Z|X}(z \mid x) = \sum_z \frac{p_{X,Y,Z}(x, y, z)}{p_{X,Z}(x, z)} \frac{p_{X,Z}(x, z)}{p_X(x)}$$
$$= \sum_z \frac{p_{X,Y,Z}(x, y, z)}{p_X(x)} = \frac{p_{X,Y}(x, y)}{p_X(x)} = p_{Y|X}(y \mid x)$$

and

$$E(E(Y \mid X, Z) \mid X) = \sum_z E(Y \mid X, Z) p_Z(z)$$
$$= \sum_z \sum_y y p_{Y|X,Z}(y \mid x, z)\, p_{Z|X}(z \mid x)$$
$$= \sum_z \sum_y y \frac{p_{X,Y,Z}(x, y, z)}{p_X(x)} = \sum_y \sum_z y \frac{p_{X,Y,Z}(x, y, z)}{p_X(x)}$$
$$= \sum_y y \frac{p_{X,Y}(x, y)}{p_X(x)} = \sum_y y p_{Y|X}(y \mid x) = E(Y \mid X).$$

(e) We have that $E(Y \mid X = 1, Z = 0) = .99, E(Y \mid X = 2, Z = 0) = .987$, $E(Y \mid X = 1, Z = 1) = .962$, and $E(Y \mid X = 2, Z = 1) = .960$.

$$\begin{aligned}
&E(Y \mid X = 1) \\
&= E(Y \mid X = 1, Z = 0) p_{Z|X}(0 \mid 1) + E(Y \mid X = 1, Z = 1) p_{Z|X}(1 \mid 1) \\
&= .99(.286) + .962(.714) = .97 \\
&E(Y \mid X = 2) \\
&= E(Y \mid X = 2, Z = 0) p_{Z|X}(0 \mid 2) + E(Y \mid X = 2, Z = 1) p_{Z|X}(1 \mid 2) \\
&= .987(.75) + .960(.25) = .98,
\end{aligned}$$

so the result is verified numerically.

The paradox is resolved by noting that the conditional distributions of Z given X indicate that hospital 1 has a far greater proportion of seriously ill patients than does hospital 2.

3.5.9 Let $S = \{1, 2, 3\}$, $P(s) = 1/3$ and $X(s) = s$ for $s \in S$, and $A = \{1, 3\}$. Then $P(A) > 0$. Also, $E(X) = (1)(1/3)+(2)(1/3)+(3)(1/3) = 2$, and $E(X^2) = (1)^2(1/3) + (2)^2(1/3) + (3)^2(1/3) = 14/3$, so $\text{Var}(X) = (14/3) - (2)^2 = 2/3$. On the other hand, $E(X \,|\, A) = (1)(1/2) + (3)(1/2) = 2$, and $E(X^2 \,|\, A) = (1)^2(1/2) + (3)^2(1/2) = 5$, so $\text{Var}(X \,|\, A) = 5 - (2)^2 = 1 > 2/3$.

3.5.10 $E(X) = E(E(X \,|\, Y)) = E(\alpha/Y) = \alpha E(1/Y) = \alpha/\lambda$.

3.5.11 Using the analog of (2.7.1) we have that $X = \mu_1 + \sigma_1 Z_1, Y = \mu_2 + \sigma_2(\rho Z_1 + \sqrt{1-\rho^2}Z_2)$, where Z_1, Z_2 are i.i.d. $N(0, 1)$. Then $X = x$ is equivalent to $Z_1 = (x - \mu_1)/\sigma_1$ and Z_2 is independent of Z_1 (and so of X), so

$$\begin{aligned}
E(Y \,|\, X = x) &= E\left(\mu_2 + \sigma_2\left(\rho Z_1 + \sqrt{1-\rho^2}Z_2\right) \,|\, X = x\right) \\
&= E\left(\mu_2 + \sigma_2\left(\rho\left(\frac{X-\mu_1}{\sigma_1}\right) + \sqrt{1-\rho^2}Z_2\right) \,|\, X = x\right) \\
&= \mu_2 + \sigma_2\left(\rho\left(\frac{x-\mu_1}{\sigma_1}\right) + \sqrt{1-\rho^2}E(Z_2)\right) = \mu_2 + \rho\sigma_2\left(\frac{x-\mu_1}{\sigma_1}\right)
\end{aligned}$$

and

$$\begin{aligned}
\text{Var}(Y \,|\, X = x) &= \text{Var}(\sigma_2\sqrt{1-\rho^2}Z_2 \,|\, X = x) = \sigma_2^2\left(1-\rho^2\right)\text{Var}(Z_2 \,|\, X - x) \\
&= \sigma_2^2\left(1-\rho^2\right)\text{Var}(Z_2) = \sigma_2^2\left(1-\rho^2\right).
\end{aligned}$$

Using (2.7.1) we have that $E(X \,|\, Y = y) = \mu_1 + \rho\sigma_1(y - \mu_2)/\sigma_2$ and $\text{Var}(X \,|\, Y = y) = \sigma_1^2\left(1-\rho^2\right)$.

3.5.12 We have that $X_2 \sim \text{Binomial}(n, \theta_2)$, so the conditional probability function of X_1 given $X_2 = x_2$ is given by

$$\begin{aligned}
p_{X_1|X_2}(x_1 \,|\, x_2) &= \frac{\binom{n}{x_1\, x_2\, n-x_1-x_2}\theta_1^{x_1}\theta_2^{x_2}\left(1-\theta_1-\theta_2\right)^{n-x_1-x_2}}{\binom{n}{x_2}\theta_2^{x_2}\left(1-\theta_2\right)^{n-x_2}} \\
&= \frac{(n-x_2)!\theta_1^{x_1}\left(1-\theta_1-\theta_2\right)^{n-x_1-x_2}}{x_1!\,(n-x_1-x_2)!\left(1-\theta_2\right)^{n-x_2}} \\
&= \binom{n-x_2}{x_2}\left(\frac{\theta_1}{1-\theta_2}\right)^{x_1}\left(1-\frac{\theta_1}{1-\theta_2}\right)^{n-x_1-x_2}
\end{aligned}$$

and this is the Binomial$(n - x_2, \theta_1/(1-\theta_2))$ probability function. Therefore, $E(X_1 \,|\, X_2 = x_2) = (n - x_2)\,\theta_1/(1-\theta_2)$ and

$$\text{Var}(X_1 \,|\, X_2 = x_2) = (n - x_2)\frac{\theta_1}{1-\theta_2}\left(1 - \frac{\theta_1}{1-\theta_2}\right).$$

3.5.13 We have that the conditional density of X_1 given $X_2 = x_2$ is given by (using Problem 2.7.12)

$$f_{X_1|X_2}(x_1 \,|\, x_2) = \frac{\frac{\Gamma(\alpha_1+\alpha_2+\alpha_3)}{\Gamma(\alpha_1)\Gamma(\alpha_2)\Gamma(\alpha_3)} x_1^{\alpha_1-1} x_2^{\alpha_2-1} (1 - x_1 - x_2)^{\alpha_3-1}}{\frac{\Gamma(\alpha_1+\alpha_2+\alpha_3)}{\Gamma(\alpha_2)\Gamma(\alpha_1+\alpha_3)} x_2^{\alpha_2-1} (1 - x_2)^{\alpha_1+\alpha_3-1}}$$
$$= \frac{\Gamma(\alpha_1+\alpha_3)}{\Gamma(\alpha_1)\Gamma(\alpha_3)} \frac{x_1^{\alpha_1-1}(1 - x_1 - x_2)^{\alpha_3-1}}{(1 - x_2)^{\alpha_1+\alpha_3-1}}$$
$$= \frac{\Gamma(\alpha_1+\alpha_3)}{\Gamma(\alpha_1)\Gamma(\alpha_3)} \left(\frac{x_1}{1-x_2}\right)^{\alpha_1-1} \left(1 - \frac{x_1}{1-x_2}\right)^{\alpha_3-1} \frac{1}{1-x_2}$$

and we see that $X_1/(1-x_2)$ given $X_2 = x_2$ is distributed Beta(α_1, α_3), so (Problem 3.3.18) $E(X_1 \,|\, X_2 = x_2) = (1 - x_2)\,\alpha_1/(\alpha_1 + \alpha_3)$, $\text{Var}(X_1 \,|\, X_2 = x_2) = (1 - x_2)^2 \alpha_1\alpha_3/(\alpha_1+\alpha_3)^2(\alpha_1+\alpha_3+1)$.

3.5.14

(a) $E(X^2) = \int_{-\infty}^{\infty} x^2 f_X(x)\,dx = \int_0^4 x^2 [(9x + 1875x^5)/1280072]\,dx$ $= 1920072/160009$. Hence, $\text{Var}(X) = E(X^2) - E(X)^2 = (1920072/160009) - (3840168/1120063)^2 = 307320963528/1254541123969 \doteq 0.244967$.

(b)

$$E(E(X \,|\, Y)^2) = E(([8(7 + 768Y^4)] \,/\, [7(3 + 256Y^4)])^2)$$
$$= \int_{-\infty}^{\infty} (8(7 + 768y^4))^2/(7(3 + 256y^4))^2 f_Y(y)\,dy$$
$$= \int_0^5 (8(7 + 768y^4))^2/(7(3 + 256y^4))^2 ((18y + 1536y^5)/4000225)\,dy$$
$$\doteq 11.754808401$$

Hence,

$$\text{Var}(E(X \,|\, Y)) = E(E(X \,|\, Y)^2) - E(E(X \,|\, Y))^2 = E(E(X \,|\, Y)^2) - E(X)^2$$
$$\doteq 11.754808401 - (3.42852857)^2 \doteq 0.0000002196$$

which is extremely small.

(c) For $0 \le y \le 5$,

$$E(X^2 \,|\, Y = y) = \int_{-\infty}^{\infty} x^2 f_{X|Y}(x|y)\,dx = \int_{-\infty}^{\infty} x^2 (f_{X,Y}(x,y)/f_Y(y))\,dx$$
$$= \int_0^4 x^2 (9(xy + x^5y^5)/16000900)/((18y + 1536y^5)/4000225)\,dx$$
$$= (24 + 3072y^4)/(3 + 256y^4).$$

Hence, $E(X^2 \mid Y) = (24 + 3072Y^4)/(3 + 256Y^4)$, for $0 \le Y \le 5$. Then

$$\begin{aligned}\mathrm{Var}(X \mid Y) &= E(X^2 \mid Y) - E(X \mid Y)^2 \\ &= [(24 + 3072Y^4)/(3 + 256Y^4)] - [(8(7 + 768Y^4))/(7(3 + 256Y^4))]^2 \\ &= (8/49)(49 + 8064Y^4 + 98304Y^8)/(3 + 256Y^4)^2.\end{aligned}$$

(d) From part (c) we have that

$$\begin{aligned}&E(\mathrm{Var}(X \mid Y)) \\ &= \int_{-\infty}^{\infty} [(8/49)(49 + 8064y^4 + 98304y^8)/(3 + 256y^4)^2]\, f_Y(y)\, dy \\ &= \int_0^5 [(8/49)(49 + 8064y^4 + 98304y^8)/(3 + 256y^4)^2]\,[(18y + 1536y^5)/4000225]\, dy \\ &\doteq 0.244967.\end{aligned}$$

Then $\mathrm{Var}(E(X \mid Y)) + E(\mathrm{Var}(X \mid Y)) \doteq 0.0000002196 + 0.244967 \doteq 0.244967 \doteq \mathrm{Var}(X)$, as it should.

3.5.15 We have that $E(g(X)h(Y) \mid Z) = \sum_{x,y} g(x)h(y)p_{X,Y|Z}(x, y \mid z)$ $= \sum_{x,y} g(x)h(y)p_{X|Z}(x \mid z)\, p_{Y|Z}(y \mid z) = \sum_x g(x)p_{X|Z}(x \mid z) \sum_y h(y)p_{Y|Z}(y \mid z)$ $= E(g(X) \mid Z)\, E(h(Y) \mid Z)$.

3.6 Inequalities

Exercises

3.6.1 Since $Z \ge 0$, $P(Z \ge 7) \le E(Z)/7 = 3/7$.

3.6.2 Since $X \ge 0$, $P(X \ge 3) \le E(X)/3 = (1/5)/3 = 1/15$.

3.6.3
(a) Since $X \ge 0$, $P(X \ge 9) \le E(X)/9 = (1 - 1/2)/(1/2)/9 = 1/9$.
(b) Since $X \ge 0$, $P(X \ge 2) \le E(X)/2 = (1 - 1/2)/(1/2)/2 = 1/2$.
(c) Since $E(X) = (1 - 1/2)/(1/2) = 1$, $P(|X - 1| \ge 1) \le \mathrm{Var}(X)/1^2 = (1 - 1/2)/(1/2)^2/1^2 = 2$.
(d) The upper bound in (b) is smaller and more useful than that in (c).

3.6.4 Since $E(Z) = 5$, $P(|Z - 5| \ge 30) \le \mathrm{Var}(Z)/30^2 = 9/30^2 = 1/100$.

3.6.5 Since $E(W) = 50$, $P(|W - 50| \ge 10) \le \mathrm{Var}(W)/10^2 = 100(1/2)(1/2)/10^2 = 1/4$.

3.6.6 We have $\mathrm{Cov}(Y, Z) = \mathrm{Corr}(Y, Z)\sqrt{\mathrm{Var}(Y)\,\mathrm{Var}(Z)} = \mathrm{Corr}(Y, Z)\sqrt{(100)(80 \cdot 1/4 \cdot 3/4)} = \mathrm{Corr}(Y, Z)\sqrt{1500}$. This is largest when $\mathrm{Corr}(Y, Z) = +1$, where $\mathrm{Cov}(Y, Z) = \sqrt{1500} \doteq 38.73$. This is smallest when $\mathrm{Corr}(Y, Z) = -1$, where $\mathrm{Cov}(Y, Z) = -\sqrt{1500} \doteq -38.73$.

3.6.7
(a) By Jensen's inequality, $E(X^4) \geq E(X)^4 = [(1-1/11)/(1/11)]^4 = 10^4 = 10,000$.
(b) By Jensen's inequality, $E(X^4) \geq E(X^2)^2 = [(1-1/11)/(1/11)^2]^2 = (10 \cdot 11)^2 = 12,100$, which is larger and hence a better lower bound.

Problems

3.6.8 Here $X \sim \text{Binomial}(M, 1/2)$, so $E(X) = M/2$ and $\text{Var}(X) = M(1/2)(1-1/2) = M/$. Hence, by Chebychev's inequality, since $E(X/M) = 1/2$, $P(|(X/M)-(1/2)| \geq \delta) \leq \text{Var}(X/M)/\delta^2 = \text{Var}(X)/M^2\delta^2 = (M/4)/M^2\delta^2 = 1/4M\delta^2$. This is $\leq \epsilon$ provided $M \geq 1/4\delta^2\epsilon$.

3.6.9 Let $a = \sigma$ and let $P(X = \mu - a) = P(X = \mu + a) = 1/2$. Then $E(X) = \mu$, $\text{Var}(X) = \sigma^2$, and $P(|X - \mu| \geq a) = 1 = \sigma^2/a^2$.

3.6.10
(a) and (b)

$$E(X) = \sum_{i=1}^{n} x_i P(X = x_i) = \sum_{i=1}^{n} x_i \frac{1}{n} = \bar{x},$$

$$\text{Var}(X) = \sum_{i=1}^{n} (x_i - \bar{x})^2 P(X = x_i) = \sum_{i=1}^{n} (x_i - \bar{x})^2 \frac{1}{n} = \hat{s}_X^2,$$

$$\begin{aligned}\text{Cov}(X, Y) &= \sum_{i=1}^{n} (x_i - \bar{x})(y_i - \bar{y}) P(X = x_i, Y = y_i) \\ &= \sum_{i=1}^{n} (x_i - \bar{x})(y_i - \bar{y}) \frac{1}{n} = \hat{s}_{XY}\end{aligned}$$

Therefore r_{XY} is as stated.
(c) Let $x_1^*, \ldots, x_{n^*}^*$ be the distinct values in $x_1, \ldots, x_n$ and let f_i denote the frequency of x_i^* in $x_1, \ldots, x_n$. Then

$$E(X) = \sum_{i=1}^{n^*} x_i^* P(X = x_i) = \sum_{i=1}^{n^*} x_i^* \frac{f_i}{n} = \frac{1}{n}\sum_{i=1}^{n} x_i = \bar{x},$$

$$\text{Var}(X) = \sum_{i=1}^{n^*} (x_i^* - \bar{x})^2 P(X = x_i) = \sum_{i=1}^{n^*} (x_i^* - \bar{x})^2 \frac{f_i}{n} = \frac{1}{n}\sum_{i=1}^{n} (x_i - \bar{x})^2 = \hat{s}_X^2$$

and, similarly, all the other expectations remain the same.
(d) Since r_{XY} is a correlation coefficient we immediately have, from the correlation inequality, that $-1 \leq r_{XY} \leq 1$ and $r_{XY} = \pm 1$ if and only if $x_i - \bar{x} = \hat{s}_{XY}(y_i - \bar{y})/\hat{s}_X^2$ for $i = 1, \ldots, n$.

3.6.11 From Chebyshev's inequality we have that

$$P(X \notin (\bar{x} - 2\hat{s}, \bar{x} + 2\hat{s})) = P(|X - \bar{x}| \geq 2\hat{s}_X) \leq \frac{\hat{s}_X^2}{(2\hat{s}_X)^2} = \frac{1}{4},$$

so the largest possible proportion is 1/4.

3.6.12
(a) We have that $\int_1^\infty (2/x^3)\,dx = -x^{-2}\big|_1^\infty = 1$, so f_X is a density.
(b) $E(X) = \int_1^\infty x\,(2/x^3)\,dx = \int_1^\infty (2/x^2)\,dx = -2x^{-1}\big|_1^\infty = 2$.
(c) Markov's inequality says that $P(X \geq k) \leq E(X)/k = 2/k$, while the precise value is $P(X \geq k) = \int_k^\infty (2/x^3)\,dx = -x^{-2}\big|_k^\infty = 1/k^2$, and we see that the tail probability declines quadratically, while Markov's inequality only declines linearly.
(d) We have that $E(X^2) = \int_1^\infty x^2\,(2/x^3)\,dx = \int_1^\infty (2/x)\,dx = -2\ln x\big|_1^\infty = \infty$. Therefore, $\text{Var}(X) = \infty$ and Chebyshev's inequality does not provide a useful bound in this case.

3.6.13
(a) For $0 < \lambda < 1$ and $x < y$, $g(\lambda x + (1-\lambda)y) = \max(-\lambda x - (1-\lambda)y,\ -10)$. If $x, y \geq 10$, then $g(x) = g(y) = g(\lambda x + (1-\lambda)y) = -10$. If $x, y \leq 10$, then $g(x) = -x$, $g(y) = -y$, and $g(\lambda x + (1-\lambda)y) = -(\lambda x + (1-\lambda)y) = \lambda g(x) - (1-\lambda)g(y)$. Finally, if $x < 10 < y$, then $g(x) = x$ and $g(y) = -10$, so $\lambda g(x)+(1-\lambda)g(y) = \lambda(-x)-(1-\lambda)(-10) = \lambda(-x)+(1-\lambda)(-y)+(1-\lambda)(y-10)$, while $g(\lambda x+(1-\lambda)y) \leq \lambda(-x)+(1-\lambda)(-y)+(\lambda x+(1-\lambda)y-10) \leq \lambda(-x)+(1-\lambda)(-y)+(1-\lambda)(y-10)$.
(b) $E(g(Z)) = E(\max(-Z,-10)) \geq g(E(Z)) = g(1/5) = \max(-1/5,-10) = -1/5$.

3.6.14
(a) $f'(x) = px^{p-1}, f''(x) = p(p-1)x^{p-2} \geq 0$ for all $x \geq 0$ since $p \geq 1$. Therefore, f is convex on $(0,\infty)$.
(b) By Jensen's inequality we have that $E(f(X)) \geq f(E(X))$, so $E(|X|^p) \geq (|E(X)|)^p$ and $(E(|X|^p))^{1/p} \geq |E(X)|$.
(c) We have that $\text{Var}(X) = E(X^2) - (E(X))^2$ and $E(X^2) - (E(X))^2 = 0$ if and only if $E(X^2) = (E(X))^2$ and this true (by Jensen's inequality) if and only if $X^2 = a + bX$ for some constants a and b. The only way this can happen is if X is degenerate at a point, say $c, a = 0$ and $b = c$.

Challenge

3.6.15 If f and $-f$ are convex, then for all $x < y$ and $0 < \lambda < 1$, $f(\lambda x + (1-\lambda)y) \geq \lambda f(x) + (1-\lambda)f(y)$ and $-f(\lambda x + (1-\lambda)y) \geq -\lambda f(x) - (1-\lambda)f(y)$, so $f(\lambda x+(1-\lambda)y) \leq \lambda f(x)+(1-\lambda)f(y)$ and $f(\lambda x+(1-\lambda)y) = \lambda f(x)+(1-\lambda)f(y)$. Hence, the graph of f from x to y is a straight line, so f must be a *linear* function, i.e., $f(x) = ax + b$ for some a and b.

3.7 General Expectations

Exercises

3.7.1 $E(X_1) = 3$, $E(X_2) = 0$, and $E(Y) = (1/5)E(X_1) + (4/5)E(X_2) = 3/5$.

3.7.2 $E(X) = (1/6)(7/2) + (5/6)(9/2) = 13/3$.

3.7.3 Here $P(X < t) = 0$ for $t < 0$, while $P(X > t) = 1$ for $0 < t < C$ and $P(X > t) = 0$ for $t > C$. Hence, $E(X) = \int_0^\infty P(X > t)\,dt - \int_{-\infty}^0 P(X < t)\,dt = \int_0^C 1\,dt + \int_C^\infty 0\,dt - \int_{-\infty}^0 0\,dt = C + 0 - 0 = C$.

3.7.4 Here $P(Z > t) = 0$ for $t > 100$. Hence, $E(Z) = \int_0^\infty P(Z > t)\,dt - \int_{-\infty}^0 P(Z < t)\,dt \le \int_0^\infty P(Z > t)\,dt = \int_0^{100} P(Z > t)\,dt \le \int_0^{100} 1\,dt = 100$.

Challenge

3.7.5 If $X > t$, then since $Y \ge X$, we also have $Y \ge X > t$. Hence, $\{X > t\} \subseteq \{Y > t\}$, so, by monotonicity, $P(X > t) \le P(Y > t)$. Similarly, $P(X < t) \ge P(Y < t)$. Then $E(X) = \int_0^\infty P(X > t)\,dx - \int_{-\infty}^0 P(X < t)\,dx \le \int_0^\infty P(Y > t)\,dx - \int_{-\infty}^0 P(Y < t)\,dx = E(Y)$, as claimed.

Chapter 4

Sampling Distributions and Limits

4.1 Sampling Distributions

Exercises

4.1.1
$P(Y_3 = 1) = (1/2)(1/2)(1/2) = 1/8$
$P(Y_3 = 2) = (1/4)(1/4)(1/4) = 1/64$
$P(Y_3 = 3) = (1/4)(1/4)(1/4) = 1/64$
$P(Y_3 = 2^{1/3}) = (1/2)(1/2)(1/4) + (1/2)(1/4)(1/2) + (1/4)(1/2)(1/2) = 3/16$
$P(Y_3 = 3^{1/3}) = (1/2)(1/2)(1/4) + (1/2)(1/4)(1/2) + (1/4)(1/2)(1/2) = 3/16$
$P(Y_3 = 4^{1/3}) = (1/2)(1/4)(1/4) + (1/4)(1/2)(1/4) + (1/4)(1/4)(1/2) = 3/32$
$P(Y_3 = 9^{1/3}) = (1/2)(1/4)(1/4) + (1/4)(1/2)(1/4) + (1/4)(1/4)(1/2) = 3/32$
$P(Y_3 = 12^{1/3}) = (1/4)(1/4)(1/4) + (1/4)(1/4)(1/4) + (1/4)(1/4)(1/4) = 3/64$
$P(Y_3 = 18^{1/3}) = (1/4)(1/4)(1/4) + (1/4)(1/4)(1/4) + (1/4)(1/4)(1/4) = 3/64$
$P(Y_3 = 6^{1/3}) = (1/2)(1/4)(1/4) + (1/4)(1/2)(1/4) + (1/4)(1/4)(1/2) +$
$(1/2)(1/4)(1/4) + (1/4)(1/2)(1/4) + (1/4)(1/4)(1/2) = 3/16$

4.1.2 If Z is the sample mean, then $P(Z = 1) = 1/36$, $P(Z = 1.5) = 2/36$, $P(Z = 2) = 3/36$, $P(Z = 2.5) = 4/36$, $P(Z = 3) = 5/36$, $P(Z = 3.5) = 6/36$, $P(Z = 4) = 5/36$, $P(Z = 4.5) = 4/36$, $P(Z = 5) = 3/36$, $P(Z = 5.5) = 2/36$, and $P(Z = 6) = 1/36$.

4.1.3 If Z is the sample mean, then $P(Z = 0) = p^2$, $P(Z = 0.5) = 2p(1-p)$, and $P(Z = 1) = (1-p)^2$.

4.1.4 If Z is the sample mean, then

$$P(Z = 0) = \frac{N}{N+M}\frac{N-1}{N+M-1}, P(Z = 0.5) = 2\frac{N}{N+M}\frac{M}{N+M-1},$$
$$P(Z = 1) = \frac{M}{N+M}\frac{M-1}{N+M-1}.$$

4.1.5 For $1 \le j \le 6$, $P(\max = j) = (j/6)^{20} - ((j-1)/6)^{20}$.

Computer Exercises

4.1.6 Using Minitab we place the values 1, 2, 3 in C1 and .5, .25, .25 in C2. Then we replace the entries in C1 by their logs and generate 1000 samples of size 50 (stored in C3-C52), calculate the mean of each of these samples, and exponentiate this (stored in C54). The mean and standard deviation of the values in C54 is what we want. We get the following results.

```
MTB > let c1=log(c1)
MTB > Random 1000 c3-c52;
SUBC> Discrete c1 c2.
MTB > RMean c3 C4 C5 C6 C7 C8 C9 C10 C11 C12 C13 C14 C15
   C16 C17 C18 C19 &
CONT> C20 C21 C22 C23 C24 C25 C26 C27 C28 C29 C30 C31 C32
   C33 C34 C35 &
CONT> C36 C37 C38 C39 C40 C41 C42 C43 C44 C45 C46 C47 C48
   C49 C50 C51 &
CONT> C52 c53.
MTB > let c54=exp(c53)
MTB > let k1=mean(c54)
MTB > let k2=stdev(c54)
MTB > print k1 k2
Data Display
K1 1.57531
K2 0.103538
```

4.1.7 Using Minitab we get the following results.

```
MTB > Random 1000 c1-c10;
SUBC> Normal 0.0 1.0.
MTB > let c11=rmax(c1-c10)
MTB > let c11=rmax(c1 c2 c3 c4 c5 c6 c7 c8 c9 c10)
MTB > RMaximum C1 C2 C3 C4 C5 C6 C7 C8 C9 C10 c11.
MTB > let k1=mean(c11)
MTB > let k2=stdev(c11)
MTB > print k1 k2
Data Display
K1 1.54637
K2 0.617012
```

Problems

4.1.8 We know that $m_Y(s) = \left(m_{X_1}(s)\right)^n = \left(e^{\lambda(e^s-1)}\right)^n = e^{n\lambda(e^s-1)}$. We recognize this as the moment generating function of Poisson$(n\lambda)$. Hence, $Y \sim$ Poisson$(n\lambda)$.

4.1.9 The density of Y, for $0 \le y \le 2$, is given by $f_Y(y) = \int_{-\infty}^{\infty} f_{X_1}(t)\, f_{X_2}(y - t)\, dt = \int_{\max(y-1,0)}^{\min(y,1)} (1)(1)\, dt = \min(y,1) - \max(y-1,0)$, which is equal to y for $0 \le y \le 1$, and to $2-y$ for $1 \le y \le 2$. Otherwise, $f_Y(y) = 0$.

4.1.10 $\ln Y = (\ln X_1 + \ln X_2)/2$. Since $X_1 \sim$ Uniform$(0,1)$, then $-\ln X_1 \sim$ Exponential(1) = Gamma$(1,1)$. Hence, $W \equiv -\ln X_1 - \ln X_2 \sim$ Gamma$(2,1)$, so $f_W(w) = we^{-w}$ for $w > 0$ (otherwise 0). Then $Y = e^{-W/2} \equiv h(W)$ and $W = -2\ln Y \equiv h^{-1}(Y)$, so the density of Y satisfies

$$f_Y(y) = f_W(-2\ln y)/|h'(h^{-1}(y)| = (-2\ln y)e^{2\ln y}/|-\frac{1}{2}e^{-(-2\ln y)/2}|$$
$$= (-2\ln y)y^2\,/\,|-y/2| = -4y\ln y$$

for $0 \le y \le 1$ (otherwise 0).

4.2 Convergence in Probability

Exercises

4.2.1 Note that $Z_n = Z$ unless $7 \le U < 7 + 1/n^2$. Hence, for any $\epsilon > 0$, $P(|Z_n - Z| \ge \epsilon) \le P(7 \le U < 7 + 1/n^2) = 1/5n^2 \to 0$ as $n \to \infty$, so $Z_n \to Z$ in probability.

4.2.2 For any $\epsilon > 0$, $P(|X_n - 0| \ge \epsilon) = P(Y^n \ge \epsilon) = P(Y \ge \epsilon^{1/n}) = 1 - \epsilon^{1/n} \to 0$ as $n \to \infty$, so $X_n \to 0$ in probability.

4.2.3 By the Weak Law of Large Numbers, since $E(W_i) = 1/3$, $\lim_{n\to\infty} P(|\frac{1}{n}(W_1 + \cdots + W_n) - \frac{1}{3}| \ge 1/6) = 0$, so there is n with $P(|\frac{1}{n}(W_1 + \cdots + W_n) - \frac{1}{3}| \ge 1/6) < 0.001$. But then $P(W_1 + \cdots + W_n < n/2) = 1 - P(W_1 + \cdots + W_n \ge n/2) \ge 1 - P(|\frac{1}{n}(W_1 + \cdots + W_n) - \frac{1}{3}| \ge 1/6) \ge 1 - 0.001 = 0.999$.

4.2.4 By the Weak Law of Large Numbers, since $E(Y_i) = 2$, $\lim_{n\to\infty} P(|\frac{1}{n}(Y_1 + \ldots + Y_n) - 2| \ge 1) = 0$, so there is n with $P(|\frac{1}{n}(Y_1 + \cdots + Y_n) - 2| \ge 1) < 0.001$. But then $P(Y_1 + \cdots + Y_n > n) = 1 - P(Y_1 + \cdots + Y_n \le n) \ge 1 - P(|\frac{1}{n}(Y_1 + \cdots + Y_n) - 2| \ge 1) \ge 1 - 0.001 = 0.999$.

4.2.5 By the Weak Law of Large Numbers, since $E(X_i) = 8$, $\lim_{n\to\infty} P(|\frac{1}{n}(X_1 + \cdots + X_n) - 8| \ge 1) = 0$, so there is n with $P(|(X_1 + \cdots + X_n)/n - 8| \ge 1) < 0.001$. But then $P(X_1 + \cdots + X_n > 9n) \le P(|(X_1 + \cdots + X_n)/n - 8| \ge 1) \le 0.001$.

Computer Exercises

4.2.6 The following results were generated using Minitab (k1 holds the proportions).

```
MTB > Random 100000 c1-c20;
SUBC> Exponential .2.
MTB > RMean C1 C2 C3 C4 C5 C7 C6 C8 C9 C10 C11 C12 C13 C14 C15
   C16 C17 C18 &
CONT> C19 C20 c21.
MTB > let c22=c21 ge .19 and c21 le .21
MTB > let k1=mean(c22)
MTB > print k1
```

```
Data Display
K1 0.176500
Random 100000 c1-c50;
SUBC> Exponential .2.
MTB > RMean C1 C2 C3 C4 C5 C7 C6 C8 C9 C10 C11 C12 C13 C14 C15
   C16 C17 C18 &
CONT> C19 C20 C21 C22 C23 C24 C25 C26 C27 C28 C29 C30 C31 C32
   C33 C34 &
CONT> C35 C36 C37 C38 C39 C40 C41 C42 C43 C44 C45 C46 C47 C48
   C49 C50 &
CONT> c51.
MTB > let c52=c51 ge .19 and c51 le .21
MTB > let k1=mean(c52)
MTB > print k1
Data Display
K1 0.276850
```

We see that about 18% of the M_{20} values are between the limits, while about 28% of the M_{50} values are between the limits. This reflects the increasing concentration of the distributions of M_n as n increases.

4.2.7 The following results were generated using Minitab (k1 holds the proportions).

```
MTB > Random 100000 c1-c20;
SUBC> Poisson 7.
MTB > RMean C1 C2 C3 C4 C5 C7 C6 C8 C9 C10 C11 C12 C13 C14 C15
   C16 C17 C18 &
CONT> C19 C20 c21.
MTB > let c22=c21 ge 6.99 and c21 le 7.01
MTB > let k1=mean(c22)
MTB > print k1
Data Display
K1 0.0328400
MTB > Random 100000 C1-c100;
SUBC> Poisson 7.
MTB > RMean C1 C2 C3 C4 C5 C7 C6 C8 C9 C10 C11 C12 C13 C14 C15
   C16 C17 C18 &
CONT> C19 C20 C21 C22 C23 C24 C25 C26 C27 C28 C29 C30 C31 C32
   C33 C34 &
CONT> C35 C36 C37 C38 C39 C40 C41 C42 C43 C44 C45 C46 C47 C48
   C49 C50 &
CONT> C51 C52 C53 C54 C55 C56 C57 C58 C59 C60 C61 C62 C63 C64
   C65 C66 &
CONT> C67 C68 C69 C70 C71 C72 C73 C74 C75 C76 C77 C78 C79 C80
   C81 C82 &
CONT> C83 C84 C85 C86 C87 C88 C89 C90 C91 C92 C93 C94 C95 C96
   C97 C98 &
```

```
CONT> C99 C100 c101.
MTB > let c102=c101 ge 6.99 and c101 le 7.01
MTB > let k1=mean(c102)
MTB > print k1
Data Display
K1 0.0463600
```

We see that about 3.2% of the M_{20} values are between the limits, while about 4.6% of the M_{50} values are between the limits. This reflects the increasing concentration of the distributions of M_n as n increases, although it is not highly concentrated yet.

Problems

4.2.8 Let $P(X_n = n) = 1/n$ and $P(X_n = 0) = 1-1/n$. Then $E(X_n) = n(1/n)+0(1-1/n) = 1$. But for any $\epsilon > 0$, $P(|X_n - 0| \geq \epsilon) \leq P(X_n = n) = 1/n \to 0$ as $n \to \infty$, so $X_n \to 0$ in probability.

4.2.9 $|X_n| \xrightarrow{P} 0$ if and only if, for any $\epsilon > 0$, $P(||X_n| - 0| \geq \epsilon) \to 0$ as $n \to \infty$. But $P(||X_n| - 0| \geq \epsilon) = P(||X_n|| \geq \epsilon) = P(|X_n| \geq \epsilon) = P(|X_n - 0| \geq \epsilon)$, so this holds if and only if $X_n \xrightarrow{P} 0$.

4.2.10 This is false. For example, suppose $X_n = -5$ for all n. Then for $0 < \epsilon < 10$, $P(|X_n - 5| \geq \epsilon) = 1 \not\to 0$ as $n \to \infty$, so $X_n \not\to 5$ in probability. On the other hand, $|X_n| = 5$ for all n, so of course $|X_n| \to 5$ in probability.

4.3 Convergence with Probability 1

Exercises

4.3.1 Note that $Z_n = Z$ unless $7 \leq U < 7 + 1/n^2$. Hence, if $U < 7$ then $Z_n = Z$ for all n, so of course $Z_n \to Z$. Also, if $U > 7$, then $Z_n = Z$ whenever $1/n^2 < 7 - U$, i.e., $n > 1/\sqrt{7-U}$, so again $Z_n \to Z$. Hence, $P(Z_n \to Z) \geq P(U \neq 7) = 1 - P(U = 7) = 1 - 0 = 1$, i.e., $Z_n \to Z$ with probability 1.

4.3.2 If $0 \leq y < 1$ then $\lim_{n\to\infty} y^n = 0$. Hence, $P(X_n \to 0) = P(Y^n \to 0) \geq P(0 \leq Y < 1) = 1$, i.e., $Y_n \to 0$ with probability 1.

4.3.3 Since $E(W_i) = 1/3$, by the strong law of large numbers, $P((W_1 + \ldots + W_n)/n \to 1/3) = 1$. But $\{(W_1 + \cdots + W_n)/n \to 1/3\} \subseteq \{\exists n;\ (W_1 + \cdots + W_n)/n < 1/2\} = \{\exists n;\ W_1 + \cdots + W_n < n/2\}$, so also $P(\exists n;\ W_1 + \cdots + W_n < n/2) = 1$.

4.3.4 Since $E(Y_i) = 2$, by the strong law of large numbers, $P(\frac{1}{n}(Y_1 + \ldots + Y_n) \to 2) = 1$. But $\{(Y_1 + \cdots + Y_n)/n \to 2\} \subseteq \{\exists n;\ (Y_1 + \cdots + Y_n)/n > 1\} = \{\exists n;\ Y_1 + \cdots + Y_n > n\}$, so also $P(\exists n;\ Y_1 + \cdots + Y_n > n) = 1$.

4.3.5 By subadditivity, $P(X_n \to X \text{ and } Y_n \to Y) = 1 - P(X_n \not\to X \text{ or } Y_n \not\to Y) \geq 1 - P(X_n \not\to X) - P(Y_n \not\to Y) = 1 - 0 - 0 = 1$.

4.3.6
(a) False, e.g., if Z_i are continuous, then $P(M_n = a) = 0$ for any a.
(b) True, by the strong law of large numbers.
(c) True, by the strong law of large numbers (the given property is implied by the fact that $\lim_{n\to\infty} A_n = m$).
(d) False, e.g., if $x < m - 0.02$ and $M_n = m$ for all n, then this will not occur.

Computer Exercises

4.3.7 The sample means are converging to $1/5 = .2$.

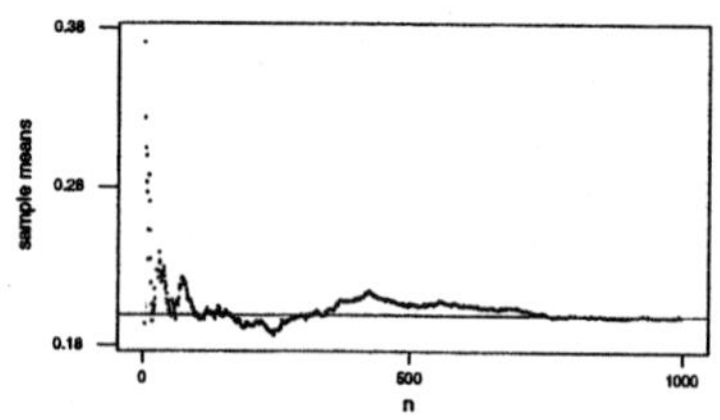

4.3.8 The sample means are converging (slowly) to 7.

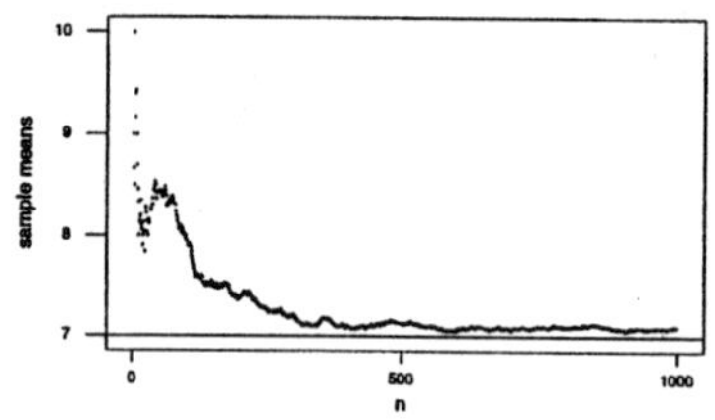

4.3.9 The sample means are converging (slowly) to -4.

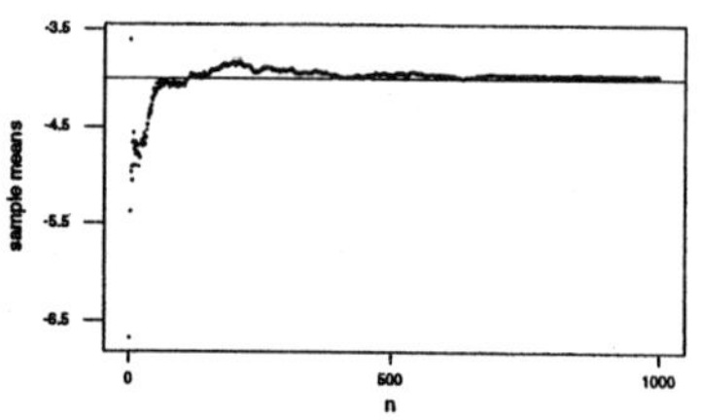

Problems

4.3.10 By countable subadditivity, $P(\lim_{n\to\infty} X_{n,k} = W_k \text{ for all } k) = 1 - P(\exists k;\ \lim_{n\to\infty} X_{n,k} \neq W_k) \geq 1 - \sum_k P(\lim_{n\to\infty} X_{n,k} \neq W_k) = 1 - \sum_k 0 = 1$.

4.3.11 Note that $x_n \to 0$ if and only if for all $\epsilon > 0$, $|x_n - 0| < \epsilon$ for all but finitely many n. But "$|x_n - 0| < \epsilon$" is the same as "$|x_n| < \epsilon$" is the same as "$||x_n| - 0| < \epsilon$." Hence, $x_n \to 0$ if and only if $|x_n| \to 0$. Thus, $P(X_n \to 0) =$

$P(|X_n| \to 0)$. Therefore, $X_n \to 0$ with probability 1 if and only if $|X_n| \to 0$ with probability 1.

4.3.12 This is false. For example, if $X_n = -5$ for all n, then $P(X_n \to 5) = 0$ but $P(|X_n| \to 5) = 1$.

Challenges

4.3.13 This is false. For example, let $U \sim$ Uniform$[0,1]$. Let $W_r = 0$ for all r and let $X_r = 0$ if $U \not r$, with $X_r = 1$ if $U = r$. Then $P(X_r \to 0) = P(U \neq r) = 1$, but $P(X_r \to 0 \text{ for all } r) = P(U \notin [0,1]) = 0$.

4.3.14 Let $S = \{1, 2, 3, \ldots\}$, with $P(s) = 2^{-s}$. Let $X_n(s) = 2^n$ for $s = n$, otherwise $X_n(s) = 0$. Then $E(X_n) = (2^n)(2^{-n}) = 1$. However, $X_n(s) = 0$ whenever $n > s$, so $P(X_n \to 0) = 1$.

4.4 Convergence in Distribution

Exercises

4.4.1 Here $\lim_{n\to\infty} P(X_n = i) = 1/3 = P(X = i)$ for $i = 1, 2, 3$, so $\lim_{n\to\infty} P(X_n \leq x) = P(X \leq x)$ for all x, so $X_n \to X$ in distribution.

4.4.2 We have that $\lim_{n\to\infty} P(Y_n = k) = 1/2^{k+1}$ for $k = 0, 1, \ldots$ so $\lim_{n\to\infty} P(Y_n \leq y) = P(Y \leq y)$.

4.4.3 Here $P(Z_n \leq 1) = 1$, and for $0 \leq z < 1$, $P(Z_n \leq z) = \int_0^z (n+1)x^n dx = z^{n+1} \to 0$ as $n \to \infty$. Also, $P(Z \leq z) = 1$ for $z \geq 1$, and 0 for $z < 1$. Hence, $\lim_{n\to\infty} P(Z_n \leq z) = P(Z \leq z)$ for all z, so $Z_n \to Z$ in distribution.

4.4.4 For $0 < w < 1$, $P(W_n \leq w) = \int_0^w (1 + x/n)/(1 + 1/2n) dx = (w + w^2/2n)/(1+1/2n) \to w$ as $n \to \infty$. Also. $P(W \leq w) = w$. Hence, $\lim_{n\to\infty} P(W_n \leq w) = P(W \leq w)$ for all w, so $W_n \to W$ in distribution.

4.4.5 Let $S = Y_1 + Y_2 + \ldots + Y_{1600}$. Then S has mean $1600/3$ and variance $1600/9$. Hence,

$$\begin{aligned}
&P(S \leq 540) = P((S - 1600/3)/\sqrt{1600/9} \leq (540 - 1600/3)/\sqrt{1600/9}) \\
&\approx \Phi((540 - 1600/3)/(40/3)) = \Phi(1/2) = 1 - \Phi(-1/2) = 1 - \Phi(-0.5) \\
&\approx 1 - 0.3085 = 0.6915.
\end{aligned}$$

4.4.6 Let $S = Z_1 + Z_2 + \ldots + Z_{900}$. Then S has mean $900(-5) = -4500$ and variance $900(30^2/12)$. Hence,

$$\begin{aligned}
&P(S \geq -4470) = P((S - (-4500))/\sqrt{900(30^2/12)} \\
&\geq (-4470 - (-4500))/\sqrt{900(30^2/12)}) \\
&\approx 1 - \Phi(-4470 - (-4500))/\sqrt{900(30^2/12)}) \\
&\approx 1 - \Phi(0.11547) = \Phi(-0.11547) \approx \Phi(-0.12) \approx 0.4522.
\end{aligned}$$

4.4.7 Let $S = X_1+X_2+\ldots+X_{800}$. Then S has mean $800(1-1/4)/(1/4) = 2400$ and variance $800(1-1/4)/(1/4)^2 = 9600$. Hence,

$$P(S \geq 2450) = P((S-2400)/\sqrt{9600} \geq (2450-2400)/\sqrt{9600})$$
$$\approx 1 - \Phi(2450-2400)/\sqrt{9600}) \approx 1 - \Phi(0.51) = \Phi(-0.51) \approx 0.3050.$$

4.4.8 Yes, $\{X_n\}$ converges in distribution to 0, since for $x < 0$, $P(X_n \leq x) = \Phi(\sqrt{n}\,x) \to 0$, while for $x > 0$, $P(X_n \leq x) = \Phi(\sqrt{n}\,x) \to 1$.

Computer Exercises

4.4.9 Using Minitab we obtain the following.

```
MTB > Random 10000 c1-c20;
SUBC> Exponential .333333333.
MTB > RMean C1 C2 C3 C4 C5 C6 C7 C8 C9 C10 C11 C12 C13 C14 C15
   C16 C17 C18 &
CONT> C19 C20 c21.
MTB > let k2=1/6
MTB > let k3=1/2
MTB > let c22 = c21 ge k2 and c21 le k3
MTB > let k1=mean(c22)
MTB > print k1
Data Display
K1 0.974600
```

And so we record 97.4% of the averages between 1/6 and 1/2. The central limit theorem gives the approximation

$$P(1/6 \leq M_{20} \leq 1/2)$$
$$= P\left(\sqrt{20}\frac{(1/6-1/3)}{1/3} \leq \sqrt{20}\frac{(M_{20}-1/3)}{1/3} \leq \sqrt{20}\frac{(1/2-1/3)}{1/3}\right)$$
$$= P(-2.236\,1 \leq \sqrt{20}\frac{(M_{20}-1/3)}{1/3} \leq 2.236\,1) \approx P(-2.236\,1 \leq Z \leq 2.236\,1)$$
$$= \Phi\,(2.2361) - \Phi\,(-2.2361) = 0.9873 - 0.01270 = 0.9746$$

and this is close to the observed proportion.

4.4.10 Using Minitab we obtain the following.

```
MTB > Random 10000 c1-c30;
SUBC> Uniform -20 10.
MTB > RMean C1 C2 C3 C4 C5 C6 C7 C8 C9 C10 C11 C12 C13 C14 C15
   C16 C17 C18 &
CONT> C19 C20 C21 C22 C23 C24 C25 C26 C27 C28 C29 C30 c31.
MTB > let c32=c31 le -5
MTB > let k1=mean(c32)
MTB > print k1
Data Display
```

```
K1 0.500500
```

And so we record 50.0% of the averages less than −5. The central limit theorem gives the approximation

$$P(M_{30} \leq -5) = P(\sqrt{30}\frac{(M_{20}+5)}{30\sqrt{1/12}} \leq \sqrt{30}\frac{(-5+5)}{30\sqrt{1/12}})$$
$$= P(\sqrt{30}\frac{(M_{20}+5)}{30\sqrt{1/12}} \leq 0) \approx P(Z \leq 0) = \Phi(0) = .5$$

and this is close to the observed proportion.

4.4.11 Using Minitab we obtain the following.

```
MTB > Random 10000 c1-c20;
SUBC> Uniform 0 1.
MTB > Let c1 = CEIL(loge(c1)/loge(.75))
MTB > Let c2 = CEIL(loge(c2)/loge(.75))
MTB > Let c3 = CEIL(loge(c3)/loge(.75))
MTB > Let c4 = CEIL(loge(c4)/loge(.75))
MTB > Let c5 = CEIL(loge(c5)/loge(.75))
MTB > Let c6 = CEIL(loge(c6)/loge(.75))
MTB > Let c7 = CEIL(loge(c7)/loge(.75))
MTB > Let c8 = CEIL(loge(c8)/loge(.75))
MTB > Let c9 = CEIL(loge(c9)/loge(.75))
MTB > Let c10 = CEIL(loge(c10)/loge(.75))
MTB > Let c11 = CEIL(loge(c11)/loge(.75))
MTB > Let c12 = CEIL(loge(c12)/loge(.75))
MTB > Let c13 = CEIL(loge(c13)/loge(.75))
MTB > Let c14 = CEIL(loge(c14)/loge(.75))
MTB > Let c15 = CEIL(loge(c15)/loge(.75))
MTB > Let c16 = CEIL(loge(c16)/loge(.75))
MTB > Let c17 = CEIL(loge(c17)/loge(.75))
MTB > Let c18 = CEIL(loge(c18)/loge(.75))
MTB > Let c19 = CEIL(loge(c19)/loge(.75))
MTB > Let c20 = CEIL(loge(c20)/loge(.75))
MTB > RMean C1 C2 C3 C4 C5 C6 C7 C8 C9 C10 C11 C12 C13 C14 C15 C16
C17 C18 &
CONT> C19 C20 c21.
MTB > let c22= c21 ge 2.5 and c21 le 3.3
MTB > let k1=mean(c22)
MTB > print k1
Data Display
K1 0.183700
```

And so we record 18.4% of the averages between 2.5 and 3.3. The central limit theorem gives the approximation (mean of Geometric(1/4) is $(1-.25)/.25 = 3$

and variance is $(1-.25)/(.25)^2 = 12.0$

$$P(2.5 \le M_{20} \le 3.3) = P(\sqrt{20}\frac{(2.5-3)}{\sqrt{12}} \le \sqrt{20}\frac{(M_{20}-3)}{\sqrt{12}} \le \sqrt{20}\frac{(3.3-3)}{\sqrt{12}})$$
$$= P(-0.64550 \le \sqrt{20}\frac{(M_{20}-1/3)}{1/3} \le 0.38730) \approx P(-0.64550 \le Z \le 0.38730)$$
$$= \Phi(0.38730) - \Phi(-0.64550) = .6507 - .2593 = 0.3914$$

and this is not close to the observed proportion because the Geometric(1/4) is a skewed distribution.

4.4.12 Using Minitab we obtain the following. Note that the histogram looks a lot like a normal density.

```
MTB > Random 10000 c1-c20;
SUBC> Gamma 4 1.
MTB > Let c1 = loge(c1)
MTB > Let c2 = loge(c2)
MTB > Let c3 = loge(c3)
MTB > Let c4 = loge(c4)
MTB > Let c5 = loge(c5)
MTB > Let c6 = loge(c6)
MTB > Let c7 = loge(c7)
MTB > Let c8 = loge(c8)
MTB > Let c9 = loge(c9)
MTB > Let c10 = loge(c10)
MTB > Let c11 = loge(c11)
MTB > Let c12 = loge(c12)
MTB > Let c13 = loge(c13)
MTB > Let c14 = loge(c14)
MTB > Let c15 = loge(c15)
MTB > Let c16 = loge(c16)
MTB > Let c17 = loge(c17)
MTB > Let c18 = loge(c18)
MTB > Let c19 = loge(c19)
MTB > Let c20 = loge(c20)
MTB > RMean C1 C2 C3 C4 C5 C6 C7 C8 C9 C10 C11 C12 C13 C14 C15 C16
C17 C18 &
CONT> C19 C20 c21.
MTB > Histogram C21;
SUBC> Density;
SUBC> CutPoint;
SUBC> Bar;
SUBC> ScFrame;
SUBC> ScAnnotation.
```

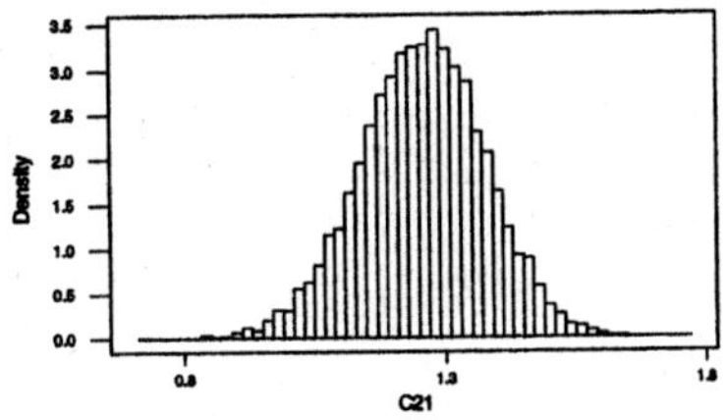

4.4.13 Using Minitab we obtain the following. Note that the histogram does not look a lot like a normal density. So a larger sample size is required for the CLT approximation to apply.

```
MTB > Random 10000 c1-c20;
SUBC> Binomial 10 .01.
MTB > RMean C1 C2 C3 C4 C5 C6 C7 C8 C9 C10 C11 C12 C13 C14 C15 C16
C17 C18 &
CONT> C19 C20 c21.
MTB > Histogram C21;
SUBC> Density;
SUBC> CutPoint;
SUBC> Bar;
SUBC> ScFrame;
SUBC> ScAnnotation.
```

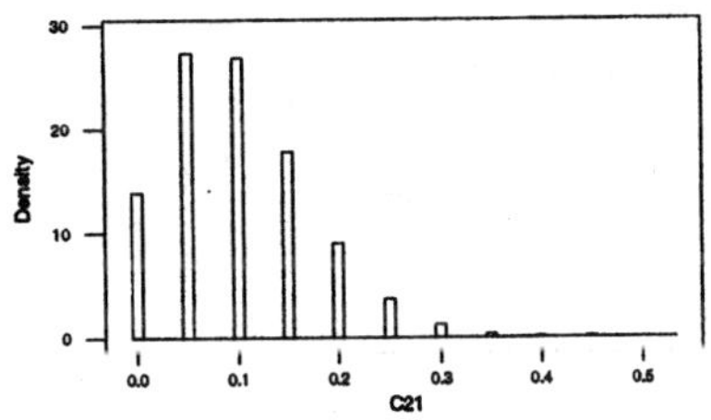

Problems

4.4.14 For example, let $Z_{j,n} \sim$ Normal$(j, 1/n)$, and let $P(Y = i) = a_i$ for positive integers i, with Y independent of the $\{Z_j\}$. Then let $X_n = Z_{Y,n}$, i.e., $X_n = Z_{j,n}$ whenever $Y = j$. Then X_n is absolutely continuous since each $Z_{j,n}$ is. Also, if $P(X = x) = 0$, then $P(X_n \le x) = \sum_i a_i P(Z_{i,n} \le x) \to \sum_{i<x} a_i$ as $n \to \infty$, so $X_n \to X$ in distribution.

4.4.15 Here $P(X_n \le x) = \{\sum_{i<nx} f(i/n)\} / \{\sum_{i=1}^{n} f(i/n)\}$. We recognize this as a *Riemann sum* (from Calculus) for the integral $\int_0^x f(x)\,dx$. Hence, since f is continuous, $P(X_n \le x) \to \int_0^x f(x)\,dx = P(X \le x)$, so $X_n \to X$ in distribution.

4.4.16 We have that $E\left(X^3\right) = 0$, so (putting $y = x^2/2, x = \sqrt{2y}, dx =$

$dy/\sqrt{2y}$)

$$\begin{aligned}\text{Var}\left(X^3\right) &= E\left(X^6\right) = \int_{-\infty}^{\infty} x^6 \frac{1}{\sqrt{2\pi}} e^{-x^2/2}\,dx = 2\int_0^{\infty} x^6 \frac{1}{\sqrt{2\pi}} e^{-x^2/2}\,dx \\ &= \frac{2}{\sqrt{2\pi}} \int_0^{\infty} \left(\sqrt{2y}\right)^5 e^{-y}\,dy = \frac{2^3}{\sqrt{\pi}} \int_0^{\infty} y^{5/2} e^{-y}\,dy = \frac{2^3}{\sqrt{\pi}} \Gamma\left(7/2\right) \\ &= \frac{2^3}{\sqrt{\pi}} \frac{5}{2}\frac{3}{2}\frac{1}{2} \Gamma\left(1/2\right) = 15.\end{aligned}$$

Therefore,

$$P(M_n \le m) = P(\frac{\sqrt{n}\,(M_n - 0)}{\sqrt{15}} \le \frac{\sqrt{n}\,(m-0)}{\sqrt{15}}) \approx P(Z \le \frac{\sqrt{n}\,(m-0)}{\sqrt{15}})$$

where $Z \sim N(0,1)$.

4.4.17 We have that

$$\begin{aligned}E\left(Y\right) &= \int_0^1 \cos\left(2\pi u\right)\,du = \left.\frac{\sin\left(2\pi u\right)}{2\pi}\right|_0^1 = 0 \\ \text{Var}(Y) &= E\left(Y^2\right) = \int_0^1 \cos^2\left(2\pi u\right)\,du \\ &= \int_0^1 \frac{1+\cos\left(4\pi u\right)}{2}\,du = \frac{1}{4} + \frac{1}{2}\left.\frac{\sin\left(4\pi u\right)}{4\pi}\right|_0^1 = \frac{1}{4}.\end{aligned}$$

Therefore,

$$P(M_n \le m) = P(\frac{\sqrt{n}\,(M_n - 0)}{\sqrt{1/4}} \le \frac{\sqrt{n}\,(m-0)}{\sqrt{1/4}}) \approx P(Z \le \frac{\sqrt{n}\,(m-0)}{\sqrt{1/4}})$$

where $Z \sim N(0,1)$.

Computer Problems

4.4.18 Using Minitab we obtain the following results.

```
MTB > Random 10000 c1;
SUBC> Normal 0.0 1.0.
MTB > let c2=c1**3
MTB > let c3=c2 le 1
MTB > let k1=mean(c3)
MTB > let k2=3*sqrt(k1*(1-k1)/10000)
MTB > let k3=k1-k2
MTB > let k4=k1+k2
MTB > print k1 k3 k4
Data Display
K1 0.837500
K3 0.826433
```

```
K4 0.848567
```

So our estimate of $P(Y \leq 1)$ is 0.837500, and the true value of this quantity lies in the interval $(0.826433, 0.848567)$ with virtual certainty. So we know the value of $P(Y \leq 1)$ with considerable accuracy. This probability can be evaluated exactly as $P(Y \leq 1) = P(X^3 \leq 1) = P(X \leq 1) = \Phi(1) = 0.8413$.

4.4.19 Using Minitab we obtain the following results.

```
MTB > Random 10000 c1;
SUBC> Normal 0.0 1.0.
MTB > let c2=cos(c1**3)
MTB > let k1=mean(c2)
MTB > let k2=stdev(c2)/sqrt(10000)
MTB > let k3=k1-3*k2
MTB > let k4=k1+3*k2
MTB > print k1 k3 k4
Data Display
K1 0.588037
K3 0.569203
K4 0.606872
```

So our estimate of $E(Y)$ is 0.588037 and the true value of this quantity lies in the interval $(0.569203, 0.606872)$ with virtual certainty.

Challenge

4.4.20 If $X_n \to C$ in distribution, then $P(X_n \leq x) \to 0$ for $x < C$, and $P(X_n \leq x) \to 1$ for $x > C$. Then for all $\epsilon > 0$, $P(|X_n - C| \geq \epsilon) = P(X_n \geq C+\epsilon) + P(X_n \leq C-\epsilon) = 1 - P(X_n < C+\epsilon) + P(X_n \leq C-\epsilon) \to 1-1+0 = 0$ as $n \to \infty$. Hence, $X_n \to C$ in probability.

4.5 Monte Carlo Approximations

Exercises

4.5.1 This integral equals $\sqrt{2\pi}E(\cos^2(Z))$, where $Z \sim N(0,1)$. Hence, let $\{U_i\}$ be i.i.d. $\sim$ Uniform[0, 1] for $1 \leq i \leq 2n$. Let $Z_i = 2\ln(1/U_{2i-1})\cos(2\pi U_{2i})$, so that $Z_i \sim N(0,1)$. Then let $I = (\sqrt{2\pi}/n)\sum_{i=1}^{n}\cos^2(Z_i)$. For large n, I is a good approximation to the integral.

4.5.2 Note that this sum equals $E(Z^6)$, where $Z \sim$ Bernoulli(2/3). Hence, let $\{U_{ij}\}$ be i.i.d. Uniform[0, 1] for $1 \leq i \leq n$ and $1 \leq j \leq m$. Let $B_{ij} = 1$ if $U_{ij} < 2/3$, otherwise $B_{ij} = 0$, so that $B_{ij} \sim$ Bernoulli(2/3). Let $Z_i = B_{i1} + B_{i2} + \ldots + B_{im}$, so that $Z_i \sim$ Binomial$(m, 2/3)$. Then let $S = (1/n)\sum_{i=1}^{n}(Z_i)^6$. For large n, S is a good approximation to the sum.

4.5.3 This integral equals $(1/5)E(e^{-14Z^2})$, where $Z \sim$ Exponential(5). Hence, let $\{U_i\}$ be i.i.d. $\sim$ Uniform[0, 1] for $1 \leq i \leq n$. Let $Z_i = \ln(1/U_i)/5$, so that $Z_i \sim$ Exponential(5). Then let $I = (1/5n)\sum_{i=1}^{n}e^{-14Z_i^2}$. For large n, I is a good approximation to the integral.

4.5.4 M_n has mean λ and variance λ/n. So the interval $M_n \pm 3\sqrt{\lambda/n}$ will contain the true value of λ with virtual certainty. But this implies that $M_n \pm 3\sqrt{10/n}$ will contain the true value of λ with virtual certainty. Therefore, the error criterion will be satisfied whenever $3\sqrt{10/n} \le .1$ or $n \ge 9\,(10)\,/(.1)^2 = 9000.0$.

4.5.5 This sum is approximately equal to $e^5 E(\sin(Z^2))$, where $Z \sim$ Poisson(5). Hence, let $Z_1, Z_2, \ldots, Z_n$ be i.i.d. with distribution Poisson(5) (perhaps generated using computer software). Then let $S = (e^5/n)\sum_{i=1}^n \sin(Z_i^2)$. For large n, S is a good approximation to the sum.

4.5.6 This integral is equal to $10E(e^{-Z^4})$, where $Z \sim$ Uniform$[0,10]$. Hence, let $\{U_i\}$ be i.i.d. $\sim$ Uniform$[0,1]$ for $1 \le i \le n$. Let $Z_i = 10U_i$, so that $Z_i \sim$ Uniform$[0,10]$. Then let $I = (10/n)\sum_{i=1}^n e^{-Z_i^4}$. For large n, I is a good approximation to the integral.

Computer Exercises

4.5.7 Using Minitab we obtain the following results.

```
MTB > Random 100000 c1;
SUBC> Uniform 0.0 1.0.
MTB > let c2=cos(c1**3)*sin(c1**4)
MTB > let k1=mean(c2)
MTB > let k2=stdev(c2)/sqrt(100000)
MTB > let k3=k1-3*k2
MTB > let k4=k1+3*k2
MTB > print k1 k3 k4
Data Display
K1 0.147770
K3 0.146163
K4 0.149378
```

So the estimate is 0.147770, and the true value of the integral lies in $(0.146163, 0.149378)$ with virtual certainty.

4.5.8 Using Minitab we obtain the following results.

```
MTB > Random 100000 c1;
SUBC> Exponential .25.
MTB > let c2=cos(c1**4)
MTB > let k1=mean(c2)
MTB > let k2=stdev(c2)/sqrt(100000)
MTB > let k3=k1-3*k2
MTB > let k4=k1+3*k2
MTB > print k1 k3 k4
Data Display
K1 0.973201
K3 0.971596
K4 0.974806
```

So the estimate is 0.973201, and the true value of the integral lies in $(0.971596, 0.974806)$ with virtual certainty.

4.5.9 Using Minitab we obtain the following results.

```
MTB > Random 100000 c1;
SUBC> Uniform 0.0 1.0.
MTB > let c2=floor(loge(c1)/loge(4/5))
MTB > let c2=(5/4)*(c2**2 +3)**(-5)
MTB > let k1=mean(c2)
MTB > let k2=stdev(c2)/sqrt(100000)
MTB > let k3=k1-3*k2
MTB > let k4=k1+3*k2
MTB > print k1 k3 k4
Data Display
K1 0.00124565
K3 0.00122658
K4 0.00126473
```

So the estimate is 0.00124565, and the true value of the integral lies in $(0.00122658, 0.00126473)$ with virtual certainty.

4.5.10 Using Minitab we obtain the following results.

```
MTB > Random 100000 c1;
SUBC> Normal 0.0 1.0.
MTB > let c2=c1**2-3*c1+2
MTB > let c3= c2 ge 0
MTB > let k1=mean(c3)
MTB > let k2=sqrt(k1*(1-k1))/sqrt(100000)
MTB > let k3=k1-3*k2
MTB > let k4=k1+3*k2
MTB > print k1 k3 k4
Data Display
K1 0.863930
K3 0.860677
K4 0.867183
```

So the estimate is 0.863930, and the true value of the integral lies in $(0.860677, 0.867183)$ with virtual certainty.

Problems

4.5.11 This requires that we determine n so that $3\sqrt{M_n(1-M_n)/n} \le \delta$. We have that $3\sqrt{M_n(1-M_n)/n} \le 3\sqrt{(1/2)(1-1/2)/n} \le \delta$ if and only if $n \ge 9/(4\delta^2)$.

4.5.12 This requires that we determine n so that $3\sigma_0/\sqrt{n} \le \delta$ or $n \ge 9\sigma_0^2/\delta^2$.

4.5.13

(a) Here for $0 \le z \le \theta$, $P(Z_n \le z) = (z/\theta)^n$. Hence, $X_{(n)}$ has density function $f(z) = nz^{n-1}\theta^{-n}$. Then

$$E(X_{(n)}) = \int_0^\theta znz^{n-1}\theta^{-n}\,dz = n\theta^{-n}\int_0^\theta z^n\,dz = \frac{n\theta^{-n}\theta^{n+1}}{n+1} = \frac{n\theta}{n+1}$$

and so $E(Z_n) = \theta$. Then

$$\begin{aligned}\operatorname{Var}(Z_n) &= E(Z_n^2) - \theta^2 = \left(\frac{n+1}{n}\right)^2 \int_0^\theta z^2 n z^{n-1} \theta^{-n}\, dz - \theta^2 \\ &= n\theta^{-n} \left(\frac{n+1}{n}\right)^2 \int_0^\theta z^{n+1}\, dz - \theta^2 = \frac{n}{n+2}\left(\frac{n+1}{n}\right)^2 \theta^{-n}\theta^{n+2} - \theta^2 \\ &= \left(\frac{(n+1)^2}{n\,(n+2)} - 1\right)\theta^2 = \frac{\theta^2}{n\,(n+2)}.\end{aligned}$$

(b) By Chebyshev's inequality we have that $P(|Z_n - \theta| \geq \epsilon) \leq \theta^2 / \left(\epsilon^2 n\,(n+2)\right) \to 0$ as $n \to \infty$.

(c) We have that $E(2M_n) = \theta$ and $\operatorname{Var}(2M_n) = 4\theta^2/(12n) = \theta^2/(3n)$. Now $n\,(n+2) \geq 3n$ for every n, so $\operatorname{Var}(2M_n) \geq \operatorname{Var}(Z_n)$. This implies that the estimator Z_n will be more accurate as the intervals given by the estimator plus/minus three standard deviations will be shorter.

4.5.14

(a) When $X \sim f$ we have that $E\left(\frac{g(X)}{f(X)}\right) = \int_a^b \frac{g(x)}{f(x)} f(x)\, dx = \int_a^b g(x)\, dx$, so $E(M_n(f)) = \int_a^b g(x)\, dx$.

(b) When $X \sim f$ then $E\left(\left(\frac{g(X)}{f(X)}\right)^2\right) = \int_a^b \frac{g^2(x)}{f^2(x)} f(x)\, dx = \int_a^b \frac{g^2(x)}{f(x)}\, dx$, so

$$\begin{aligned}\operatorname{Var}(M_n(f)) &= \frac{1}{n}\left\{E\left(\left(\frac{g(X)}{f(X)}\right)^2\right) - \left(E\left(\frac{g(X)}{f(X)}\right)\right)^2\right\} \\ &= \frac{1}{n}\left\{\int_a^b \frac{g^2(x)}{f(x)}\, dx - \left(\int_a^b g(x)\, dx\right)^2\right\}\end{aligned}$$

(d) Put $q(x) = |g(x)| / \int_a^b |g(x)|\, dx$. We have that

$$\begin{aligned}\operatorname{Var}(M_n(f)) &= \frac{1}{n}\left\{\int_a^b \frac{g^2(x)}{f(x)}\, dx - \left(\int_a^b g(x)\, dx\right)^2\right\} \\ &= \frac{1}{n}\left(\int_a^b |g(x)|\, dx\right)^2 \left\{\int_a^b \frac{q^2(x)}{f(x)}\, dx - \frac{\left(\int_a^b g(x)\, dx\right)^2}{\left(\int_a^b |g(x)|\, dx\right)^2}\right\}\end{aligned}$$

and $\int_a^b \frac{q^2(x)}{f(x)}\, dx = \int_a^b \frac{(q(x)-f(x))^2}{f(x)}\, dx + 1$, and this is minimized by taking $f = q$ and the minimum variance is $\left(\int_a^b |g(x)|\, dx\right)^2 - \left(\int_a^b g(x)\, dx\right)^2$. This is 0 when g is nonnegative.

The optimum importance sampler is not feasible because it requires that we be able to compute $\int_a^b |g(x)|\, dx$, which is typically as least as hard to evaluate as the original integral.

(e) We have that $\int_a^b (g^2(x)/f(x))\,dx \le \int_a^b (|g(x)|\,cf(x)/f(x))\,dx =$
$c\int_a^b |g(x)|\,dx < \infty$.
(f) The standard error of $M_n(f)$ is given by

$$S = \left\{ \frac{1}{n-1} \left(\sum_{i=1}^{n} \frac{g^2(X_i)}{f^2(X_i)} - n \left(\frac{1}{n} \sum_{i=1}^{n} \frac{g(X_i)}{f(X_i)} \right)^2 \right) \right\}^{1/2}$$

divided by $\sqrt{n}$. The CLT then implies that the true value of the integral is in the interval $M_n(f) \pm 3S/\sqrt{n}$ with virtual certainty when n is large.

Computer Problems

4.5.15 Using Minitab we obtain the following results.

```
MTB > Random 100000 c1;
SUBC> Normal 1 2.
MTB > Random 100000 c2;
SUBC> Gamma 1 1.
MTB > let c1=c1**3
MTB > let c2=c2**3
MTB > let c3=c1+c2
MTB > let c4=c3 le 3
MTB > let k1=mean(c4)
MTB > let k2=sqrt(k1*(1-k1))/sqrt(100000)
MTB > let k3=k1-3*k2
MTB > let k4=k1+3*k2
MTB > print k1 k3 k4
Data Display
K1 0.451620
K3 0.446899
K4 0.456341
```

So the estimate is 0.451620, and the true value of the probability lies in (0.446899, 0.456341) with virtual certainty.

By Problem 4.5.11 we must have $n \ge 9/\left(4(.01)^2\right) = 22500.0$.

4.5.16 Using Minitab we obtain the following results.

```
MTB > Random 100000 c1;
SUBC> Normal 1 2.
MTB > Random 100000 c2;
SUBC> Gamma 1 1.
MTB > let c1=c1**3
MTB > let c2=c2**3
MTB > let c3=c1+c2
MTB > let k1=mean(c3)
MTB > let k2=stdev(c3)/sqrt(100000)
MTB > let k3=k1-3*k2
```

```
MTB > let k4=k1+3*k2
MTB > print k1 k3 k4
Data Display
K1 18.9665
K3 18.5143
K4 19.4186
```

So the estimate is 18.9665, and the true value of the expectation lies in $(18.5143, 19.4186)$ with virtual certainty.

4.5.17 Using Minitab we obtain the following results for the algorithm based on generating from the Exponential(5) distribution.

```
MTB > Random 100000 c1;
SUBC> Exponential .2.
MTB > let c2=(exp(-14*c1*c1))/4
MTB > let k1=mean(c2)
MTB > let k2=stdev(c2)/sqrt(100000)
MTB > print k1 k2
Data Display
K1 0.159487
K2 0.000274517
```

Using Minitab we obtain the following results for the algorithm based on generating from the $N(0, 1/7)$ distribution.

```
MTB > let k1=sqrt(1/7)
MTB > print k1
Data Display
K1 0.377964
MTB > let k2=k1*sqrt(2*3.1415926)
MTB > Random 100000 c1;
SUBC> Normal 0 .377964.
MTB > let c2=k2*exp(-5*c1)
MTB > let c3=c1>0
MTB > let c2=c2*c3
MTB > let k1=mean(c2)
MTB > let k2=stdev(c2)/sqrt(100000)
MTB > print k1 k2
Data Display
K1 0.165965
K2 0.000788127
```

Notice that while the estimates 0.159487 and 0.165965 are similar, the standard error for the Exponential(5) algorithm is 0.000274517 and the standard error for the $N(0, 1/7)$ algorithm is 0.000788127. So the Exponential(5) algorithm is substantially more accurate.

Challenge

4.5.18

(a) Let D and A be as in the hint, and let L be the distance between the lines.

Then $D \sim \text{Uniform}[0, L]$, and $A \sim \text{Uniform}[0, \pi]$. Also, the needle will touch the line just below it if and only if $L\sin(A) \geq D$. This happens with probability

$$(1/L)\int_0^L (1/\pi)\int_0^\pi I_{L\sin(A)\geq D}\, dA\, dD = (1/\pi)\int_0^\pi (1/L)\int_0^L I_{D\leq L\sin(A)}\, dD\, dA$$
$$= (1/\pi)\int_0^\pi (1/L)(L\sin(A))\, dA = (1/\pi)\int_0^\pi \sin(A)\, dA$$
$$= (1/\pi)[-\cos(\pi) + \cos(0)] = 2/\pi.$$

(b) Repeat the experiment a large number N of times. Let M be the number of times the needle is touching a line. Then by the strong law of large numbers, for large N, we should have $M/N \approx 2/\pi$, so that $\pi \approx 2N/M$. Hence, for large N, the quantity $2N/M$ is a good Monte Carlo estimate of π.

4.6 Normal Distribution Theory

Exercises

4.6.1
(a) $U \sim \text{Normal}(1(3) - 5(-8), 1^2(2^2) + 5^2(5^2)) = \text{Normal}(44, 629)$. $V \sim \text{Normal}(-6(3) + C(-8), 6^2(2^2) + C^2(5^2)) = \text{Normal}(-18 - 8C, 144 + 25C^2)$.
(b) $\text{Cov}(U, V) = (1)(-6)(2^2) + (-5)(C)(5^2) = -24 - 125C$. Hence, U and V are independent if and only if $\text{Cov}(U, V) = 0$ if and only if $C = -24/125$.

4.6.2
(a) $Z \sim \text{Normal}(4(3) - (1/3)(-7), 4^2(5) + (1/3)^2(2)) = \text{Normal}(43/3, 722/9)$.
(a) $\text{Cov}(X, Z) = 4\,\text{Var}(X) = 20$.

4.6.3 $C_1 = 1/\sqrt{5}$. $C_2 = -3$. $C_3 = 1/\sqrt{2}$. $C_4 = 7$. $C_5 = 2$.

4.6.4 Since $X \sim \chi^2(n)$, we can find $Z_1, \ldots, Z_n \sim N(0, 1)$, which are i.i.d., with $X = (Z_1)^2 + \ldots + (Z_n)^2$. Then $X + Y^2 = (Z_1)^2 + \ldots + (Z_n)^2 + Y^2 \sim \chi^2(n+1)$ since it is the sum of squares of $n + 1$ independent standard normal random variables.

4.6.5 Since $X \sim \chi^2(n)$ and $Y \sim \chi^2(m)$, we can find $Z_1, \ldots, Z_n, W_1, \ldots, W_m \sim N(0, 1)$, which are i.i.d., with $X = (Z_1)^2 + \ldots + (Z_n)^2$ and $Y = (W_1)^2 + \ldots + (W_n)^2$. Then $X + Y = (Z_1)^2 + \ldots + (Z_n)^2 + (W_1)^2 + \ldots + (W_m)^2 \sim \chi^2(n+m)$ since it is the sum of squares of $n + m$ independent standard normal random variables.

4.6.6 $C = (1/n) / (1/3n) = 3$.

4.6.7 $C = 1 / (1/\sqrt{n}) = \sqrt{n}$.

4.6.8 $C_1 = \sqrt{2/5}$. $C_2 = -3$. $C_3 = 1$. $C_4 = 7$. $C_5 = 1$. $C_6 = 1$.

4.6.9 $C_1 = 2/5$. $C_2 = -3$. $C_3 = 2$. $C_4 = 7$. $C_5 = 2$. $C_6 = 1$. $C_7 = 1$.

Problems

4.6.10

(a) Note that $P(X \le z) = P(X \ge -z) = P(-X \le z) = \Phi(z)$. Hence, $P(Z \le z) = P(XY \le z) = P(XY \le z,\ Y = 1) + P(XY \le z,\ Y = -1) = P(X \le z,\ Y = 1) + P(-X \le z,\ Y = -1) = P(X \le z)P(Y = 1) + P(-X \le z)P(Y = -1) = \Phi(z)P(Y = 1) + \Phi(z)P(Y = -1) = \Phi(z)$, so $Z \sim N(0, 1)$.

(b) $\mathrm{Cov}(X, Z) = E(XZ) = E(X(XY)) = E(X^2)E(Y) = (1)(0) = 0$.

(c) For example, $P(X < -10,\ Z < -10) = P(X < -10,\ Y = 1) = \Phi(-10)/2$, while $P(X < -10)P(Z < -10) = \Phi(-10)^2 \ne \Phi(-10)/2$, so X and Z and not independent.

(d) Here X and Z do not arise as linear combinations of the same collection of independent normal random variables.

4.6.11 We see that $f_Z(-z) = \Gamma((n+1)/2)(1+(-z)^2/n)^{-(n+1)/2}/\Gamma(n/2)\sqrt{\pi n} = \Gamma((n+1)/2)(1+z^2/n)^{-(n+1)/2}/\Gamma(n/2)\sqrt{\pi n} = f_Z(z)$. Then using the substitution $s = -t$, we have $P(Z < -x) = \int_{-\infty}^{-x} f_Z(t)\,dt = -\int_x^\infty f_Z(-s)\,(-ds) = \int_x^\infty f_Z(s)\,ds = P(Z > x)$.

4.6.12 If $X_n \sim F(n, 2n)$, then we can find $X_1, \ldots, X_{3n}$ i.i.d. $\sim N(0, 1)$, with $X_n = (((X_1)^2 + \ldots + (X_n)^2)/n)/(((X_{n+1})^2 + \ldots + (X_{3n})^2)/2n)$. But as $n \to \infty$, by the strong law of large numbers, since $E((X_i)^2) = 1$, $((X_1)^2 + \ldots + (X_n)^2)/n \to 1$ $((X_{n+1})^2 + \ldots + (X_{3n})^2)/2n \to 1$ with probability 1. Hence, $X_n \to 1/1 = 1$ with probability 1, and hence also in probability.

4.6.13 The Gamma($\alpha/2$, 1/2) distribution has density function $g(x) = (1/2)^{\alpha/2}x^{\alpha/2-1}e^{-x/2}/\Gamma(\alpha/2)$. By inspection, $g(z) = f(z)$, i.e., the $\chi^2(\alpha)$ distribution corresponds to the Gamma($\alpha/2$, 1/2) distribution and is thus a well-defined probability distribution on $(0, \infty)$.

4.6.14 Just replace n by α throughout in the proof of Theorem 4.6.7.

4.6.15 Just replace n by α throughout in the proof of Theorem 4.6.9.

4.6.16 When $\alpha > 1$, we have that

$$E(X) = \int_{-\infty}^{\infty} x\left(1+\frac{x^2}{\alpha}\right)^{-\frac{\alpha+1}{2}} dx = -\frac{2\alpha}{2(\alpha-1)}\left(1+\frac{x^2}{\alpha}\right)^{-\frac{\alpha-1}{2}}\Bigg|_{-\infty}^{\infty} = 0.$$

When $\alpha > 2$ we can write (using $X \sim t(\alpha)$ implies that $Y = X^2 \sim F(1,\alpha)$)

$$\begin{aligned}
E\left(X^2\right) &= \frac{\Gamma\left(\frac{1+\alpha}{2}\right)}{\Gamma\left(\frac{1}{2}\right)\Gamma\left(\frac{\alpha}{2}\right)}\alpha\int_0^\infty u^{\frac{1}{2}}\left(1+u\right)^{-\frac{\alpha+1}{2}}du \\
&= \frac{\Gamma\left(\frac{1+\alpha}{2}\right)}{\Gamma\left(\frac{1}{2}\right)\Gamma\left(\frac{\alpha}{2}\right)}\alpha\int_0^\infty u^{\frac{3}{2}-1}\left(1+u\right)^{-\frac{3+\alpha-2}{2}}du \\
&= \frac{\Gamma\left(\frac{1+\alpha}{2}\right)}{\Gamma\left(\frac{1}{2}\right)\Gamma\left(\frac{\alpha}{2}\right)}\alpha\int_0^\infty \left(\frac{3}{\alpha-2}v\right)^{\frac{3}{2}-1}\left(1+\frac{3}{\alpha-2}v\right)^{-\frac{3+\alpha-2}{2}}\frac{3}{\alpha-2}dv \\
&= \frac{\Gamma\left(\frac{1+\alpha}{2}\right)}{\Gamma\left(\frac{1}{2}\right)\Gamma\left(\frac{\alpha}{2}\right)}\frac{\Gamma\left(\frac{3}{2}\right)\Gamma\left(\frac{\alpha-2}{2}\right)}{\Gamma\left(\frac{1+\alpha}{2}\right)}\alpha = \frac{1}{2}\frac{1}{\alpha/2-1}\alpha = \frac{\alpha}{\alpha-2}.
\end{aligned}$$

4.6.17 Making the transformation $v = \left(\alpha\left(\beta-2\right)/\beta\left(\alpha+2\right)\right)u$, we have that

$$\begin{aligned}
E(X) &= \frac{\Gamma\left(\frac{\alpha+\beta}{2}\right)}{\Gamma\left(\frac{\alpha}{2}\right)\Gamma\left(\frac{\beta}{2}\right)}\frac{\beta}{\alpha}\int_0^\infty \left(\frac{\alpha}{\beta}u\right)^{\frac{\alpha+2}{2}-1}\left(1+\frac{\alpha}{\beta}u\right)^{-\frac{\alpha+\beta}{2}}\frac{\alpha}{\beta}du \\
&= \frac{\Gamma\left(\frac{\alpha+\beta}{2}\right)}{\Gamma\left(\frac{\alpha}{2}\right)\Gamma\left(\frac{\beta}{2}\right)}\frac{\beta}{\alpha}\int_0^\infty \left(\frac{\alpha+2}{\beta-2}u\right)^{\frac{\alpha+2}{2}-1}\left(1+\frac{\alpha+2}{\beta-2}u\right)^{-\frac{\alpha+\beta}{2}}\frac{\alpha+2}{\beta-2}du \\
&= \frac{\Gamma\left(\frac{\alpha+\beta}{2}\right)}{\Gamma\left(\frac{\alpha}{2}\right)\Gamma\left(\frac{\beta}{2}\right)}\frac{\beta}{\alpha}\frac{\Gamma\left(\frac{\alpha+2}{2}\right)\Gamma\left(\frac{\beta-2}{2}\right)}{\Gamma\left(\frac{\alpha+\beta}{2}\right)} = \frac{\beta}{\alpha}\frac{\alpha/2}{\left(\beta-2\right)/2} = \frac{\beta}{\beta-2},
\end{aligned}$$

$$\begin{aligned}
E(X^2) &= \frac{\Gamma\left(\frac{\alpha+\beta}{2}\right)}{\Gamma\left(\frac{\alpha}{2}\right)\Gamma\left(\frac{\beta}{2}\right)}\left(\frac{\beta}{\alpha}\right)^2\int_0^\infty \left(\frac{\alpha}{\beta}u\right)^{\frac{\alpha+4}{2}-1}\left(1+\frac{\alpha}{\beta}u\right)^{-\frac{\alpha+\beta}{2}}\frac{\alpha}{\beta}du \\
&= \frac{\Gamma\left(\frac{\alpha+\beta}{2}\right)}{\Gamma\left(\frac{\alpha}{2}\right)\Gamma\left(\frac{\beta}{2}\right)}\left(\frac{\beta}{\alpha}\right)^2\frac{\Gamma\left(\frac{\alpha+4}{2}\right)\Gamma\left(\frac{\beta-4}{2}\right)}{\Gamma\left(\frac{\alpha+\beta}{2}\right)} = \left(\frac{\beta}{\alpha}\right)^2\frac{(\alpha/2)\left((\alpha/2)+1\right)}{\left(\left(\beta-2\right)/2\right)\left(\left(\beta-4\right)/2\right)}.
\end{aligned}$$

Therefore,

$$\begin{aligned}
\operatorname{Var}(X) &= \left(\frac{\beta}{\alpha}\right)^2\frac{(\alpha)(\alpha+2)}{(\beta-2)(\beta-4)} - \frac{\beta^2}{(\beta-2)^2} \\
&= \frac{\beta^2}{\alpha(\beta-2)^2(\beta-4)}\left\{(\alpha+2)(\beta-2)-\alpha(\beta-4)\right\} = \frac{2\beta^2(\alpha+\beta-2)}{\alpha(\beta-2)^2(\beta-4)}.
\end{aligned}$$

Challenges

4.6.18 Suppose that $0 < \alpha \le 1$. We have that

$$\int_0^\infty x\left(1+\frac{x^2}{\alpha}\right)^{-\frac{\alpha+1}{2}}dx \ge \int_0^\infty x\left(1+\frac{x^2}{\alpha}\right)^{-1}dx = \frac{\alpha}{2}\ln\left(1+\frac{x^2}{\alpha}\right)\Bigg|_0^\infty = \infty$$

and, similarly, $\int_{-\infty}^{0} x\left(1+\frac{x^2}{\alpha}\right)^{-\frac{\alpha+1}{2}} dx = -\infty$, which implies that the mean doesn't exist.

Consider now the second moment. Since $X \sim t(\alpha)$ implies that $Y = X^2 \sim F(1, \alpha)$, we have that

$$E(X^2) = \frac{\Gamma\left(\frac{1+\alpha}{2}\right)}{\Gamma\left(\frac{1}{2}\right)\Gamma\left(\frac{\alpha}{2}\right)} \int_0^\infty \left(\frac{y}{\alpha}\right)^{\frac{1}{2}} \left(1+\frac{y}{\alpha}\right)^{-\frac{\alpha+1}{2}} dy$$
$$= \frac{\Gamma\left(\frac{1+\alpha}{2}\right)}{\Gamma\left(\frac{1}{2}\right)\Gamma\left(\frac{\alpha}{2}\right)} \alpha \int_0^\infty u^{\frac{1}{2}} (1+u)^{-\frac{\alpha+1}{2}} du.$$

Now if $0 < \alpha \leq 2$, we have that

$$\int_0^\infty u^{\frac{1}{2}} (1+u)^{-\frac{\alpha+1}{2}} du \geq \int_0^\infty u^{\frac{1}{2}} (1+u)^{-\frac{3}{2}} du.$$

Since $\lim_{u\to\infty} u^{\frac{1}{2}} (1+u)^{-\frac{1}{2}} = 1$, we have that $u^{\frac{1}{2}} (1+u)^{-\frac{1}{2}} \geq 1-\epsilon$ for a specified $\epsilon > 0$ whenever $u > c_\epsilon$. Therefore,

$$\int_0^\infty u^{\frac{1}{2}} (1+u)^{-\frac{3}{2}} du \geq \int_{c_\epsilon}^\infty u^{\frac{1}{2}} (1+u)^{-\frac{3}{2}} du$$
$$\geq (1-\epsilon) \int_{c_\epsilon}^\infty (1+u)^{-1} du = (1-\epsilon) \ln(1+u)\big|_{c_\epsilon}^\infty = \infty.$$

Obviously, the variance is undefined when the mean doesn't exist, as when $0 < \alpha \leq 1$, and the above shows that it is infinite when $1 < \alpha \leq 2$.

4.6.19 We use induction on n. For $n = 1$ both sides are 0, so the equation holds. Assume now that it holds for some value of n. We shall prove that it holds for $n+1$. Multiplying through by σ^2, it suffices to take $\sigma = 1$. Let $\bar{X}_m = (1/m)(X_1 + \ldots + X_m)$ for any m. Then $\bar{X}_{n+1} = (n\bar{X}_n + X_{n+1})/(n+1)$, so that $\bar{X}_{n+1} - \bar{X}_n = (X_{n+1} - \bar{X}_n)/(n+1)$. Hence, for $1 \leq i \leq n$, $(X_i - \bar{X}_{n+1})^2 = (X_i - \bar{X}_n + (\bar{X}_n - \bar{X}_{n+1}))^2 = (X_i - \bar{X}_n)^2 + (\bar{X}_n - \bar{X}_{n+1})^2 + 2(X_i - \bar{X}_n)(\bar{X}_n - \bar{X}_{n+1})$. Now, $\sum_{i=1}^n (X_i - \bar{X}_n)^2$ equals the right-hand side of (4.7.1) by the induction assumption. Also, $\sum_{i=1}^n (\bar{X}_n - \bar{X}_{n+1})^2 = n(\bar{X}_n - \bar{X}_{n+1})^2 = (n/(n+1)^2)(\bar{X}_n - X_{n+1})^2$. Also, $\sum_{i=1}^n (X_i - \bar{X}_n)(\bar{X}_n - \bar{X}_{n+1}) = (\bar{X}_n - \bar{X}_{n+1}) \sum_{i=1}^n (X_i - \bar{X}_n) = 0$ by definition of $\bar{X}_n$. Hence, $\sum_{i=1}^{n+1} (X_i - \bar{X}_n)^2$ equals the right-hand side of (4.7.1) plus $n(\bar{X}_n - X_{n+1})^2$ plus $(X_{n+1} - \bar{X}_{n+1})^2$. But $(n/(n+1)^2)(\bar{X}_n - X_{n+1})^2 + (X_{n+1} - \bar{X}_{n+1})^2 = (n/(n+1)^2)((1/n)(X_1 + \ldots + X_n) - X_{n+1})^2 + (X_{n+1} - (1/(n+1))(X_1 + \ldots + X_{n+1})^2 = (1/n(n+1)^2)((X_1 + \ldots + X_n) - nX_{n+1})^2 + (1/(n+1))^2(X_1 + \ldots + X_{n+1} - (n+1)X_{n+1})^2 = (1/n(n+1)^2)((X_1 + \ldots + X_n) - nX_{n+1})^2 + (1/(n+1))^2(X_1 + \ldots + X_n - nX_{n+1})^2 = ((1/n(n+1)^2) + (1/(n+1))^2)(X_1 + \ldots + X_n - nX_{n+1})^2 = (1/n(n+1))(X_1 + \ldots + X_n - nX_{n+1})^2 = ((X_1 + \ldots + X_n - nX_{n+1})/\sqrt{n(n+1)})^2$. Hence, $\sum_{i=1}^{n+1} (X_i - \bar{X}_n)^2$ equals the right-hand side of (4.7.1) plus $((X_1 + \ldots + X_n - nX_{n+1})/\sqrt{n(n+1)})^2$. The result follows by induction.

Chapter 5

Statistical Inference

5.1 Why Do We Need Statistics?

Exercises

5.1.1 The mean survival times for the control group and the treatment group are 93.2 days and 356.2 days respectively. As we can see, there is a big difference between the two means, which might suggest that the treatment is indeed effective, but we can't base our conclusions about the effectiveness of the treatment based only on these numbers. We have to consider sampling variability as well.

5.1.2 In the control group there are two unusual observations, namely, observations 11 and 30, and these tend to make the mean for this group much larger. In the treatment group there wouldn't appear to be any unusual observations.

5.1.3 For those who are still alive, their survival times will be longer than the recorded values, so these data values are incomplete.

5.1.4 We could construct a probability distribution based on the database of marks. For example, recording the proportion of students receiving marks greater than 80, etc. Then for a student randomly selected from the database, this proportion is the probability that the student will have a mark greater than 80.

5.1.5 We use the sample average $\bar{x} = -0.1375$. We base this on the weak law of large numbers because we know that $\bar{x}$ will be close to μ when n is large.

Problem

5.1.6 We note that patients who didn't receive transplants could have been much unhealthier than those that did. It is not clear what factors influenced which group a patient wound up in and these factors could have a profound impact on the survival times.

Computer Problem

5.1.7 A good method would be to generate a large sample (say $n = 1000$) from the $N(0,1)$ distribution, calculate the values of Y, and then record the empirical distribution function of Y. This allows us to estimate the probability $P(Y \in A)$ for any interval A. For example, the following Minitab commands do this and record the estimate .021 for $P(Y \in (1,2))$. As we know from the weak law of large numbers, the proportion of Y values in this interval will converge to this probability as $n \to \infty$.

```
MTB > random 1000 c1;
SUBC> normal 0 1.
MTB > let c2=c1**4+2*c1**3-3
MTB > let c3=c2>1 and c2<2
MTB > let k1=mean(c3)
MTB > print k1
Data Display
K1 0.0210000
```

Statistical methodology is relevant to determine if n is large enough to accurately estimate the probability and how accurate this estimate is.

5.2 Inference Using a Probability Model

Exercises

5.2.1 In Example 5.2.1 the lifelength in years of a machine was known to be $X \sim$ Exponential(1), so the mode is given by 0. In Example 5.2.2 the conditional density is given by $e^{-(x-1)}$ fro $x > 1$. The mode of this density is 1.

In both cases the mode is at the extreme left end of the distribution and so doesn't seem like a very good predictor.

5.2.2 Using the mean of a distribution to predict a future response, the mean squared error of this predictor is $E(X-1)^2 = Var(X) = 1$, where X is the future response and 1 is the mean of the distribution.

5.2.3 The density of the distribution obtained as a mixture of a $N(-4,1)$ and a $N(4,1)$ with mixture probability .5 has density given by $\frac{.5}{\sqrt{2\pi}} \exp\left\{-\frac{(x+4)^2}{2}\right\} + \frac{.5}{\sqrt{2\pi}} \exp\left\{-\frac{(x-4)^2}{2}\right\}$ for $-\infty < x < \infty$. This is plotted below.

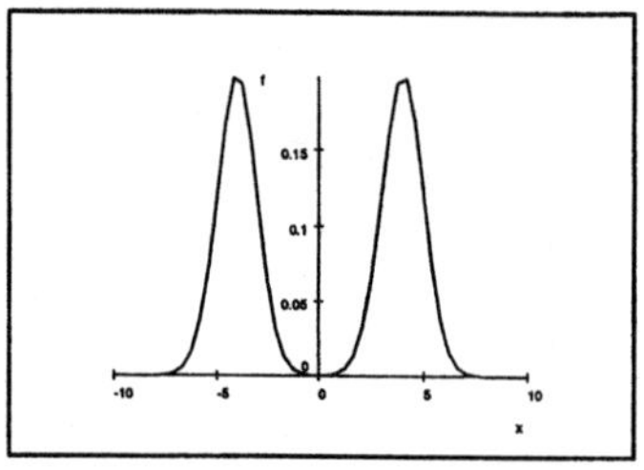

5.2.4 First, if $X \sim$Uniform(0,1), then the density of $Y = 10X$ is given by $f_Y(y) = 1/10$ for $0 \le y \le 10$, i.e., $Y \sim$ Uniform$(0, 10)$, so $E(Y) = 5$ years.

The smallest interval containing 95% of the probability for Y is an interval (a, b), where a and b satisfy $0 \le a \le b \le 10$ and $0.95 = \int_a^b \frac{1}{10} dy = \frac{1}{10}(b - a)$ or $b - a = 9.5$. We thus see that any subinterval of $(0, 10)$ of length 9.5 will work, e.g., $(.5, 10)$.

Next, if we want to assess whether or not $x_0 = 5$ is a plausible lifelength for a new machine, we need to compute the tail probability $P(Y \ge 5) = \int_5^{10} \frac{1}{10} dy = \frac{5}{10} = 0.5$, which in this case is quite high and therefore indicates that $x_0 = 5$ is a plausible lifelength for that new machine.

Now, the density of the conditional distribution of Y, given $Y > 1$, is given by $f_Y(y \,|\, Y > 1) = \frac{1}{9}$ for $1 \le y \le 10$. So the predicted lifelength is now $E(Y \,|\, Y > 1) = \int_1^{10} \frac{y}{9} dy = \frac{1}{9}\left(\frac{100}{2} - \frac{1}{2}\right) = 5.5$.

The tail probability measuring the plausibility of the value $x_0 = 5$ is given by $P(Y > 5 \,|\, Y > 1) = \int_5^{10} \frac{1}{9} dy = \frac{5}{9} = 0.555555$, which indicates that $x_0 = 5$ is slightly more plausible now.

Finally, the shortest interval containing 0.95 of the conditional probability is of the form (c, d), where c,and d satisfy $1 \le c \le d \le 10$ and $0.95 = \int_c^d \frac{1}{9} dy = \frac{1}{9}(d - c)$ or $d - c = (.95)9 = 8.55$. We thus see that any subinterval of $(1, 10)$ of length 8.5 will work, e.g., $(1.45, 10)$.

Problems

5.2.5 Let X be the number of heads in 10 tosses of a fair coin. Then $s \sim$Binomial$(10, 0.5)$

(a) The expected value of the response is $E(X) = 10 \cdot 0.5 = 5$.

(b) The probability function of X looks like the graph below.

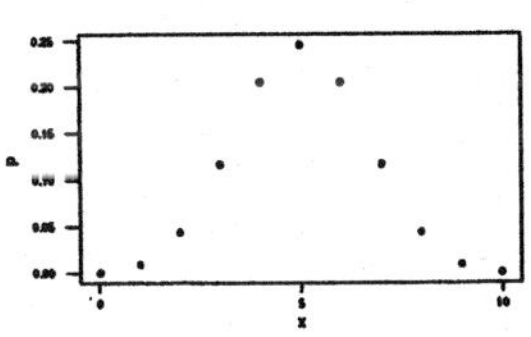

From this we can see that the shortest interval containing 0.95 of the probability is symmetric about 5. So c satisfying $2(P(X = 0) + \cdots + P(X = c)) \le .05$ and $2(P(X = 0) + \cdots + P(X = c + 1)) > .05$ gives the shortest interval as $(c, n - c)$. In this case $c = 2$ since $2(P(X = 0) + \cdots + P(X = 2)) = 0.02148$ and $2(P(X = 0) + \cdots + P(X = 3)) = 0.10937$.

(c) As we can see, the probability function of X is symmetric, so to assess whether or not a value x is a possible future value we would use the probability of obtaining a value whose probability of occurrence was as small or smaller than that of x. In this case it is the probability $2(P(X = 0) + \cdots + P(X = \min(x, n - x)))$. When $x = 8$ this probability is given by $2(P(X = 0) + \cdots + P(X = 2)) = 0.02148$. It seems small, so we have evidence against the coin being fair. Note that it is also plausible to use the left or right tail alone, but the two-tailed approach seems more sensible.

5.2.6 We have that the probability of an interval (a, b) is given by $e^{-a} - e^{-b}$, and we want a and b such that $e^{-a} - e^{-b} = .95$ and $b - a$ is smallest. From the graph of the density e^{-x}, we see that for two intervals of the same length the one closest to 0 has the most probability. So taking $a = 0$ means that choosing b appropriately will give the shortest interval.

5.2.7
(a) The condition implies that $X \in C = \{0, 2, 4, 6, 8, 10\}$ and this has probability

$$P(C) = \left\{\binom{10}{0} + \binom{10}{2} + \binom{10}{4} + \binom{10}{6} + \binom{10}{8} + \binom{10}{0}\right\}\left(\frac{1}{2}\right)^{10} = \frac{1}{2}$$

The conditional distribution of X, given C, is

$$P(X = 0 \,|\, C) = 2\binom{10}{0}\left(\frac{1}{2}\right)^{10} = 1.9531 \times 10^{-3}$$
$$P(X = 2 \,|\, C) = 2\binom{10}{2}\left(\frac{1}{2}\right)^{10} = 8.7891 \times 10^{-2}$$
$$P(X = 4 \,|\, C) = 2\binom{10}{4}\left(\frac{1}{2}\right)^{10} = 0.41016$$
$$P(X = 6 \,|\, C) = 2\binom{10}{6}\left(\frac{1}{2}\right)^{10} = 0.41016$$
$$P(X = 8 \,|\, C) = 2\binom{10}{8}\left(\frac{1}{2}\right)^{10} = 8.7891 \times 10^{-2}$$
$$P(X = 10 \,|\, C) = 2\binom{10}{10}\left(\frac{1}{2}\right)^{10} = 1.9531 \times 10^{-3}$$

so the conditional expectation of X is $\mu = 0\left(1.9531 \times 10^{-3}\right) +$
$2\left(8.7891 \times 10^{-2}\right) + 4\,(0.41016) + 6\,(0.41016) + 8\left(8.7891 \times 10^{-2}\right) +$
$10\left(1.9531 \times 10^{-3}\right) = 5.0$.
(b) The shortest interval containing at least 0.95 of the probability for X is $(2, 8)$.
(c) We assess $x = 8$ by computing $2\,(P(X = 0) + P(X = 2)) = 2(1.9531 \times 10^{-3} + 8.7891 \times 10^{-2}) = 0.17969$, and we see that we now don't have any evidence against 8 as a plausible value.

5.2.8 Suppose that $X \sim \text{Beta}(a, b)$. We have that

$$E(X) = \int_0^1 x \frac{\Gamma(a+b)}{\Gamma(a)\Gamma(b)} x^{a-1}(1-x)^{b-1}\,dx = \frac{\Gamma(a+b)}{\Gamma(a)\Gamma(b)} \int_0^1 x^a (1-x)^{b-1}\,dx$$
$$= \frac{\Gamma(a+b)}{\Gamma(a)\Gamma(b)} \frac{\Gamma(a+1)\Gamma(b)}{\Gamma(a+b+1)} = \frac{a}{a+b}.$$

and the mean-squared error of this predictor is just

$$E\left(\left(X-\frac{a}{a+b}\right)^2\right)=Var(X)=\frac{ab}{(a+b+1)(a+b)^2}.$$

We have that

$$\begin{aligned}E\left(X^2\right)&=\int_0^1 x^2\frac{\Gamma(a+b)}{\Gamma(a)\Gamma(b)}x^{a-1}(1-x)^{b-1}\,dx\\&=\frac{\Gamma(a+b)}{\Gamma(a)\Gamma(b)}\int_0^1 x^{a+1}(1-x)^{b-1}\,dx=\frac{\Gamma(a+b)}{\Gamma(a)\Gamma(b)}\frac{\Gamma(a+2)\Gamma(b)}{\Gamma(a+b+2)}\\&=\frac{a(a+1)}{(a+b)(a+b+1)}\end{aligned}$$

so $Var(X)=ab/(a+b+1)(a+b)^2$.

To obtain the mode, we need to maximize $x^{a-1}(1-x)^{b-1}$ or equivalently $(a-1)\ln x+(b-1)\ln(1-x)$, which has derivative $(a-1)/x-(b-1)/(1-x)$, and setting this equal to 0 yields the solution $\hat{x}=(a-1)/(a+b-2)$. The second derivative is given by $-(a-1)/x^2-(b-1)/(1-x)^2$, which is always less than or equal to 0 so $\hat{x}$ is the mode. Then

$$\begin{aligned}&E\left(\left(X-\frac{a-1}{a+b-2}\right)^2\right)=E\left(\left(X-\frac{a}{a+b}+\frac{a}{a+b}-\frac{a-1}{a+b-2}\right)^2\right)\\&=Var(X)+\left(\frac{a}{a+b}-\frac{a-1}{a+b-2}\right)^2\\&=\frac{ab}{(a+b+1)(a+b)^2}+\left(\frac{a}{a+b}-\frac{a-1}{a+b-2}\right)^2\geq E\left(\left(X-\frac{a}{a+b}\right)^2\right).\end{aligned}$$

Therefore, the mean is a better predictor.

5.2.9 Suppose $X\sim N(0,1)$ and that we use the mean of the distribution to predict a future value. Then:
(a) $E(X)=0$ is a prediction for a future X.
(b) If $Y=X^2$, then $Y\sim\chi^2_{(1)}$ and $E(Y)=1$.
(c) We notice that we predict X by 0 but do not predict X^2 by 0^2.

5.3 Statistical Models

Exercises

5.3.1 Let θ denote the type of coin being selected, then $\theta\in\Omega=\{1,2,3\}$, where coin 1 is the fair one, coin 2 has probability 1/3 of yielding a head, and coin 3 has probability 2/3 of yielding a head. So the statistical model for a single response consists of three probability functions $\{f_1,f_2,f_3\}$, where f_1 is the probability function for the Bernoulli(1/2) distribution, f_1 is the probability function for the Bernoulli(1/3) distribution, and f_3 is the probability function for the Bernoulli(2/3) distribution. Then $(x_1, x_2, \ldots, x_5)$ is a sample from one of these Bernoulli(θ) distributions.

5.3.2 There are 6 possible distributions in the model as given in the following table. Here p_i denotes the distribution relevant when the face with i pips is duplicated.

	1	2	3	4	5	6
p_1	1/3	0	1/6	1/6	1/6	1/6
p_2	0	1/3	1/6	1/6	1/6	1/6
p_3	0	1/6	1/3	1/6	1/6	1/6
p_4	0	1/6	1/6	1/3	1/6	1/6
p_5	0	1/6	1/6	1/6	1/3	1/6
p_6	0	1/6	1/6	1/6	1/6	1/3

5.3.3 The sample $(X_1, \ldots, X_n)$ is a sample from $N(\mu, \sigma^2)$ distribution, where $\theta = (\mu, \sigma^2) \in \Omega = \{(10, 2), (8, 3)\}$. We could parameterize this model by the population mean or by the population variance as both of these quantities uniquely identify the two populations. For example, if we know the mean of the distribution is 10, then we know that we are sampling from population I (and similarly if we know the variance is 2).

5.3.4 We cannot parameterize the model by the population mean since the two populations have the same mean, but we can parameterize by the population variance, as this is unique.

5.3.5 A single observation is from an Exponential(θ) distribution, where $\theta \in \Omega = [0, \infty)$. We can parameterize this model by the mean $1/\theta$ since the mean is a 1-1 function of θ. We can also parameterize this model by the variance, since it is a 1-1 transformation of $\theta \geq 0$. The coefficient of variation is given by $\theta^{-1}/\sqrt{\theta^{-2}} = 1$. This quantity is free of θ, and so we cannot use this quantity to parameterize the model.

5.3.6 The first quartile c of the Uniform$[0, \beta]$ distribution satisfies $0.25 = \int_0^c \frac{1}{\beta} dx = \frac{c}{\beta}$, so $c = 0.25\beta$. Since c is a 1-1 transformation of β, we can parameterize this model by the first quartile.

Problems

5.3.7 We have that $\exp(\psi) = \theta/(1-\theta)$, so $1 + \exp(\psi) = 1 + \theta/(1-\theta) = 1/(1-\theta)$, giving that $\theta = \exp(\psi) / (1 + \exp(\psi))$. Then the probability function for X_i is given by

$$\left(\frac{\exp(\psi)}{1 + \exp(\psi)}\right)^{x_i} \left(\frac{1}{1 + \exp(\psi)}\right)^{1-x_i}$$

for $x_i \in \{0, 1\}$ with $\psi \in [0, \infty]$ (note $\psi = \infty$ when $\theta = 1$). The probability function for the sample $(X_1, \ldots, X_n)$ is given by the product of these individual probability functions, and the parameter is ψ, which takes values in the parameter space $[0, \infty]$.

5.3.8 We have that $\psi = \ln \sigma$, so $\sigma = \exp(\psi)$. The density function for X_i is then given by

$$\frac{\exp\left(-\frac{\psi}{2}\right)}{\sqrt{2\pi}} \exp\left\{\frac{\exp(-2\psi)}{2}(x_i - \mu)^2\right\}$$

and $(\mu, \psi) \in R^2$ so that the parameter space is now R^2. The density function for the sample $(X_1, \ldots, X_n)$ is given by the product of these individual probability functions and the parameter is (μ, ψ), which takes values in the parameter space R^2.

5.3.9 $\sum_{i=1}^n (x_i - \mu)^2 = \sum_{i=1}^n (x_i - \bar{x} + \bar{x} - \mu)^2 = \sum_{i=1}^n ((x_i - \bar{x})^2 + 2(x_i - \bar{x})(\bar{x} - \mu) + (\mu - \bar{x})^2) = \sum_{i=1}^n (x_i - \bar{x})^2 - 2(\mu - \bar{x})\sum_{i=1}^n (x_i - \bar{x}) + \sum_{i=1}^n (\mu - \bar{x})^2 = \sum_{i=1}^n (x_i - \bar{x})^2 + n(\mu - \bar{x})^2$ since $\sum_{i=1}^n (x_i - \bar{x}) = \sum_{i=1}^n x_i - n\bar{x} = 0$.

5.3.10 We know that $T \sim \text{Binomial}(n, \theta)$, where $\theta \in [0, 1]$ is unknown. Therefore, the probability function for T is given by $f_\theta(t) = \binom{n}{t}\theta^t(1-\theta)^{n-t}$ for $t \in \{0, \ldots, n\}$. The parameter is θ and the parameter space is $[0, 1]$.

5.3.11 The first quartile c,of a $N(\mu, \sigma^2)$ distribution satisfies

$$0.25 = \int_{-\infty}^{c} \frac{1}{\sqrt{2\pi}\sigma} \exp\left\{-\frac{(x-\mu)^2}{2\sigma^2}\right\} dx = \Phi\left(\frac{c-\mu}{\sigma}\right).$$

Therefore, $c = \mu + \sigma z_{.25}$, where $z_{.25}$ is the first quartile of the $N(0, 1)$ distribution, i.e., $\Phi(z_{.25}) = .25$. But we see from this that several different values of (μ, σ^2) can give the same first quartile, e.g., $(\mu, \sigma^2) = (0, 1)$ and $(\mu, \sigma^2) = (z_{.25}/2, 1/4)$ both give rise to normal distributions whose first quartile equals $z_{.25}$. Therefore, we cannot parameterize this model by the first quartile.

5.3.12 The statistical model for (X, Y) is given by the densities

$$\begin{aligned}
f(x, y \mid \sigma^2, \delta^2) &= f(x \mid \sigma^2, y) f(y \mid \delta^2) \\
&= \frac{1}{\sqrt{2\pi}\sigma} \exp\left\{-\frac{(x-y)^2}{2\sigma^2}\right\} \frac{1}{\sqrt{2\pi}\delta} \exp\left\{-\frac{1}{2\delta^2}y^2\right\} \\
&= \frac{1}{2\pi\sigma\delta} \exp\left\{-\frac{1}{2}\left(\frac{x^2}{\sigma^2} + \frac{2xy}{\sigma^2} - \left(\frac{1}{\sigma^2} + \frac{1}{\delta^2}\right)y^2\right)\right\} \\
&= \frac{1}{2\pi\sigma\delta} \exp\left\{-\frac{1}{2}\left(\frac{x^2}{\sigma^2} + \frac{2xy}{\sigma^2} - \frac{\delta^2 + \sigma^2}{\sigma^2\delta^2}y^2\right)\right\} \\
&= \frac{1}{2\pi\sigma\delta} \exp\left\{-\frac{\delta^2 + \sigma^2}{2\sigma^2}\left(\frac{x^2}{\delta^2 + \sigma^2} + \frac{2xy}{\delta^2 + \sigma^2} - \frac{1}{\delta^2}y^2\right)\right\}
\end{aligned}$$

where the parameter (σ^2, δ^2) ranges in the parameter space $(\sigma^2, \delta^2) \times (\sigma^2, \delta^2)$. From Example 2.7.8 we see that this is the density of a bivariate normal $0, 0, \delta^2 + \sigma^2, \delta^2, \rho$ distribution, where $\rho = \sqrt{\frac{\delta^2}{\delta^2 + \sigma^2}}$. Using Problem 2.7.8 we have immediately that $X \sim N(0, \sigma^2 + \delta^2)$. Therefore, the statistical model for X alone is given by the collection of all $N(0, \tau^2)$ distributions, where the parameter

τ^2 is any value greater than 0. Alternatively, this result can be obtained by integrating out y in the joint density to obtain

$$\begin{aligned} f(x\,|\,\sigma^2,\delta^2) &= \int_{-\infty}^{\infty} \frac{1}{2\pi\sigma\delta} \exp\left\{ -\frac{x^2}{2\sigma^2} + \frac{2xy}{2\sigma^2} - (\frac{1}{2\sigma^2} + \frac{1}{2\delta^2})y^2 \right\} dy \\ &= \frac{1}{\sqrt{2\pi(\sigma^2+\delta^2)}} \exp\left\{ -\frac{x^2}{2(\delta^2+\sigma^2)} \right\}. \end{aligned}$$

5.4 Data Collection

Exercises

5.4.1 We have that

$$F_X(x) = \begin{cases} 0 & x < 1 \\ \frac{4}{10} & 1 \le x < 2 \\ \frac{7}{10} & 2 \le x < 3 \\ \frac{9}{10} & 3 \le x < 4 \\ 1 & 4 \le x \end{cases}, \quad f_X(x) = \begin{cases} \frac{4}{10} & x = 1 \\ \frac{3}{10} & x = 2 \\ \frac{2}{10} & x = 3 \\ \frac{1}{10} & x = 4 \end{cases}$$

and $\mu_X = \sum_{x=1}^{4} x f_X(x) = 2, \sigma_X^2 = \left(\sum_{x=1}^{4} x^2 f_X(x)\right) - 2^2 = 1.$

5.4.2
(a) We can't consider this as an approximate i.i.d. sample from the population distribution since the size of the population is small and the sample size is large relative to the population size.
(b) Place ten chips in a bowl. Each chip should have a unique number on it from 1 to 10. Thoroughly mix the chips and draw three of them without replacement. The numbers on the selected chips correspond to the individuals to be selected from the population. Alternatively, we can use Table D.1 by selecting a row and reading off the first three single numbers (treat 0 in the table as a 10).
(c) Using row 108 of Table D.1 (treating 0 as 10)we get:
First sample — we obtain random numbers $6, 0, 9$ and so compute $(X(\pi_6) + X(\pi_{10})$
$+X(\pi_9))/3 = (3+4+2)/3 = 3.0$
Second sample — we obtain random numbers $4, 0, 7$ and so compute $(1+4+3)/3$ $= 2.6667$
Third sample — we obtain random numbers $2, 0, 4$ (note we had to skip the second 2) and so compute $(1+4+1)/3 = 2.0$.

5.4.3
(a) We can consider this as an exact i.i.d. sample from the population distribution since it's a sample with replacement, so each individual has the same chance to be chosen on each draw.
(b) Place ten chips in a bowl. Each chip should have a unique number on it from 1 to 10. Thoroughly mix the chips and draw three of them with replacement.

The numbers on the selected chips correspond to the individuals to be selected from the population. Alternatively, we can use Table D.1 by selecting a row and reading off the first three single numbers (treat 0 in the table as a 10).
(c) Using row 108 of Table D.1 (treating 0 as 10)we get:
First sample — we obtain random numbers 6, 0, 9 and so compute $(X(\pi_6) + X(\pi_{10})$
$+X(\pi_9))/3 = (3+4+2)/3 = 3.0$
Second sample — we obtain random numbers 4, 0, 7 and so compute $(1+4+3)/3 = 2.6667$
Third sample — we obtain random numbers 2, 0, 2 (note we do not skip the second 2) and so compute $(1+4+1)/3 = 2.0$.

5.4.4
(a) $f_X(0) = a/N, f_X(1) = (N-a)/N$. This is a Bernoulli$((N-a)/N)$ distribution.
(b) $P\left(\hat{f}_X(0) = f_X(0)\right) = P\left(n\hat{f}_X(0) = nf_X(0)\right) = P$(number of 0's in the sample equals $nf_X(0)) = \binom{a}{n-nf_X(0)}\binom{N-a}{nf_X(0)}/\binom{N}{n}$ since $n\hat{f}_X(0) \sim$ Hypergeometric(N, a, n).
(c) We have that $n\hat{f}_X(0) \sim$ Binomial$(n, a/N)$, so $P(\hat{f}_X(0) = f_X(0)) = P(n\hat{f}_X(0) = nf_X(0)) = P$(number of 0's in the sample equals $nf_X(0)) = \binom{n}{nf_X(0)}\left(\frac{a}{N}\right)^{nf_X(0)} \times \left(1-\frac{a}{N}\right)^{n-nf_X(0)}$.

5.4.5
(a)

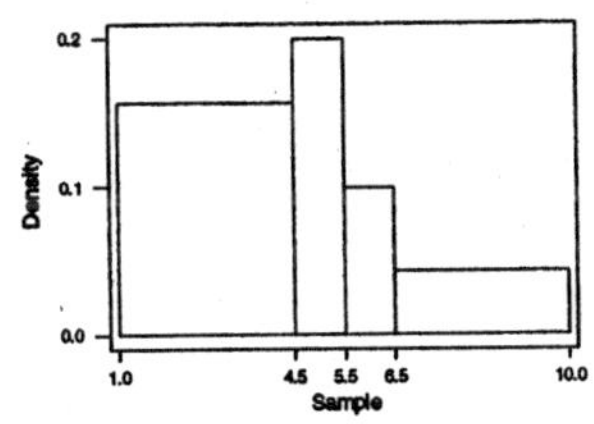

(b)

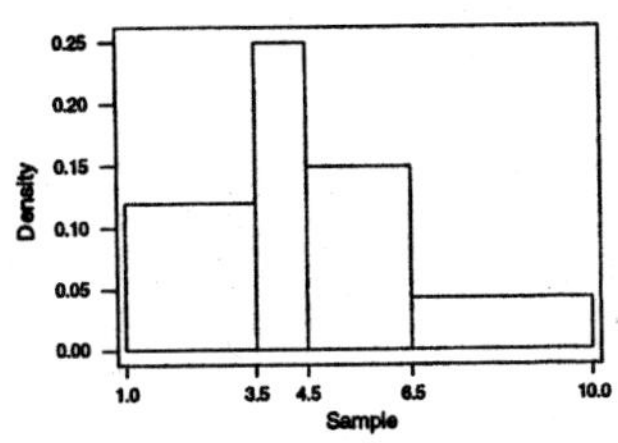

(c) The shape of a histogram depends on the intervals being used.

5.4.6
(a) Through a census of the population.
(b) We can't represent the population distribution of X by F_X since X is a categorical variable.
(c)

$$f_X(x) = \begin{cases} 0.35 & x = A \\ 0.55 & x = B \\ 0.1 & x = C \end{cases}$$

(d) We can select a simple random sample from the population, record the political opinion of each student, and compute the sample proportions for each party.
(e) It doesn't allow for those who do not have a preference.

Computer Exercises

5.4.7
(a) After generating the sample $(x_1, \ldots, x_{1000})$, you need to sort it to obtain the order statistics $(x_{(1)}, \ldots, x_{(n)})$ and then record the proportion of data values less than or equal to each value. Then $F_X(x)$ equals the largest value i/n, such that $x_{(i)} \leq x$.
(b)

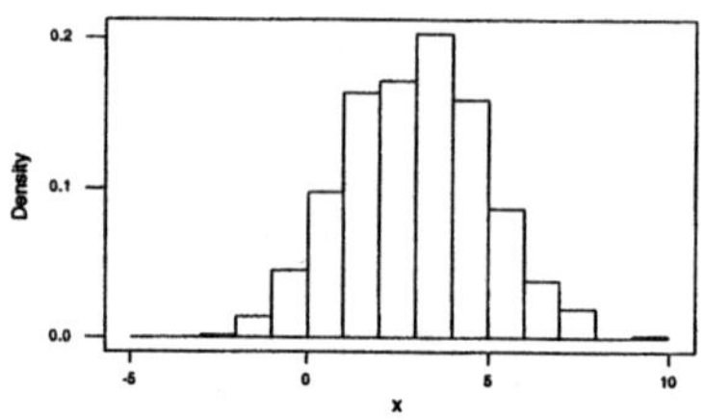

(c)

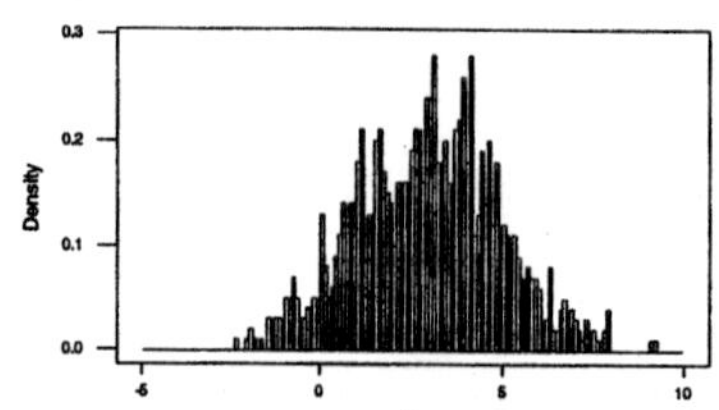

(d) The histogram in (c) is much more erratic than that in (b). Some of this is due to sampling error.
(e) If we make the lengths of the intervals too short, then there will inevitably only be one or a few data points per interval, and the histogram will not have any kind of recognizable shape. This is sometimes called over-fitting, as the erratic shape is caused by making the intervals too small.

5.4.8 Using Minitab this can be carried out by placing the numbers 1 through 10,000 in a column and then using the `Sample from columns` command, with the subcommand to carry out sampling with replacement.

Problems

5.4.9
(a) $f_X(0) = a/N, f_X(1) = b/N, f_X(2) = (N - a - b)/N.$
(b) Assuming f_1, f_2, and f_3 are nonnegative integers summing to n (otherwise probability is 0), the probability is $\binom{a}{f_0}\binom{b}{f_1}\binom{N-a-b}{f_2}/\binom{N}{n}$.
(c) The probability that $\hat{f}_X(0) = f_0, \hat{f}_X(1) = f_1$and $\hat{f}_X(2) = f_2$ is

$$\binom{a}{f_0\, f_1\, f_2}\left(\frac{a}{N}\right)^{f_0}\left(\frac{b}{N}\right)^{f_1}\left(\frac{N-a-b}{N}\right)^{f_2}$$

since each sequence of f_0 zeros, f_1 ones, and f_2 twos has probability

$$\left(\frac{a}{N}\right)^{f_0}\left(\frac{b}{N}\right)^{f_1}\left(\frac{N-a-b}{N}\right)^{f_2}$$

of occurring, and there are $\binom{a}{f_0\, f_1\, f_2}$ such sequences.

5.4.10
(a) The population mean is given by $\mu_X = \frac{1}{N}\sum_{i=1}^N X(\pi_i) = \sum_x x f_X(x)$ since $f_X(x) =$ (the number of population elements with $X(\pi_i) = x)/N$.
(b) The population variance is given by

$$\begin{aligned}\sigma_X^2 &= \frac{1}{N}\sum_{i=1}^N (X(\pi_i) - \mu_X)^2 = \frac{1}{N}\sum_{i=1}^N X^2(\pi_i) - \frac{2}{N}\sum_{i=1}^N X(\pi_i)\mu_X + \mu_X^2 \\ &= \sum_x x^2 f_X(x) - 2\mu_X^2 + \mu_X^2 = \sum_x x^2 f_X(x) - \mu_X^2 = \sum_x (x - \mu_X)^2 f_X(x).\end{aligned}$$

5.4.11
(a) First, note that $\hat{f}_X(0) = \frac{1}{n}\sum_{i=1}^{n} I_{\{0\}}(X(\pi_i))$, so

$$\begin{aligned}E\left(\hat{f}_X(0)\right) &= \frac{1}{n}E\left(\sum_{i=1}^n I_{\{0\}}(X(\pi_i))\right) = \frac{1}{n}\sum_{i=1}^n E\left(I_{\{0\}}(X(\pi_i))\right) \\ &= \frac{1}{n}\sum_{i=1}^n P(X(\pi_i) = 0) = \frac{1}{n}\sum_{i=1}^n f_X(0) = f_X(0).\end{aligned}$$

(b) We have that

$$\begin{aligned}
\text{Var}\left(\hat{f}_X(0)\right) &= \frac{1}{n^2} Var \left(\sum_{i=1}^{n} I_{\{0\}}\left(X(\pi_i)\right)\right) \\
&= \frac{1}{n^2}\sum_{i=1}^{n} \text{Var}\left(I_{\{0\}}(X(\pi_i))\right) + \frac{2}{n^2}\sum_{i<j} \text{Cov}\left(I_{\{0\}}(X(\pi_i), I_{\{0\}}(X(\pi_j))\right) \\
&= \frac{1}{n}\text{Var}\left(I_{\{0\}}(X(\pi_i))\right) + \frac{2}{n^2}\frac{n(n-1)}{2}\text{Cov}\left(I_{\{0\}}(X(\pi_1), I_{\{0\}}(X(\pi_2))\right) \\
&= \frac{f_X(0)\left(1 - f_X(0)\right)}{n} + \frac{n-1}{n}\text{Cov}\left(I_{\{0\}}(X(\pi_1), I_{\{0\}}(X(\pi_2))\right)
\end{aligned}$$

and

$$\begin{aligned}
&\text{Cov}\left(I_{\{0\}}(X(\pi_1), I_{\{0\}}(X(\pi_2))\right) = E\left(I_{\{0\}}(X(\pi_1) I_{\{0\}}(X(\pi_2))\right) - \left(f_X(0)\right)^2 \\
&= P\left(X(\pi_{1})) = 0, X(\pi_2) = 0\right) - \left(f_X(0)\right)^2 = f_X(0)\left(\frac{N f_X(0) - 1}{N-1}\right) - \left(f_X(0)\right)^2 \\
&= f_X(0)\left\{\frac{N f_X(0) - 1 - N f_X(0) + f_X(0)}{N-1}\right\} = -\frac{f_X(0)\left(1 - f_X(0)\right)}{N-1}
\end{aligned}$$

Therefore,

$$\begin{aligned}
Var\left(\hat{f}_X(0)\right) &= \frac{f_X(0)\left(1 - f_X(0)\right)}{n} - \frac{n-1}{n}\frac{f_X(0)\left(1 - f_X(0)\right)}{N-1} \\
&= \frac{f_X(0)\left(1 - f_X(0)\right)}{n}\frac{N-n}{N-1}.
\end{aligned}$$

(c) If we take a sample with replacement, then we can assume this is an i.i.d. sample, so $n\hat{f}_X(0) \sim \text{Binomial}(n, f_X(0))$. Therefore,

$$\begin{aligned}
E\left(\hat{f}_X(0)\right) &= \frac{1}{n}E\left(n\hat{f}_X(0)\right) = \frac{1}{n}n f_X(0) = f_X(0), \\
\text{Var}\left(\hat{f}_X(0)\right) &= \frac{1}{n^2}\text{Var}\left(n\hat{f}_X(0)\right) = \frac{n f_X(0)\left(1 - f_X(0)\right)}{n^2} = \frac{f_X(0)\left(1 - f_X(0)\right)}{n}.
\end{aligned}$$

(d) Because this factor is the only difference with the variance for sampling with and without replacement. Note that when n is small relative to N, then this factor is approximately 1.

5.4.12 When $f_X(0) = a/N$ is unknown, then we estimate it by $\hat{f}_X(0)$. Now $N = a/f_X(0)$, so we can estimate N by setting $\hat{N} = a/\hat{f}_X(0)$, provided $\hat{f}_X(0) \neq 0$.

5.4.13 If we knew N but not T, then, based on a sample $X(\pi_1), \ldots, X(\pi_n)$, we would estimate T/N by $\bar{X} = \frac{1}{n}\sum_{i=1} X(\pi_i)$. Therefore, when we know T and don't know N, we can estimate N by $T/\bar{X}$ provided $\bar{X} \neq 0$.

5.4.14 We have that $\bar{X} = \frac{1}{n}\sum_{i=1}^{n} X(\pi_i)$, so $E\left(\bar{X}\right) = \frac{1}{n}\sum_{i=1}^{n} E\left(X(\pi_i)\right)$. Since each $X(\pi_i) \sim f_X$, we have that $E\left(X(\pi_i)\right) = \sum_x x f_X(x) = \mu_X$, so $E\left(\bar{X}\right) = \mu_X$.

Under the assumption of i.i.d. sampling, each $X(\pi_i)$ has the same variance $\sigma_X^2 = \sum_x (x - \mu_X)^2 f_X(x)$. So we get $\text{Var}(\bar{X})] = \sigma_X^2 / n$.

5.4.15 Note that $\hat{f}_X(x) = \frac{1}{n}\sum_{i=1}^n I_{\{x\}}(X(\pi_i))$, so it is an average of i.i.d. terms and $E\left(I_{\{x\}}(X(\pi_i))\right) = f_X(x)$. Then by the weak law of large numbers $\hat{f}_X(x) \xrightarrow{P} f_X(x)$ as $n \to \infty$.

Challenge

5.4.16
(a)

$$\begin{aligned}
&f_X(x) \\
&= \frac{|\{\pi \in \Pi : X(\pi) = x\}|}{|\Pi|} = \frac{|\{\pi \in \Pi_1 : X(\pi) = x\}| + |\{\pi \in \Pi_2 : X(\pi) = x\}|}{|\Pi|} \\
&= \frac{|\Pi_1|}{|\Pi|}\frac{|\{\pi \in \Pi_1 : X(\pi) = x\}|}{|\Pi_1|} + \frac{|\Pi_2|}{|\Pi|}\frac{|\{\pi \in \Pi_2 : X(\pi) = x\}|}{|\Pi_2|} \\
&= pf_{1X}(x) + (1-p)f_{2X}(x)
\end{aligned}$$

(b)

$$\begin{aligned}
\mu_X &= \sum_x x f_X(x) = \frac{1}{|\Pi|}\sum_{\pi\in\Pi} X(\pi) = \frac{|\Pi_1|}{|\Pi|}\frac{1}{|\Pi_1|}\sum_{\pi\in\Pi_1} X(\pi) \\
&+ \frac{|\Pi_2|}{|\Pi|}\frac{1}{|\Pi_2|}\sum_{\pi\in\Pi_2} X(\pi) \\
&= p\mu_{1X} + (1-p)\mu_{2X}
\end{aligned}$$

(c) Using

$$\sum_{\pi\in\Pi_1} (X(\pi) - \mu_{1X}) = \sum_{\pi\in\Pi_2} (X(\pi) - \mu_{2X}) = 0$$

we have that

$$\begin{aligned}
\sigma_X^2 &= \frac{1}{|\Pi|}\sum_{\pi\in\Pi}^{N} (X(\pi) - \mu_X)^2 \\
&= \frac{|\Pi_1|}{|\Pi|}\frac{1}{|\Pi_1|}\sum_{\pi\in\Pi_1} (X(\pi) - \mu_X)^2 + \frac{|\Pi_2|}{|\Pi|}\frac{1}{|\Pi_2|}\sum_{\pi\in\Pi_2} (X(\pi) - \mu_X)^2 \\
&= p\frac{1}{|\Pi_1|}\sum_{\pi\in\Pi_1} (X(\pi) - p\mu_{1X} - (1-p)\mu_{2X})^2 + \\
&\quad (1-p)\frac{1}{|\Pi_2|}\sum_{\pi\in\Pi_2} (X(\pi) - p\mu_{1X} + (1-p)\mu_{2X})^2
\end{aligned}$$

$$= p\frac{1}{|\Pi_1|}\sum_{\pi\in\Pi_1}\left((X(\pi)-\mu_{1X})+(1-p)(\mu_{1X}-\mu_{2X})\right)^2+$$
$$(1-p)\frac{1}{|\Pi_2|}\sum_{\pi\in\Pi_2}\left((X(\pi)-\mu_{2X})-p(\mu_{1X}-\mu_{2X})\right)^2$$
$$= p\sigma_{1X}^2+p(1-p)^2(\mu_{1X}-\mu_{2X})^2+(1-p)\sigma_{2X}^2+(1-p)p^2(\mu_{1X}-\mu_{2X})^2$$
$$= p\sigma_{1X}^2+(1-p)\sigma_{2X}^2+p(1-p)(\mu_{1X}-\mu_{2X})^2.$$

(d) Under the assumption of i.i.d. sampling and using Problem 5.4.14
$$E\left(p\bar{X}_1+(1-p)\bar{X}_2\right)=pE\left(\bar{X}_1\right)+(1-p)E\left(\bar{X}_2\right)=p\mu_{1X}+(1-p)\mu_{2X}=\mu_X$$
and
$$Var\left(p\bar{X}_1+(1-p)\bar{X}_2\right)=p^2Var\left(\bar{X}_1\right)+(1-p)^2Var\left(\bar{X}_2\right)=p^2\frac{\sigma_{1X}^2}{n_1}+(1-p)^2\frac{\sigma_{2X}^2}{n_2}$$

(e) Again, under the assumption of i.i.d. sampling and by Problem 5.4.14, part (c), and using $n_1=pn, n_2=(1-p)n$ we have
$$Var\left(\bar{X}\right)=\frac{\sigma_X^2}{n}=p\frac{\sigma_{1X}^2}{n}+(1-p)\frac{\sigma_{2X}^2}{n}+\frac{p(1-p)(\mu_{1X}-\mu_{2X})^2}{n}$$
$$=p^2\frac{\sigma_{1X}^2}{n_1}+(1-p)^2\frac{\sigma_{2X}^2}{n_2}+\frac{p(1-p)(\mu_{1X}-\mu_{2X})^2}{n}$$
so $Var\left(p\bar{X}_1+(1-p)\bar{X}_2\right)\le Var\left(\bar{X}\right)$.

(f) If $\mu_{1X}=\mu_{2X}$, then there are no benefits as the two estimators have the same variances. When the means μ_{1X} and μ_{2X} are quite different, then there will be a big improvement through the use of stratified sampling. This indicates that the populations Π_1 and Π_2 are quite different with respect to the measurement X.

5.5 Some Basic Inferences

Exercises

5.5.1

(a) $\hat{f}_X(0)=.2667, \hat{f}_X(1)=.2, \hat{f}_X(2)=.2667, \hat{f}_X(3)=\hat{f}_X(4)=.1333.$

(b) $\hat{F}_X(0)=.2667, \hat{F}_X(1)=.4667, \hat{F}_X(2)=.7333, \hat{F}_X(3)=.8667, \hat{F}_X(4)=1.000$

(c) A plot of $\hat{f}_X$ is given below.

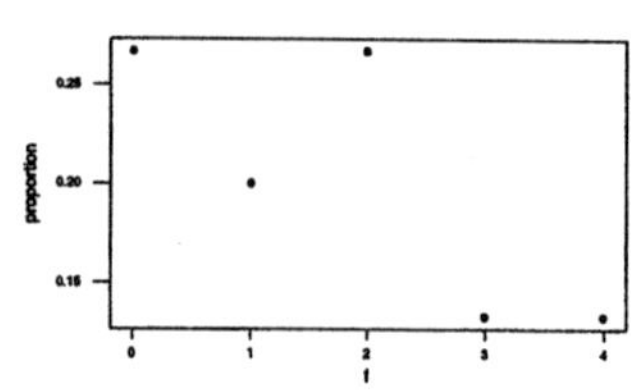

(d) The mean $\bar{x} = 15$ and the variance $s^2 = 1.952$.
(e) The median is 2 and the $IQR = 3$. The boxplot is plotted below. According to the 1.5 IQR rule, there are no outliers.

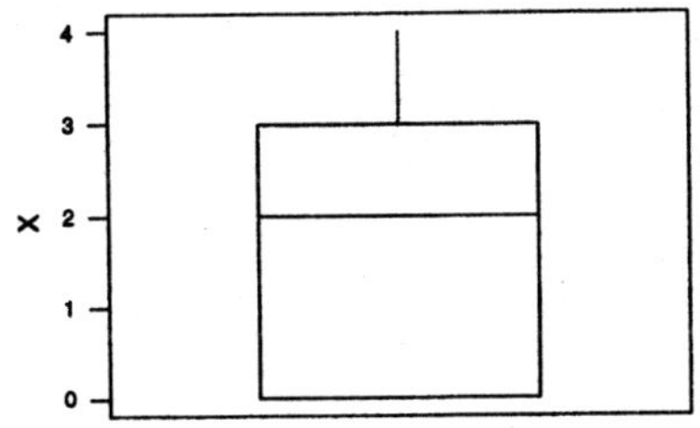

5.5.2
(a) The empirical distribution function is given by

$$\hat{F}_X(x) = \begin{cases} 0 & x < 0 \\ \frac{1}{16} & 0 \le x < 1 \\ \frac{3}{16} & 1 \le x < 2 \\ \frac{5}{16} & 2 \le x < 3 \\ \frac{8}{16} & 3 \le x < 4 \\ \frac{11}{16} & 4 \le x < 5 \\ \frac{14}{16} & 5 \le x < 10 \\ \frac{15}{16} & 10 \le x < 15 \\ 1 & 15 \le x. \end{cases}$$

(b) A plot of $\hat{f}_X$ is given below.

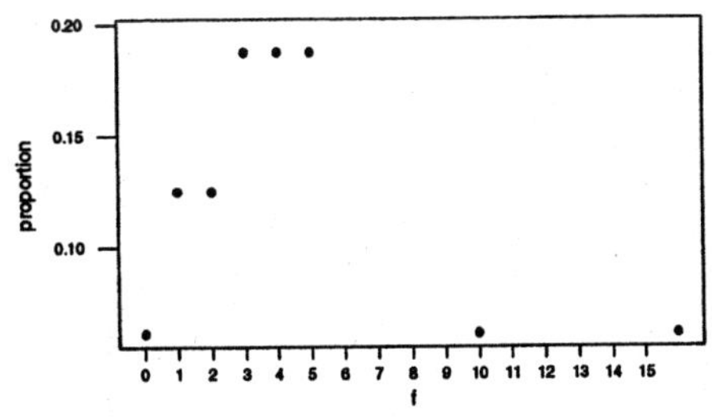

(c) The mean is $\bar{x} = 4.188$, the variance is $s^2 = 13.63$.
(d) The median is 3.5 and the $IQR = 3$. A boxplot is provided as follows.

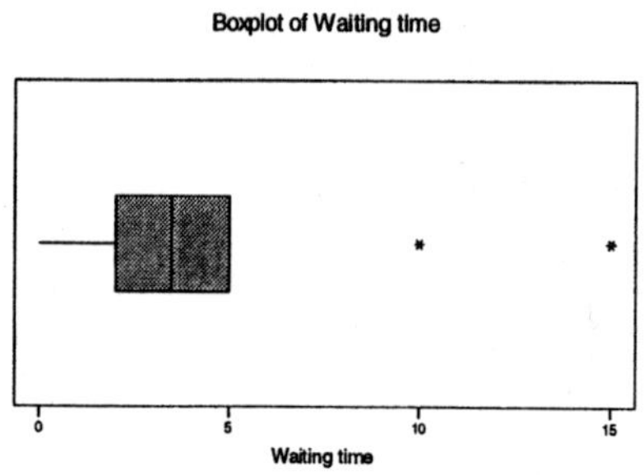

According to the 1.5 IQR rule, there are two outliers, namely 10 and 15.

5.5.3
(a) $\hat{f}_X(1) = 25/82, \hat{f}_X(2) = 35/82, \hat{f}_X(3) = 22/82.$
(b) It doesn't make sense to estimate $F_X(i)$ since this is a categorical variable.
(c) A bar chart is given below.

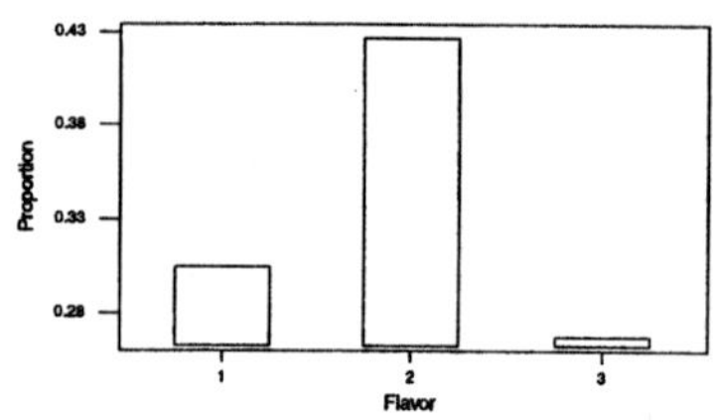

5.5.4 It means that 90% of all students got a score equal to his or lower and only 10% got a higher score.

5.5.5 A plot of the empirical distribution function is given below (we have joined consecutive points by line segments).

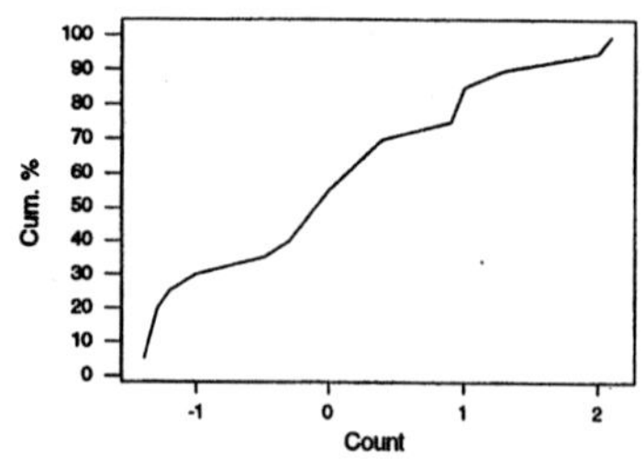

The sample median is 0, first quartile is -1.150, third quartile is 0.975, and the $IQR = 2.125$. We estimate $F_X(1)$ by $\hat{F}_X(1) = 17/20 = 0.85$.

5.5.6 Since the shape of the distribution is asymmetric, we should choose the median as a measure of location and the IQR as a measure of spread. This is because the distribution is skewed to the right.

5.5.7 We have that $\psi(\mu) = x_{0.25} = \mu + \sigma_0 z_{0.25}$, where $z_{0.25}$ satisfies $\Phi(z_{0.25}) = .25$.

5.5.8 First, recall that the third moment of the distribution is $E_\mu(X^3)$. So $\psi(\mu)$ is given by

$$\begin{aligned}\psi(\mu) &= E_\mu(X^3) = E_\mu((X-\mu+\mu)^3)\\ &= E_\mu((X-\mu)^3) + 3\mu E_\mu((X-\mu)^2) + 3\mu^2 E_\mu((X-\mu)) + \mu^3\\ &= 0 + 3\mu\sigma_0^2 + 0 + \mu^3 = 3\mu\sigma_0^2 + \mu^3.\end{aligned}$$

5.5.9 We have that $\psi(\mu) = F_\mu(3) = P_\mu(X \le (3-\mu)/\sigma_0) = \Phi((3-\mu)/\sigma_0)$.

5.5.10 We have that $\psi(\mu, \sigma^2) = x_{0.25} = \mu + \sigma z_{0.25}$, where $\Phi(z_{0.25}) = .25$.

5.5.11 We have that $\psi(\mu, \sigma^2) = F_{(\mu,\sigma^2)}(3) = P_{(\mu,\sigma^2)}(X \le 3) = P(Z \le (3-\mu)/\sigma) = \Phi((3-\mu)/\sigma)$.

5.5.12 We have that $\psi(\theta) = (1-\theta)^2 + \theta^2$.

5.5.13 We have that $\psi(\theta) = 2\theta(1-\theta)$.

5.5.14 First, recall that the coefficient of variation is given by σ_X/μ_X. So $\psi(\theta) = \sqrt{\theta^2/12}/(\theta/2) = 1/\sqrt{3}$. So we know $\psi(\theta)$ exactly and do not require data to make inference about this quantity.

5.5.15 We have that $\psi(\theta) = \alpha_0/\beta^2$.

Computer Exercises

5.5.16
(a) The order statistics are given by $x_{(1)} = 1.2$, $x_{(2)} = 1.8$, $x_{(3)} = 2.3$, $x_{(4)} = 2.5$, $x_{(5)} = 3.1$, $x_{(6)} = 3.4$, $x_{(7)} = 3.7$, $x_{(8)} = 3.9$, $x_{(9)} = 4.3$, $x_{(10)} = 4.4$, $x_{(11)} = 4.5$, $x_{(12)} = 4.8$, $x_{(13)} = 5.6$, $x_{(14)} = 5.8$, $x_{(15)} = 6.9$, $x_{(16)} = 7.2$, and $x_{(17)} = 8.5$.
(b) $\hat{F}_X(x_{(i)}) = i/n$ (there are no ties).
(c) The sample mean $\bar{x} = 4.345$ and the sample variance $s^2 = 3.345$.
(d) The sample median is 4.350 and the $IQR = 2.225$.
(e) Since the distribution looks somewhat skewed, the descriptive statistics in part (c) are appropriate for measuring location and spread.
(f) The sample mean $\bar{x} = 4.845$ and the sample variance $s^2 = 7.874$, while the sample median is 4.450 and the $IQR = 2.575$. As we can see, the sample mean and sample variance changed quite a lot, while the sample median and the IQR have hardly changed. This suggests that the median and the IQR are more resistant to extreme observations.

5.5.17
(a) The order statistics are given by $x_{(1)} = 59.8$, $x_{(2)} = 60.9$, $x_{(3)} = 61.4$, $x_{(4)} = 61.5$, $x_{(5)} = 61.6$, $x_{(6)} = 61.9$, $x_{(7)} = 62.5$, $x_{(8)} = 63.1$, $x_{(9)} = 63.4$,

$x_{(10)} = 63.6$, $x_{(11)} = 64.0$, $x_{(12)} = 64.2$, $x_{(13)} = 64.3$, $x_{(14)} = 64.3$, $x_{(15)} = 64.4$, $x_{(16)} = 64.9$, $x_{(17)} = 64.9$, $x_{(18)} = 65.0$, $x_{(19)} = 65.0$, $x_{(20)} = 65.1$, $x_{(21)} = 65.8$, $x_{(22)} = 65.8$, $x_{(23)} = 66.3$, $x_{(24)} = 66.3$, $x_{(25)} = 66.4$, $x_{(26)} = 66.5$, $x_{(27)} = 66.6$, $x_{(28)} = 66.8$, $x_{(29)} = 67.8$, and $x_{(30)} = 71.4$.

(b) The graph empirical distribution function is plotted as follows. Note that there are two values at 64.3, two values at 64.9, two values at 65.0, two values at 65.8, and two values at 66.3, so the empirical cdf jumps by 2/30 at these points. Otherwise, the jump is 1/30 at a data point.

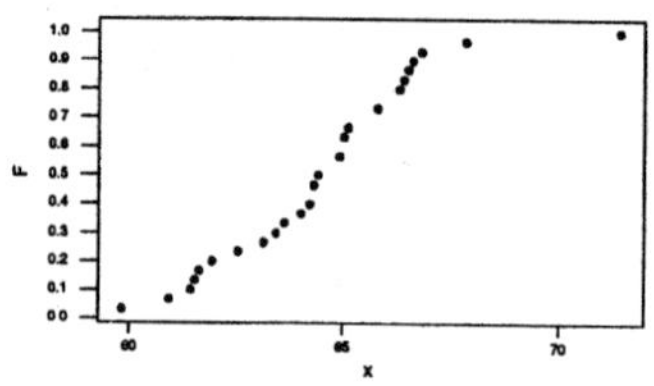

(c) The sample median is 64.650 and the sample $IQR = 66.300 - 62.950 = 3.35$. The boxplot is given below and there is one outlier, namely 71.4.

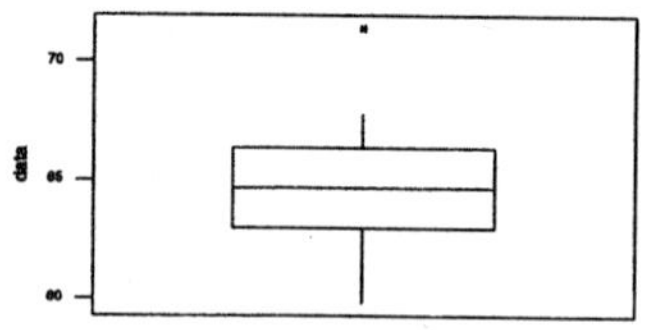

(d) Since the shape of the distribution is somewhat skewed to the left, the median and the IQR are the appropriate descriptive statistics for the location and spread.

(e) The sample median is still 64.650 and the sample $IQR = 66.325 - 62.950 = 3.375$, so these values barely change. The boxplot is given below and identifies two outliers, 71.4 and 84.9.

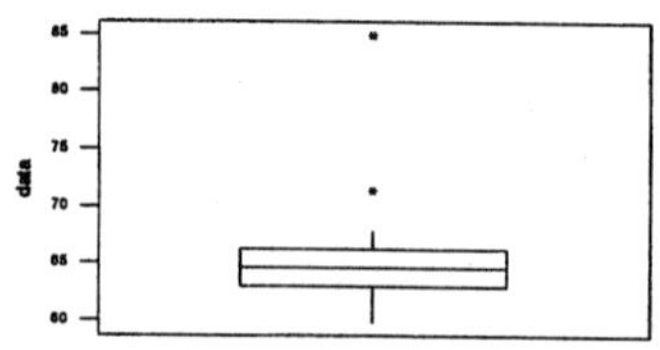

Problems

5.5.18 Using (5.5.3), we have that $\tilde{x}_{.5} = x_{(i-1)} + n\left(x_{(i)} - x_{(i-1)}\right)\left(.5 - \frac{i-1}{n}\right)$, where $(i-1)/n < 1/2 \le i/n$. Now $i - 1 < n/2 \le i$ implies that $i = n/2$ when n is even and $i = \lceil n/2 \rceil$ when n is odd. So we have that

$$\tilde{x}_{.5} = \begin{cases} x_{(n/2)} & n \text{ even} \\ x_{(\lceil n/2 \rceil - 1)} + n\left(x_{(\lceil n/2 \rceil)} - x_{(\lceil n/2 \rceil - 1)}\right)\left(.5 - \frac{i-1}{n}\right) & n \text{ odd.} \end{cases}$$

5.5.19
(a) We have that $\tilde{F}(x_{(i)}) = \hat{F}(x_{(i)}) = i/n, \tilde{F}(x_{(i+1)}) = \hat{F}(x_{(i+1)}) = (i+1)/n$, and

$$\frac{\hat{F}(x_{(i+1)}) - \hat{F}(x_{(i)})}{x_{(i+1)} - x_{(i)}} \ge 0$$

shows that $\tilde{F}(x)$ is an increasing function from 0 to 1.
(b) Since $\tilde{F}$ is linear on each interval $(x_{(i)}, x_{(i+1)}]$ it is continuous there. Therefore, $\tilde{F}$ is continuous on $(x_{(1)}, \infty)$. It is also continuous on $\left(-\infty, x_{(1)}\right)$ and right-continuous at $x_{(1)}$. Therefore, $\tilde{F}$ is right-continuous everywhere.
(c) From (a) there is an i such that $(i-1)/n < p \le i/n$ and then

$$\begin{aligned} p &= \hat{F}(x_{(i-1)}) + \frac{\hat{F}(x_{(i)}) - \hat{F}(x_{(i-1)})}{x_{(i)} - x_{(i-1)}}\left(\tilde{x}_p - x_{(i-1)}\right) \\ &= \frac{i-1}{n} + \frac{1}{n}\frac{1}{x_{(i)} - x_{(i-1)}}\left(\tilde{x}_p - x_{(i-1)}\right) \end{aligned}$$

so

$$\tilde{x}_p = x_{(i-1)} + n\left(x_{(i)} - x_{(i-1)}\right)\left(p - \frac{i-1}{n}\right).$$

Chapter 6

Likelihood Inference

6.1 The Likelihood Function

Exercises

6.1.1 The appropriate statistical model is the Binomial(n, θ), where $\theta \in \Omega = [0,1]$ is the probability of having this antibody in the blood (we can also think of θ as the unknown proportion of the population who have this antibody in their blood). The likelihood function is given by $L(\theta\,|\,s) = \binom{n}{s}\theta^s(1-\theta)^{n-s}$, where s is the number of people whose result was positive. The likelihood function for $n = 10$ people and $s = 3$ is given by $L(\theta\,|\,3) = \binom{10}{3}\theta^3(1-\theta)^7$, and the graph of this function is given below.

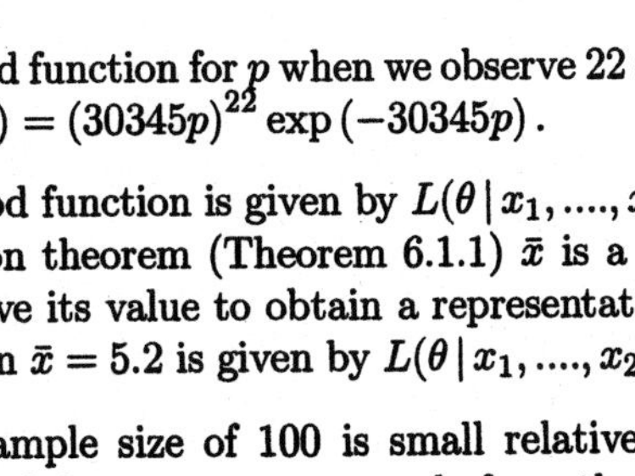

6.1.2 The likelihood function for p when we observe 22 suicides with $N = 30,345$ is given by $L(p\,|\,22) = (30345p)^{22}\exp(-30345p)$.

6.1.3 The likelihood function is given by $L(\theta\,|\,x_1, ..., x_{20}) = \theta^{20}\exp(-(20\bar{x})\,\theta)$. By the factorization theorem (Theorem 6.1.1) $\bar{x}$ is a sufficient statistic, so we only need to observe its value to obtain a representative likelihood. The likelihood function when $\bar{x} = 5.2$ is given by $L(\theta\,|\,x_1, ..., x_{20}) = \theta^{20}\exp(-20\,(5.2)\,\theta)$.

6.1.4 Since the sample size of 100 is small relative to the total population size, we can think of the counts as a sample from the Multinomial$(1, \theta_1, \theta_2, \theta_3)$

distribution. The likelihood function is then given by $L(\theta_1, \theta_2, \theta_3 \,|\, 34, 44, 22) = \theta_1^{34}\theta_2^{44}\theta_3^{22}$.

6.1.5 If we denote the likelihood in Example 6.1.2 by $L_1(\theta \,|\, 4)$ and the likelihood in Example 6.1.3 by $L_2(\theta \,|\, 4)$, then $L_1(\theta \,|\, 4) = cL_2(\theta \,|\, 4)$, where $c = \binom{10}{4}/\binom{9}{3}$.

6.1.6 The likelihood function is given by

$$L(\theta \,|\, x_1, ..., x_n) = \prod_{i=1}^{n} \theta^{x_i}(1-\theta)^{1-x_i} = \theta^{\sum x_i}(1-\theta)^{n-\sum x_i} = \theta^{n\bar{x}}(1-\theta)^{n(1-\bar{x})}.$$

By the factorization theorem $\bar{x}$ is a sufficient statistic. If we differentiate $\ln L(\theta \,|\, x_1, ..., x_n) = n\bar{x}\ln\theta + n(1-\bar{x})\ln(1-\theta)$, we get

$$(\ln L(\theta \,|\, x_1, ..., x_n))' = \frac{n\bar{x}}{\theta} - \frac{n(1-\bar{x})}{1-\theta}$$

and setting this equal to 0 gives the solution $\theta = \bar{x}$. Therefore, we can obtain $\bar{x}$ from the likelihood and we conclude that it is a minimal sufficient statistic.

6.1.7 The likelihood function is given by

$$L(\theta \,|\, x_1, ..., x_n) = \prod_{i=1}^{n} \frac{\theta^{x_i}e^{-\theta}}{x_i!} = \frac{\theta^{n\bar{x}}e^{-n\theta}}{\prod x_i!}$$

By the factorization theorem $\bar{x}$ is a sufficient statistic. If we differentiate $\ln L(\theta \,|\, x_1, ..., x_n) = -\ln\prod x_i! + n\bar{x}\ln\theta - n\theta$, we get

$$(\ln L(\theta \,|\, x_1, ..., x_n))' = \frac{n\bar{x}}{\theta} - n$$

and setting this equal to 0 gives the solution $\theta = \bar{x}$. Therefore, we can obtain $\bar{x}$ from the likelihood and we conclude that it is a minimal sufficient statistic.

Problems

6.1.8 Example 6.1.6 showed that $T : S \to \{0, 1\}$ given by $T(1) = 0$ and $T(2) = T(3) = T(4) = 1$ is a sufficient statistic. Now given the likelihood $L(\cdot \,|\, s)$, we know whether or not the likelihood ratio of a to b is 2 or 4/6, so can identify whether or not $T(s)$ takes the value 0 or 1. Therefore, T is minimal sufficient.

6.1.9 We see that $L(\cdot \,|\, 2) = L(\cdot \,|\, 3)$, so the data values in $\{2, 3\}$ all give the same likelihood ratios. Therefore, $T : S \to \{0, 1\}$ given by $T(1) = 0, T(2) = T(3) = 1$, and $T(4) = 3$ is a sufficient statistic. We also see that once we know the likelihood function $L(\cdot \,|\, s)$, we can determine all the likelihood ratios and so determine if $s = 1, s \in \{2, 3\}$ or $s = 4$ has occurred.

The minimal sufficient statistic in Example 6.1.6 is not sufficient for this model since the data value $s = 4$ does not give the same likelihood ratios when $s = 2$ or $s = 3$.

6.1.10 The likelihood function is given by $L(\mu \,|\, x_1, ..., x_n) = \exp\left(-n(\bar{x} - \mu)^2/2\sigma_0^2\right)$. A likelihood interval has the form

$$\{\mu : \exp\left(-n(\bar{x}-\mu)^2/2\sigma_0^2\right) > c\} = \{\mu : -n(\bar{x}-\mu)^2/2\sigma_0^2 > \ln c\}$$
$$= \left\{\mu : \bar{x} - \frac{\sigma_0}{\sqrt{2n}}\ln c < \mu < \bar{x} + \frac{\sigma_0}{\sqrt{2n}}\ln c\right\} = \left(\bar{x} - \frac{\sigma_0}{\sqrt{2n}}\ln c, \bar{x} + \frac{\sigma_0}{\sqrt{2n}}\ln c\right).$$

So for any constant a, the interval $(\bar{x}-a, \bar{x}+a)$ is a likelihood interval for this model.

6.1.11 We have that the likelihood function is given by $L(\theta \,|\, x_1,...,x_n) = \prod_{i=1}^{n} f_\theta(x_i) = \prod_{i=1}^{n} f_\theta\left(x_{(i)}\right)$, so once we know the order statistics, we know the likelihood function and so they are sufficient.

6.1.12 The likelihood function is given by

$$L(\theta \,|\, x_1,...,x_n) = \prod_{i=1}^{n} \frac{1}{\Gamma(\alpha_0)} (\theta x_i)^{\alpha_0 - 1} \exp\{-\theta x_i\}\, \theta$$
$$= \Gamma^{-n}(\alpha_0) \left(\prod x_i\right)^{\alpha-1} \theta^{n\alpha_0} \exp(-\theta n\bar{x}).$$

By the factorization theorem $\bar{x}$ is a sufficient statistic. The logarithm of the likelihood is given by $\ln L(\theta \,|\, x_1,...,x_n) = \ln\{\Gamma^{-n}(\alpha_0)\left(\prod x_i\right)^{\alpha_0-1}\} + n\alpha \ln\theta - \theta n\bar{x}$. Differentiating this and setting it equal to 0, we obtain $\theta = \alpha/\bar{x}$. So given a likelihood function, we can determine $\bar{x}$ and this proves that $\bar{x}$ is minimal sufficient.

6.1.13 The likelihood function is given by $L(\theta \,|\, x_1,...,x_n) = \theta^{-n} I_{[x_{(n)},\infty)}(\theta)$ when $\theta > 0$. By the factorization theorem $x_{(n)}$ is a sufficient statistic. Now notice that the likelihood function is 0 to the left of $x_{(n)}$ and positive to the right. So given the likelihood, we can determine $x_{(n)}$ and it is minimal sufficient.

6.1.14 The likelihood function is given by

$$L(\theta_1, \theta_2 \,|\, x_1,...,x_n) = \left(\frac{1}{\theta_2 - \theta_1}\right)^n I_{[-\infty, x_{(1)})}(\theta_1)\, I_{[x_{(n)},\infty)}(\theta_2)$$

By the factorization theorem $\left(x_{(1)}, x_{(n)}\right)$ is a sufficient statistic. Now given the likelihood function, we see that the likelihood becomes 0 at $x_{(1)}$ on the left and at $x_{(n)}$ on the right. So given the likelihood, we can determine these points. This implies that $\left(x_{(1)}, x_{(n)}\right)$ is a minimal sufficient statistic.

6.1.15 From the argument in Example 6.1.8 we have that $L(\bar{x}, \sigma^2 \,|\, x_1,...,x_n) \geq L(\mu, \sigma^2 \,|\, x_1,...,x_n)$ for every μ. Further, the argument there shows that, as a function of σ^2, the function $\ln L(\bar{x}, \sigma^2 \,|\, x_1,...,x_n)$ has a critical point at $\hat{\sigma}^2$. The second derivative of this function at $\hat{\sigma}^2$ is given by

$$\left.\frac{\partial^2 \ln L\left((\bar{x},\sigma^2) \,|\, x\right)}{\partial\left(\sigma^2\right)^2}\right|_{\sigma^2=\hat{\sigma}^2} = \left.\frac{\partial}{\partial\sigma^2}\left(-\frac{n}{2\sigma^2} + \frac{n-1}{2\sigma^4}s^2\right)\right|_{\sigma^2=\hat{\sigma}^2}$$
$$= \left.\frac{1}{\sigma^4}\left(\frac{n}{2} - \frac{n-1}{\sigma^2}s^2\right)\right|_{\sigma^2=\hat{\sigma}^2} = \frac{1}{\hat{\sigma}^4}\left(\frac{n}{2} - n\right) < 0$$

\so $L(\bar{x}, \hat{\sigma}^2 \,|\, x_1, ..., x_n) \geq L(\mu, \sigma^2 \,|\, x_1, ..., x_n)$ for every μ and σ^2.

6.1.16 The likelihood function is given by $L(\theta \,|\, x_1, ..., x_n) = \theta^{n\bar{x}}(1-\theta)^{n(1-\bar{x})}$ for $\theta \in [0, .5]$. The factorization theorem establishes that $\bar{x}$ is sufficient. Just as with the full Bernoulli(θ) model, we can determine $\bar{x}$ as the point where this function is maximized — provided that $\bar{x} \in [0, .5]$ — otherwise we do not know the form of this function outside of $[0, .5]$. In general, the maximum value of this likelihood function is attained at $\min\{.5, \bar{x}\}$. When the maximum occurs at .5, we only know that $.5 \leq \bar{x} \leq 1$. But the second derivative of the log of the likelihood is given by

$$-\frac{n\bar{x}}{\theta^2} - \frac{n - n\bar{x}}{(1-\theta)^2} = \bar{x}\left(\frac{n}{(1-\theta)^2} + \frac{n}{\theta^2}\right) - \frac{n}{(1-\theta)^2}$$

so we can determine $\bar{x}$ from this value at any specified $\theta \in (0, .5)$ (since specifying θ allows us to compute $n/(1-\theta)^2, n/\theta^2$ and then knowing the value of the right-hand side allows us to compute $\bar{x}$). Therefore, $\bar{x}$ is minimal sufficient.

6.1.17 The likelihood function is given by

$$L(\theta \,|\, x_1, ..., x_n) = \theta^{x_1}(2\theta)^{x_2}(1-3\theta)^{x_3} = 2^{x_2}\theta^{x_1+x_2}(1-3\theta)^{n-(x_1+x_2)}$$

for $\theta \in [0, 1/6]$. By the factorization theorem $x_1 + x_2$ is sufficient. The log of the likelihood function is given by$(x_1 + x_2)\ln\theta + x_3\ln(1-3\theta)$, which has derivative $(x_1 + x_2)/\theta - 3x_3/(1-3\theta)$. Setting this equal to 0, we obtain $(x_1 + x_2)(1-3\theta) = 3x_3\theta$ and so $\theta = (x_1 + x_2)/3n$. Just as in Problem 6.1.17, this says we can determine $x_1 + x_2$ from the likelihood provided $0 \leq (x_1 + x_2)/3n \leq 1/6$. But taking the second derivative we obtain

$$-\frac{(x_1 + x_2)}{\theta^2} - \frac{9x_3}{(1-3\theta)^2} = (x_1 + x_2)\left(-\frac{1}{\theta^2} + \frac{9n}{(1-3\theta)^2}\right) - \frac{9n}{(1-3\theta)^2}$$

so we can calculate $x_1 + x_2$ from the likelihood. Therefore, $x_1 + x_2$ is minimal sufficient.

6.1.18 The likelihood is given by $L(i \,|\, s) = f_i(s)$ for $i \in \{1, 2\}$. Now note that when $T(s_1) = f_1(s_1)/f_2(s_1) = f_1(s_2)/f_2(s_2) = T(s_2)$

$$L(1 \,|\, s_1) = f_1(s_1) = \frac{f_2(s_1)}{f_2(s_2)} f_1(s_2) = \frac{f_2(s_1)}{f_2(s_2)} L(1 \,|\, s_2)$$

$$L(2 \,|\, s_1) = f_2(s_1) = \frac{f_1(s_1)}{f_1(s_2)} f_2(s_2) = \frac{f_2(s_1)}{f_2(s_2)} L(2 \,|\, s_2)$$

and to T is sufficient. Once we know $L(i \,|\, s)$, we can certainly compute $T(s)$ and so T is minimal sufficient.

Challenge

6.1.19 The likelihood function is given by

$$L(\mu,\sigma\,|\,x_1,...,x_n)=\left(\prod_{i=1}^{n}(x_i-\mu)\right)^{\alpha_0-1}\exp\left\{-n\frac{\bar{x}-\mu}{\sigma}\right\}\left(\frac{1}{\sigma}\right)^{n\alpha_0}$$

for $\mu > x_{(1)}, \sigma > 0$ and is 0 otherwise.

Now observe that the logarithm of the likelihood function is given by

$$\ln L(\mu,\sigma\,|\,x_1,...,x_n)=(\alpha_0-1)\sum_{i=1}^{n}\ln\left(x_{(i)}-\mu\right)-n\frac{\bar{x}-\mu}{\sigma}-n\alpha_0\ln\sigma$$

and

$$\frac{\partial}{\partial\mu}\ln L(\mu,\sigma\,|\,x_1,...,x_n)=-(\alpha_0-1)\sum_{i=1}^{n}\frac{1}{x_{(i)}-\mu}+\frac{n}{\sigma}\mu.$$

(a) When $\alpha_0 = 1$ the likelihood function is determined by $\left(x_{(1)},\bar{x}\right)$, so $\left(x_{(1)},\bar{x}\right)$ is sufficient. Given the likelihood, we can determine $x_{(1)}$ (this is the point where the likelihood becomes 0), and $\bar{x}$ is the point where the the derivative of the log of the likelihood becomes 0. Therefore, we can determine $\left(x_{(1)},\bar{x}\right)$ from the likelihood and it is minimal sufficient.

(b) When $\alpha_0 \neq 1$ this derivative is infinite at each order statistic and nowhere else. So when $\alpha_0 \neq 1$ we can calculate the order statistic from the likelihood by determining every point where the log of the likelihood has an infinite derivative. Also, by Problem 6.1.12 the order statistic is sufficient. Therefore, the order statistic is minimal sufficient in this case.

6.2 Maximum Likelihood Estimation

Exercises

6.2.1 The MLEs are $\hat{\theta}(1)=a, \hat{\theta}(2)=b,\ \hat{\theta}(3)=b, \hat{\theta}(4)=a.$

6.2.2 The likelihood function is given by $L(\theta\,|\,x_1,...,x_n)=\theta^{n\bar{x}}(1-\theta)^{n(1-\bar{x})}$. The log-likelihood function is given by $l(\theta\,|\,x_1,...,x_n)=n\bar{x}\ln\theta+n(1-\bar{x})\ln(1-\theta)$. The score function is given by

$$S(\theta\,|\,x_1,...,x_n)=\frac{n\bar{x}}{\theta}-\frac{n(1-\bar{x})}{1-\theta}$$

Solving the score equation gives $\hat{\theta}(x_1,...,x_n)=\bar{x}$. Note that since $0\le\bar{x}\le1$ we have that

$$\left.\frac{\partial S(\theta\,|\,x_1,...,x_n)}{\partial\theta}\right|_{\theta=\bar{x}}=-\left.\frac{n\bar{x}}{\theta^2}-\frac{n(1-\bar{x})}{(1-\theta)^2}\right|_{\theta=\bar{x}}=-\frac{n}{\bar{x}}-\frac{n}{1-\bar{x}}<0$$

So $\bar{x}$ is indeed the MLE.

6.2.3 Since $\psi(\theta) = \theta^2$ is a 1-1 transformation of θ when θ is restricted to $[0, 1]$, we can apply Theorem 6.2.1, so the MLE is $\psi(\hat{\theta}(x_1, ..., x_n)) = \bar{x}^2$.

6.2.4 The likelihood function is given by $L(\theta \,|\, x_1, ..., x_n) = e^{-n\theta}\theta^{n\bar{x}}$, the log-likelihood function is given by $l(\theta \,|\, x_1, ..., x_n) = -n\theta + n\bar{x}\ln\theta$, and the score function is given by

$$S(\theta \,|\, x_1, ..., x_n) = -n + \frac{n\bar{x}}{\theta}.$$

Solving the score equation gives $\hat{\theta}(x_1, ..., x_n) = \bar{x}$. Note that since $\bar{x} \geq 0$, we have

$$\left.\frac{\partial S(\theta \,|\, x_1, ..., x_n)}{\partial\theta}\right|_{\theta=\bar{x}} = -\left.\frac{n\bar{x}}{\theta^2}\right|_{\theta=\bar{x}} = -\frac{n}{\bar{x}} < 0$$

so $\bar{x}$ is the MLE.

6.2.5 The likelihood function is given by $L(\theta \,|\, x_1, ..., x_n) = \theta^{n\alpha_0}\exp(-n\bar{x}\theta)$, the log-likelihood function is given by $l(\theta \,|\, x_1, ..., x_n) = n\alpha_0 \ln\theta - n\bar{x}\theta$, and the score function is given by $S(\theta \,|\, x_1, ..., x_n) = n\alpha_0/\theta - n\bar{x}$. Solving the score equation gives $\hat{\theta}(x_1, ..., x_n) = \alpha_0/\bar{x}$. Note that since $\bar{x} > 0$ we have that

$$\left.\frac{\partial S(\theta \,|\, x_1, ..., x_n)}{\partial\theta}\right|_{\theta=\frac{\alpha_0}{\bar{x}}} = -\left.\frac{n\alpha_0}{\theta^2}\right|_{\theta=\frac{\alpha_0}{\bar{x}}} = -\frac{n\bar{x}^2}{\alpha_0} < 0,$$

so $\hat{\theta} = \alpha_0/\bar{x}$ is the MLE.

6.2.6 First, note that each x_i comes from a Geometric(θ) distribution. The likelihood function is then given by $L(\theta \,|\, x_1, ..., x_n) = \theta^n (1-\theta)^{n\bar{x}}$, the log-likelihood function is given by $l(\theta \,|\, x_1, ..., x_n) = n\ln\theta + n\bar{x}\ln(1-\theta)$, and the score function is given by

$$S(\theta \,|\, x_1, ..., x_n) = \frac{n}{\theta} - \frac{n\bar{x}}{1-\theta}.$$

Solving the score equation gives $\hat{\theta}(x_1, ..., x_n) = 1/(1+\bar{x})$. Note that since $0 \leq \bar{x} \leq 1$, we have that

$$\left.\frac{\partial S(\theta \,|\, x_1, ..., x_n)}{\partial\theta}\right|_{\hat{\theta}=\frac{1}{1+\bar{x}}} = -\frac{n}{\theta^2} - \left.\frac{n\bar{x}}{(1-\theta)^2}\right|_{\hat{\theta}=\frac{1}{1+\bar{x}}} = -n\left((1+\bar{x})^2 + \frac{(1+\bar{x})^2}{\bar{x}}\right)$$
$$< 0.$$

So $\hat{\theta} = 1/(1+\bar{x})$ is the MLE.

6.2.7 The likelihood function is given by

$$L(\alpha \,|\, x_1, ..., x_n) = \left(\frac{\Gamma(\alpha+1)}{\Gamma(\alpha)}\right)^n \prod_{i=1}^n x_i^{\alpha-1} = \left(\prod_{i=1}^n x_i\right)^{\alpha-1}$$

The log-likelihood function is given by

$$l(\alpha \mid x_1, \ldots, x_n) = n \ln(\Gamma(\alpha+1)) - n \ln(\Gamma(\alpha)) + (\alpha - 1) \sum_{i=1}^{n} \ln x_i$$

The score function is given by

$$\begin{aligned} S(\alpha \mid x_1, \ldots, x_n) &= \frac{n (\Gamma(\alpha+1))'}{\Gamma(\alpha+1)} - \frac{n\Gamma'(\alpha)}{\Gamma(\alpha)} + \sum_{i=1}^{n} \ln x_i \\ &= \frac{n(\Gamma(\alpha) + \alpha\Gamma'(\alpha))}{\alpha\Gamma(\alpha)} - \frac{n\Gamma'(\alpha)}{\Gamma(\alpha)} + \sum_{i=1}^{n} \ln x_i = \frac{n}{\alpha} + \sum_{i=1}^{n} \ln x_i. \end{aligned}$$

Then the solution to the score equation is given by

$$\hat{\alpha} = -\frac{n}{\sum_{i=1}^{n} \ln x_i}.$$

The second derivative of the score at $\hat{\alpha}$ is given by

$$-\frac{n}{\alpha^2}\bigg|_{\alpha=\hat{\alpha}} = -\frac{n}{\hat{\alpha}^2} < 0$$

so $\hat{\alpha}$ is the MLE.

6.2.8 The likelihood function is given by

$$L(\beta \mid x_1, \ldots, x_n) = \beta^n \left(\prod_{i=1}^{n} x_i\right)^{\beta-1} \exp\left(-\sum_{i=1}^{n} x_i^{\beta}\right),$$

the log-likelihood function is given by

$$l(\beta \mid x_1, \ldots, x_n) = n \ln \beta + (\beta - 1)\left(\sum_{i-1}^{n} \ln x_i\right) - \sum_{i=1}^{n} x_i^{\beta},$$

and the score equation is given by

$$S(\beta \mid x_1, \ldots, x_n) = \frac{n}{\beta} + \sum_{i=1}^{n} \ln x_i - \sum_{i=1}^{n} x_i^{\beta} \ln x_i = 0.$$

6.2.9 The likelihood function is given by

$$L(\alpha \mid x_1, \ldots, x_n) = \prod_{i=1}^{n} \alpha (1 + x_i)^{-(\alpha+1)} = \alpha^n \left(\prod_{i=1}^{n} (1 + x_i)\right)^{-(\alpha+1)},$$

the log-likelihood function is given by

$$l(\alpha \mid x_1, \ldots, x_n) = n \ln \alpha - (\alpha + 1) \sum_{i=1}^{n} \ln(1 + x_i),$$

and the score function is given by

$$S(\alpha \,|\, x_1, ..., x_n) = \frac{n}{\alpha} - \sum_{i=1}^{n} \ln(1 + x_i).$$

Solving the score equation gives

$$\hat{\alpha}(x_1, ..., x_n) = \frac{n}{\sum_{i=1}^{n} \ln(1 + x_i)}.$$

Note also that $\frac{\partial}{\partial \alpha} S(\alpha \mid x_1, ..., x_n) = -\frac{n}{\alpha^2} < 0$ for every α, so $\hat{\alpha}$ is the MLE.

6.2.10 The likelihood function is given by

$$L(\tau \,|\, x_1, ..., x_n) = \left(\frac{1}{\sqrt{2\pi}\tau}\right)^n \exp\left(-\sum_{i=1}^{n} \frac{(\ln x_i)^2}{2\tau^2}\right) \prod_{i=1}^{n} \frac{1}{x_i},$$

the log-likelihood function is given by

$$l(\tau \,|\, x_1, ..., x_n) = -\frac{n}{2}\ln(2\pi) - n\ln\tau - \frac{1}{2\tau^2}\sum_{i=1}^{n}(\ln x_i)^2 + \sum_{i=1}^{n}\ln\frac{1}{x_i},$$

and the score function is given by

$$S(\tau \,|\, x_1, ..., x_n) = -\frac{n}{\tau} + \frac{1}{\tau^3}\sum_{i=1}^{n}(\ln x_i)^2.$$

Solving the score equation gives

$$\hat{\tau}(x_1, ..., x_n) = \pm\sqrt{\frac{\sum_{i=1}^{n}(\ln x_i)^2}{n}}$$

and since $\tau > 0$, we take the positive root. Now

$$\left.\frac{\partial S(\tau \,|\, x_1, ..., x_n)}{\partial \tau}\right|_{\tau=\hat{\tau}} = \left.\frac{n}{\tau^2} - \frac{3}{\tau^4}\sum_{i=1}^{n}(\ln x_i)\right|_{\tau=\hat{\tau}} = -\frac{2n^2}{\sum_{i=1}^{n}(\ln x_i)^2} < 0.$$

So $\hat{\tau}$ is the MLE.

Problems

6.2.11

(a) The counts are distributed Multinomial$\left(\theta^2, 2\theta(1-\theta), (1-\theta)^2\right)$.

(b) The likelihood function is given by

$$L(\theta \,|\, s_1, ..., s_n) = \theta^{2x_1}(2\theta(1-\theta))^{x_2}(1-\theta)^{2x_3} = 2^{x_2}\theta^{2x_1+x_2}(1-\theta)^{x_2+2x_3},$$

the log-likelihood function is given by

$$l(\theta \,|\, s_1, ..., s_n) = x_2\ln 2 + (2x_1 + x_2)\ln\theta + (x_2 + 2x_3)\ln(1-\theta),$$

and the score function is given by

$$S(\theta \mid s_1, ..., s_n) = \frac{2x_1 + x_2}{\theta} - \frac{x_2 + 2x_3}{1 - \theta}.$$

(c) Solving the score equation gives

$$\hat{\theta}(s_1, ..., s_n) = \frac{2x_1 + x_2}{2(x_1 + x_2 + x_3)}.$$

Since

$$\frac{\partial S(\theta \mid s_1, ..., s_n)}{\partial \theta} = -\frac{2x_1 + x_2}{\theta^2} - \frac{x_2 + 2x_3}{(1-\theta)^2} < 0$$

for every $\theta \in [0, 1]$ this is the MLE for θ.

6.2.12 First, recall that the MLE for μ is $\bar{x}$ (Example 6.2.2). The parameter of interest now is $\psi(\mu) = P_\mu(X < 1) = \Phi(1 - \mu)$, where Φ is the cdf of a $N(0, 1)$. Since $\Phi(1 - \mu)$ is a strictly decreasing function μ, then ψ is a 1-1 function of μ. Hence, we can apply Theorem 6.2.1 and conclude that $\hat{\psi} = \Phi(1 - \bar{x})$ is the MLE.

6.2.13 The log-likelihood function is $l(\mu \mid x_1, ..., x_n) = -n(\bar{x} - \mu)^2/2$, and, as a function of μ, its graph is a concave parabola and its maximum value occurs at $\bar{x}$. So if $\bar{x} \geq 0$, this is the MLE. If $\bar{x} < 0$, however, the maximum occurs at 0 and this is the MLE.

6.2.14 By the factorization theorem $L(\theta \mid s) = f_\theta(s) = h(s) g_\theta(T(s))$. The probability function for T is given by

$$f_{\theta T}(t) = \sum_{\{s:T(s)=t\}} f_\theta(s) = \sum_{\{s:T(s)=t\}} h(s) g_\theta(T(s)) = g_\theta(t) \sum_{\{s:T(s)=t\}} h(s).$$

So the likelihood based on the observed value $T(s) = t$ is given by$L(\theta \mid t) = g_\theta(t)$, and this is positive multiple times the likelihood based on the observed s. Therefore, the MLE based on s is the same as the MLE based on T.

6.2.15
(a) First, note that $\theta_3 = 1 - \theta_1 - \theta_2$, so the likelihood function is only a function of θ_1 and θ_2 and is given by $L(\theta_1, \theta_2 \mid x_1, x_2, x_3) = \theta_1^{x_1} \theta_2^{x_2} (1 - \theta_1 - \theta_2)^{x_3}$. The log-likelihood function is then given by $l(\theta_1, \theta_2 \mid x_1, x_2, x_3) = x_1 \ln \theta_1 + x_2 \ln \theta_2 + x_3 \ln(1 - \theta_1 - \theta_2)$. Using the methods discussed in Section 6.2.1 we obtain the score function as

$$S(\theta_1, \theta_2 \mid x_1, x_2, x_3) = \begin{pmatrix} \frac{x_1}{\theta_1} - \frac{x_3}{1-\theta_1-\theta_2} \\ \frac{x_2}{\theta_2} - \frac{x_3}{1-\theta_1-\theta_2} \end{pmatrix}$$

The score equation is given by

$$\frac{x_1}{\theta_1} - \frac{x_3}{1 - \theta_1 - \theta_2} = 0, \quad \frac{x_2}{\theta_2} - \frac{x_3}{1 - \theta_1 - \theta_2} = 0$$

so $x_1 = (x_1 + x_3)\,\theta_1 + x_1\theta_2$, and $x_2 = x_2\theta_1 + (x_2 + x_3)\,\theta_2$. The solution to this system of linear equations is given by

$$\hat{\theta}_1 = \frac{x_1}{x_1 + x_2 + x_3} = \frac{x_1}{n}, \quad \hat{\theta}_2 = \frac{x_2}{x_1 + x_2 + x_3} = \frac{x_2}{n}.$$

Also note that the matrix of second partial derivatives is given by

$$\frac{\partial S(\theta_1, \theta_2 \,|\, x_1, x_2, x_3)}{\partial \theta} = \begin{pmatrix} -\frac{x_1}{\theta_1^2} - \frac{x_3}{(1-\theta_1-\theta_2)^2} & -\frac{x_3}{(1-\theta_1-\theta_2)^2} \\ -\frac{x_3}{(1-\theta_1-\theta_2)^2} & -\frac{x_2}{\theta_2^2} - \frac{x_3}{(1-\theta_1-\theta_2)^2} \end{pmatrix}$$

and evaluated at $\left(\hat{\theta}_1, \hat{\theta}_2\right)$ this equals

$$-n^2 \begin{pmatrix} \frac{1}{x_1} + \frac{1}{x_3} & \frac{1}{x_3} \\ \frac{1}{x_3} & \frac{1}{x_2} + \frac{1}{x_3} \end{pmatrix}.$$

Now the negative of this matrix has $(1,1)$ entry greater than 0 and its determinant equals

$$\left(\frac{1}{x_1} + \frac{1}{x_3}\right)\left(\frac{1}{x_1} + \frac{1}{f_3}\right) - \left(\frac{1}{x_3}\right)^2 > 0$$

so the matrix is positive definite. This implies that the matrix of second partial derivatives of the log-likelihood evaluated at $\left(\hat{\theta}_1, \hat{\theta}_2\right)$ is negative definite. Therefore,

$$\left(\hat{\theta}_1, \hat{\theta}_2, \hat{\theta}_3\right) = \left(\frac{x_1}{n}, \frac{x_2}{n}, 1 - \frac{x_1}{n} - \frac{x_2}{n}\right) = \left(\frac{x_1}{n}, \frac{x_2}{n}, \frac{x_3}{n}\right)$$

is the MLE for $(\theta_1, \theta_2, \theta_3)$.
(b) The plug-in MLE of $\theta_1 + \theta_2^2 - \theta_3^2$ is $x_1/n + (x_2/n)^2 - (x_3/n)^2$.

6.2.16 The likelihood function is given by

$$L(\theta_1, \theta_2 \,|\, x_1, ..., x_n) = \left(\frac{1}{\theta_2 - \theta_1}\right)^n I_{[-\infty, x_{(1)})}(\theta_1)\, I_{[x_{(n)}, \infty)}(\theta_2).$$

Fixing θ_2, we see that $L(\cdot, \theta_2 \,|\, x_1, ..., x_n)$ is largest when $\theta_2 - \theta_1$ is smallest, and this occurs when $\theta_1 = x_{(1)}$. Now $L(x_{(1)}, \cdot \,|\, x_1, ..., x_n)$ is largest when $\theta_2 - x_{(1)}$ is smallest, and this occurs when $\theta_2 = x_{(n)}$. Therefore, $L(\theta_1, \theta_2 \,|\, x_1, ..., x_n) \leq L(x_{(1)}, \theta_2 \,|\, x_1, ..., x_n) \leq L(x_{(1)}, x_{(n)} \,|\, x_1, ..., x_n)$ and $\left(x_{(1)}, x_{(n)}\right)$ is the MLE.

Computer Problem

6.2.17
(a) Assuming that the individuals are independent (sample size small relative to the population size), the log-likelihood function is given by $4 \ln \theta + 16 \ln(1 - \theta)$. The plot of this function is provided here (note that it goes to $-\infty$ at 0 and 1). We can determine the MLE exactly in this case as $\hat{\theta} = 4/16 = 0.25$.

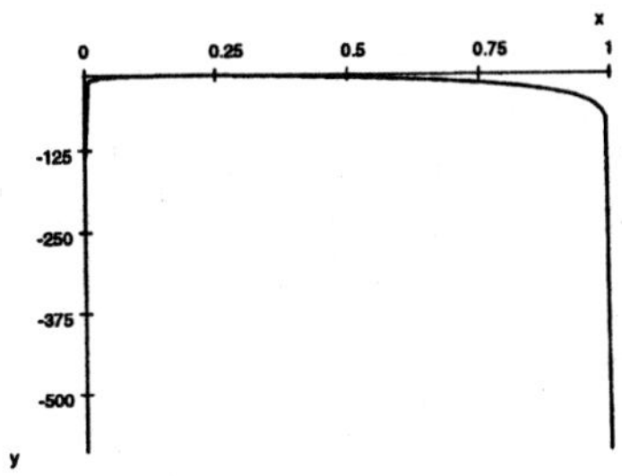

(b) Now the sample size is not small relative to the population size, so the number of left-handed individuals in the sample is distributed Hypergeometric$(50, 50\theta, 20)$. Note that θ is no longer a continuous variable but must take a value in $0, 1/5, 2/50, \ldots, 49/50, 1\}$. The log-likelihood is then given by (ignoring the denominator in the hypergeometric)

$$\begin{aligned}
&\ln \binom{50\theta}{4} + \ln \binom{50\,(1-\theta)}{16} \\
&= \ln \Gamma\,(50\theta + 1) - \ln \Gamma\,(50\theta - 4 + 1) - \ln \Gamma\,(4+1) + \\
&\ln \Gamma\,(50\,(1-\theta) + 1) - \ln \Gamma\,(50\,(1-\theta) - 16 + 1) - \ln \Gamma\,(16+1)
\end{aligned}$$

for $200\theta = 4, 5, \ldots, 34$. Ignoring the $\ln \Gamma\,(4+1)$ and $\ln \Gamma\,(16+1)$ (as they don't involve θ), we plot the log-likelihood below. From the tabulation required for this plot we obtain the MLE as $\hat{\theta} = .22$.

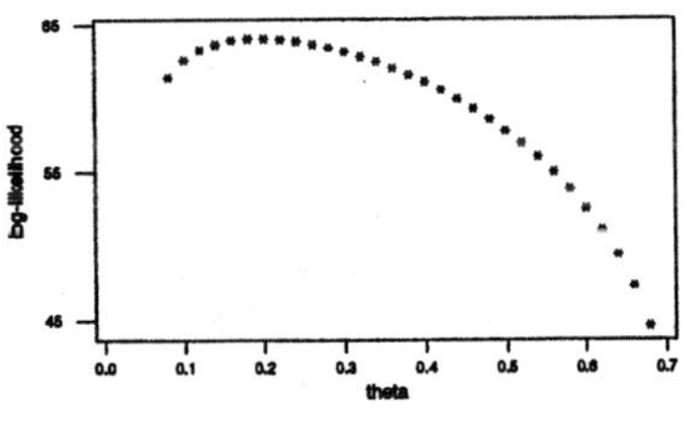

Challenge

6.2.18 First we write the density as

$$f_\theta\,(x) = \begin{cases} \frac{1}{2}\exp\,(\theta - x) & \theta \le x \\ \frac{1}{2}\exp\,(x - \theta) & \theta \ge x. \end{cases}$$

The log-likelihood function is then given by

$$l\,(\theta\,|\,x_1, \ldots, x_n) = \sum_{\theta \le x_{(j)}} \left(\theta - x_{(j)}\right) + \sum_{\theta \ge x_{(j)}} \left(x_{(j)} - \theta\right).$$

When $\theta < x_{(1)}$, $l(\theta \mid x_1, ..., x_n) = n\theta - \sum_{i=1}^n x_{(i)}$, and this is maximized by taking $\theta = x_{(1)}$, giving the value $nx_{(1)} - \sum_{i=1}^n x_{(i)} \le 0$.

When $\theta \ge x_{(n)}$, $l(\theta \mid x_1, ..., x_n) = \sum_{i=1}^n x_{(i)} - n\theta$, and this is maximized by taking $\theta = x_{(n)}$, giving the value $\sum_{i=1}^n x_{(i)} - nx_{(n)} \le 0$.

When $x_{(i)} \le \theta < x_{(i+1)}$,

$$
\begin{aligned}
l(\theta \mid x_1, ..., x_n) &= \sum_{j=1}^{i} \left(x_{(j)} - \theta\right) + \sum_{j=i+1}^{n} \left(\theta - x_{(j)}\right) \\
&= (n-2i)\theta + \sum_{j=1}^{i} x_{(j)} - \sum_{j=i+1}^{n} x_{(j)} = (n-2i)\theta + 2\sum_{j=1}^{i} x_{(j)} - \sum_{j=1}^{n} x_{(j)}
\end{aligned}
$$

and this is maximized (provided $n \ne 2i$) by taking $\theta = x_{(i+1)}$ when $i \le n/2$ and by $\theta = x_{(i)}$ when $i > n/2$. When $n = 2i$ all values in $x_{(i)} \le \theta < x_{(i+1)}$ are maximizers.

When $n = 1$, then $\hat{\theta} = x_{(1)}$. Now suppose $n > 1$. We have that $i = 1 \le n/2$ and

$$
(n-2)x_{(2)} + 2\sum_{j=1}^{1} x_{(j)} - \sum_{j=1}^{n} x_{(j)} = nx_{(2)} - \sum_{j=1}^{n} x_{(j)} \ge nx_{(1)} - \sum_{j=1}^{n} x_{(j)}.
$$

Now suppose $i < i + 1 \le n/2$. Then

$$
\begin{aligned}
&(n-2i)x_{(i+1)} + 2\sum_{j=1}^{i} x_{(j)} - \sum_{j=1}^{n} x_{(j)} \\
&= (n - 2(i+1))x_{(i+2)} + 2\sum_{j=1}^{i+1} x_{(j)} - \sum_{j=1}^{n} x_{(j)} + (n-2i)\left(x_{(i+1)} - x_{(i+2)}\right) \\
&\le (n - 2(i+1))x_{(i+2)} + 2\sum_{j=1}^{i+1} x_{(j)} - \sum_{j=1}^{n} x_{(j)}.
\end{aligned}
$$

If $n/2 \le i < i + 1$, then

$$
\begin{aligned}
&(n-2i)x_{(i)} + 2\sum_{j=1}^{i} x_{(j)} - \sum_{j=1}^{n} x_{(j)} \\
&= (n - 2(i+1))x_{(i+1)} + 2\sum_{j=1}^{i+1} x_{(j)} - \sum_{j=1}^{n} x_{(j)} + (n-2i)\left(x_{(i)} - x_{(i+1)}\right) \\
&\ge (n - 2(i+1))x_{(i+2)} + 2\sum_{j=1}^{i+1} x_{(j)} - \sum_{j=1}^{n} x_{(j)}
\end{aligned}
$$

and finally when $i = n$, then $(n-2n)x_{(n)} + 2\sum_{j=1}^n x_{(j)} - \sum_{j=1}^n x_{(j)} = \sum_{j=1}^n x_{(j)} - nx_{(n)}$.

When n is odd this argument shows that $l(\theta \mid x_1, ..., x_n)$ increases in $(-\infty, x_{(\lfloor n/2 \rfloor)})$ and decreases in $[x_{(\lfloor n/2 \rfloor)}, \infty)$, so $\hat{\theta} = x_{(\lfloor n/2 \rfloor)}$ (the middle value). When n is even this argument shows that $l(\theta \mid x_1, ..., x_n)$ increases in $(-\infty, x_{(n/2)})$, is constant in $[x_{(\lfloor n/2 \rfloor)}, x_{(\lfloor n/2 \rfloor + 1)})$, and decreases in $[x_{(\lfloor n/2 \rfloor)}, \infty)$, so any value $\hat{\theta} \in [x_{(\lfloor n/2 \rfloor)}, x_{(\lfloor n/2 \rfloor + 1)})$ is a maximizer.

6.3 Inferences Based on the MLE

Exercises

6.3.1 This is a two-sided z-test with the z statistic equal to -0.54 and the P-value equal to 0.592, which is very high. So we conclude that we don't have any evidence against H_0. A .95-confidence interval for the unknown μ is $(4.442, 5.318)$. Note that the confidence interval contains the value 5, which confirms our conclusion using the above test.

6.3.2 This is a two-sided t-test with the t statistic equal to -0.55 and the P-value equal to 0.599, which is very high. We conclude that we don't have enough evidence against H_0. A .95-confidence interval for the unknown μ is $(4.382, 5.378)$. Note that the confidence interval contains the value 5, which confirms our conclusion using the above test.

6.3.3 This is a two-sided z-test with the z statistic equal to 5.14 and the P-value equal to 0.000. So we conclude that we have enough evidence against H_0 being true. A .95-confidence interval for the unknown μ is $(63.56, 67.94)$. Note that the confidence interval does not contain the value 60, which confirms our conclusion using the above test.

6.3.4 This is a two-sided t-test with the t statistic equal to 9.12 and (using the Student(3) distribution) the P-value equals 0.452, which is not small and so we don't reject the null hypothesis. A .95-confidence interval for the unknown μ is $(44.55, 86.95)$. Note that the confidence interval contains the value 60.

6.3.5 If we assume that the population variance is known then under H_0 we have $Z = \frac{X - \mu_0}{\sigma_0} \sim N(0,1)$ and the P-value then is given by

$$P\left(Z \geq \left|\frac{x_0 - \mu_0}{\sigma_0}\right|\right) = P\left(Z \geq \left|\frac{52 - 60}{\sqrt{5}}\right|\right) = 2\left(1 - \Phi\left|\frac{52 - 60}{\sqrt{5}}\right|\right)$$
$$= 2(1 - .99983) = .00034$$

and a .95 confidence interval for μ is given by

$$[x_0 - z_{0.975}\sigma_0, x_0 + z_{0.975}\sigma_0] = \left[52 - 1.96\sqrt{5}, 52 + 1.96\sqrt{5}\right] = [47.617, 56.383]$$

Note that both the P-value and the .95 confidence interval indicate that there is evidence against H_0 being true.

If we don't assume that the population variance is known, then, since we only have a single observation the sample variance is 0, and we do not have a sensible estimate of the population variance. So we can't use the t procedures to compute the P-value and construct a confidence interval. The minimum sample size n for which inference is possible, without the assumption that the population variance is known, is 2.

6.3.6 A .99 confidence interval for μ is given by $(22.70, 29.72)$. The P-value for testing $H_0 : \mu = 24$ is 0.099, so we conclude that there isn't much evidence against H_0 being true. Note also that the .99 confidence interval for μ contains the value 24.

6.3.7 To detect if these results are statistically significant or not we need to perform a z-test for testing $H_0 : \mu = 1$. The P-value is given by

$$P\left(|Z| \geq \left|\frac{1.05-1}{\sqrt{0.1/100}}\right|\right) = 2\left[1 - \Phi\left(1.581\,1\right)\right] = 2\left(1 - 0.9431\right) = 0.1138.$$

So these results are not statistically significant at the 5% level, so we have no evidence against $H_0 : \mu = 1$. Also, the observed difference of $1.05 - 1 = .05$ is well within the range that the manufacturer thinks is of practical significance. So the test has detected a small difference that is not practically significant.

6.3.8 Based on a two-sided z-test, the z-statistic (using standard error $\sqrt{.65(.32)/250}$) equals -0.994490 and the P-value equals 0.32. So we conclude that there is no evidence against H_0 being true. A .90-confidence interval for θ is given by $(0.559832, 0.680168)$, which includes the value 0.65, so agrees with the result of the above test.

6.3.9 Based on a two-sided z-test to assess $H_0 : \theta = 0.5$, the z-statistic is equal to 0.63 and the P-value is equal to 0.527. So we conclude that there is no evidence against H_0 being true, in other words, there are not enough evidence to conclude that the coin is unfair.

6.3.10 Let θ be the probability of head on a single toss. The sample sizes required so that the margin of error (half of the length) of a $\gamma = 0.95$ confidence interval for θ is less then 0.05, 0.025, 0.005 are given by

$$n \geq \frac{1}{4}\left(\frac{z_{\frac{1+\gamma}{2}}}{\delta}\right)^2$$

So for $\delta = 0.1$ $n > 384.15, \delta = 0.05$ $n \geq 1536.6$ and $\delta = 0.01$ $n \geq 38415$.

6.3.11 Based on a two-sided z-test to assess $H_0 : \theta = \frac{1}{6}$, the z-statistic is equal to 2.45 and the P-value is equal to 0.014. So we can conclude that at the 5% significance level, there is evidence to conclude that the die is biased.

6.3.12 The sample size that will guarantee that a 0.95-confidence interval for μ is no longer than 1 is given by

$$n \geq \sigma_0^2\left(\frac{z_{\frac{1+\gamma}{2}}}{\delta}\right)^2 = 2\left(\frac{1.96}{0.5}\right)^2 = 30.732$$

So the minimum sample size is 31.

Computer Exercises

6.3.13 The sample size that will guarantee that a 0.95-confidence interval for μ is no longer than 1 is given by

$$n \geq 25 \left(\frac{t_{.975}(n-1)}{\delta} \right)^2.$$

When n is large then $t_{.975}(n-1) \approx z_{.975} = 1.96$, and in that case

$$n \geq \frac{25}{.5^2}(1.96)^2 = 384.16$$

So the minimum sample size is 385. Now when $n = 400$ we have that $t_{.975}(400) = 1.9659$ and

$$400 \geq \frac{25}{.5^2}(1.9659)^2 = 386.48$$

so $n = 400$ suffices.

6.3.14 The power function is given by ($z_{.975} = 1.96$)

$$1 - \Phi\left(\frac{.5}{\sqrt{2}/\sqrt{n}} + 1.96 \right) + \Phi\left(\frac{.5}{\sqrt{2}/\sqrt{n}} - 1.96 \right).$$

A partial tabulation of the power function (as a function of n) is given below. We see that $n = 63$ is the appropriate sample size.

60 0.78190
61 0.78853
62 0.79500
63 0.80129
64 0.80742
65 0.81339
66 0.81919
67 0.82484

6.3.15 We expect to observe approximately 950 confidence intervals containing the true value of θ. In practice, we do not observe exactly this number. The number covering will be less for sample size $n = 5$ than for sample size $n = 20$.

6.3.16 As n increases, you should observe that the proportion of intervals that actually contains 0 increases as s becomes a better estimate of $\sigma = 1$.

Problems

6.3.17 Using $c(x_1, ..., x_n) = \bar{x} + k(\sigma_0/\sqrt{n})$, we have that k satisfies

$$P(\mu \leq \bar{x} + k(\sigma_0/\sqrt{n})) = P\left(\frac{\bar{x} - \mu}{\sigma_0/\sqrt{n}} \geq -k \right) = P(Z \geq -k) \geq \gamma$$

So $k = -z_{1-\gamma} = z_\gamma$, i.e., the γ-percentile of a $N(0,1)$ distribution.

6.3.18 The P-value for testing $H_0 : \mu \leq \mu_0$ is given by

$$\begin{aligned}\max_{\mu \in H_0} P_\mu \left(\frac{\bar{X} - \mu}{\sigma_0/\sqrt{n}} > \frac{\bar{x}_o - \mu}{\sigma_0/\sqrt{n}} \right) &= \max_{\mu \in H_0} P \left(Z > \frac{\bar{x}_o - \mu}{\sigma_0/\sqrt{n}} \right) \\ &= \max_{\mu \in H_0} \left(1 - \Phi \left(\frac{\bar{x}_o - \mu}{\sigma_0/\sqrt{n}} \right) \right)\end{aligned}$$

Since $(1 - \Phi((\bar{x}_o - \mu) / (\sigma_0/\sqrt{n})))$ is an increasing function of μ, its maximum is at $\mu = \mu_0$.

6.3.19 The form of the power function associated with the above hypothesis assessment procedure is given by

$$\begin{aligned}\beta(\mu) &= P_\mu \left(1 - \Phi \left(\frac{\bar{X} - \mu_0}{\sigma_0/\sqrt{n}} \right) < \alpha \right) = P_\mu \left(\Phi \left(\frac{\bar{X} - \mu_0}{\sigma_0/\sqrt{n}} \right) > 1 - \alpha \right) \\ &= P_\mu \left(\frac{\bar{X} - \mu_0}{\sigma_0/\sqrt{n}} > z_{1-\alpha} \right) = P_\mu \left(\frac{\bar{X} - \mu}{\sigma_0/\sqrt{n}} > \frac{\mu_0 - \mu}{\sigma_0/\sqrt{n}} + z_{1-\alpha} \right) \\ &= 1 - \Phi \left(\frac{\mu_0 - \mu}{\sigma_0/\sqrt{n}} + z_{1-\alpha} \right).\end{aligned}$$

6.3.20 Using $c(x_1, ..., x_n) = \bar{x} + k(\sigma_0/\sqrt{n})$ we have that k satisfies

$$P\left(\mu \geq \bar{x} + k(\sigma_0/\sqrt{n})\right) = P\left(\frac{\bar{X} - \mu}{\sigma_0/\sqrt{n}} \leq -k \right) = P(Z \leq -k) \geq \gamma$$

So $k = -z_\gamma = z_{1-\gamma}$, i.e., the $1 - \gamma$ percentile of a $N(0, 1)$ distribution.
The P-value for testing $H_0 : \mu \geq \mu_0$ is given by

$$\max_{\mu \in H_0} P_\mu \left(\frac{\bar{X} - \mu}{\sigma_0/\sqrt{n}} < \frac{\bar{x}_o - \mu}{\sigma_0/\sqrt{n}} \right) = \max_{\mu \in H_0} P \left(Z < \frac{\bar{x}_o - \mu}{\sigma_0/\sqrt{n}} \right) = \max_{\mu \in H_0} \Phi \left(\frac{\bar{x}_o - \mu}{\sigma_0/\sqrt{n}} \right).$$

Since $\Phi((\bar{x}_o - \mu) / (\sigma_0/\sqrt{n}))$ is a decreasing function of μ, its maximum is at $\mu = \mu_0$.

6.3.21 Using $c(x_1, ..., x_n) = \bar{x} + ks/\sqrt{n}$, we have that k satisfies

$$P\left(\mu \leq \bar{X} + ks/\sqrt{n}\right) = P\left(\frac{\bar{X} - \mu}{s/\sqrt{n}} \geq -k \right) \geq \gamma.$$

So $k = -t_{1-\gamma}(n-1) = t_\gamma(n-1)$, i.e., the γ percentile of a $t(n-1)$ distribution.
The P-value for testing $H_0 : \mu \leq \mu_0$ is given by

$$\max_{\mu \in H_0} P_\mu \left(\frac{\bar{X} - \mu}{\sigma_0/\sqrt{n}} > \frac{\bar{x}_o - \mu}{\sigma_0/\sqrt{n}} \right) = \max_{\mu \in H_0} \left(1 - G \left(\frac{\bar{x}_o - \mu}{\sigma_0/\sqrt{n}}; n - 1 \right) \right).$$

Since $(1 - G((\bar{x}_o - \mu) / (\sigma_0/\sqrt{n}); n - 1))$ is an increasing function of μ, its maximum is at $\mu = \mu_0$.

6.3.22 Using $c(x_1, ..., x_n) = Ks^2$ we have that k satisfies $P(\sigma^2 \le kS^2) = P\left(\frac{(n-1)S^2}{\sigma^2} \ge \frac{n-1}{k}\right) \ge \gamma$. So $k = (n-1)/\chi^2_{1-\gamma}(n-1)$.

6.3.23 The P-value for testing $H_0 : \sigma^2 \le \sigma_0^2$ is given by

$$\max_{(\mu,\sigma^2)\in H_0} P_\mu\left(S^2 > s^2\right) = \max_{(\mu,\sigma^2)\in H_0} P_\mu\left(\frac{(n-1)S^2}{\sigma^2} > \frac{(n-1)s^2}{\sigma^2}\right)$$
$$= \max_{(\mu,\sigma^2)\in H_0}\left(1 - H\left(\frac{(n-1)s^2}{\sigma^2}; n-1\right)\right) = \left(1 - H\left(\frac{(n-1)s^2}{\sigma_0^2}; n-1\right)\right)$$

since $\left(1 - H\left((n-1)s_0^2/\sigma^2; n-1\right)\right)$ is an increasing function of σ^2.

6.3.24 We have that

$$\beta\left(\mu,\sigma^2\right) = P_{(\mu,\sigma^2)}\left(1 - H\left(\frac{(n-1)S^2}{\sigma_0^2}; n-1\right) < \alpha\right).$$
$$= P_{(\mu,\sigma^2)}\left(H\left(\frac{(n-1)S^2}{\sigma_0^2}; n-1\right) > 1-\alpha\right)$$
$$= P_{(\mu,\sigma^2)}\left(\frac{(n-1)S^2}{\sigma_0^2} > \chi^2_{1-\alpha}(n-1)\right)$$
$$= P_{(\mu,\sigma^2)}\left(\frac{(n-1)S^2}{\sigma^2} > \frac{\sigma_0^2}{\sigma^2}\chi^2_{1-\alpha}(n-1)\right) = 1 - H\left(\frac{\sigma_0^2}{\sigma^2}\chi^2_{1-\alpha}(n-1); n-1\right).$$

6.3.25 To detect if these results are statistically significant or not we need to perform a t-test for testing $H_0 : \mu = 1$. The P-value is given by

$$P\left(|T| \ge \left|\frac{1.05 - 1}{\sqrt{0.083/100}}\right|\right) = 2\left[1 - G(1.7355; 99)\right] = 2(1 - .95712) = 0.08576.$$

Since the P-value is greater then 5%, these result are not statistically significant at the 5% level, so we have no evidence against $H_0 : \mu = 1$.

The P-value for testing $H_0 : \sigma^2 \le \sigma_0^2$ is given by $\left(1 - H\left((n-1)s^2/\sigma_0^2; n-1\right)\right)$. Using $\sigma_0^2 = 0.01, s^2 = 0.083, n = 100$ we obtain P-value equal to 0. So we have enough evidence against H_0, i.e., the result is statistically significant and we have evidence that the process is not under control.

Challenges

6.3.26 Equation (6.3.11) is given by

$$\left[\varphi\left(\frac{\mu_0 - \mu}{\sigma_0/\sqrt{n}} - z_{1-\frac{\alpha}{2}}\right) - \varphi\left(\frac{\mu_0 - \mu}{\sigma_0/\sqrt{n}} + z_{1-\frac{\alpha}{2}}\right)\right]\frac{\mu_0 - \mu}{\sigma_0}.$$

Put $x = \sqrt{n}(\mu_0 - \mu)/\sigma_0$. If $x < 0$, then $x - z_{1-\frac{\alpha}{2}} < x + z_{1-\frac{\alpha}{2}} < -x + z_{1-\frac{\alpha}{2}} = -\left(x - z_{1-\frac{\alpha}{2}}\right)$. Since $\varphi\left(x - z_{1-\frac{\alpha}{2}}\right) = \varphi\left(-\left(x - z_{1-\frac{\alpha}{2}}\right)\right)$ and $\varphi(z)$ increases to the left of 0 and decreases on the right, this implies that (6.3.11) is nonnegative.

If $x > 0$, then $-\left(x + z_{1-\frac{\alpha}{2}}\right) < x - z_{1-\frac{\alpha}{2}} < x + z_{1-\frac{\alpha}{2}}$ and again (6.3.11) is nonnegative.

6.3.27 Equation (6.3.12) is given by

$$\left[\varphi\left(\frac{\mu_0 - \mu}{\sigma_0/\sqrt{n}} + z_{1-\frac{\alpha}{2}}\right) - \varphi\left(\frac{\mu_0 - \mu}{\sigma_0/\sqrt{n}} - z_{1-\frac{\alpha}{2}}\right)\right]\frac{\sqrt{n}}{\sigma_0}.$$

Put $x = \sqrt{n}\,(\mu_0 - \mu)\,/\sigma_0$. Then if $x < 0$, we have that $x - z_{1-\frac{\alpha}{2}} < x + z_{1-\frac{\alpha}{2}} < -x + z_{1-\frac{\alpha}{2}} = -\left(x - z_{1-\frac{\alpha}{2}}\right)$. Since $\varphi\left(x - z_{1-\frac{\alpha}{2}}\right) = \varphi\left(-\left(x - z_{1-\frac{\alpha}{2}}\right)\right)$ and $\varphi(z)$ increases to the left of 0 and decreases on the right, this implies that (6.3.12) is positive when $\mu > \mu_0$. When $x = 0$, clearly (6.3.12) equals 0. When $x > 0$ then $-\left(x + z_{1-\frac{\alpha}{2}}\right) < x - z_{1-\frac{\alpha}{2}} < x + z_{1-\frac{\alpha}{2}}$, and this implies that (6.3.12) is less than 0 when $\mu < \mu_0$.

6.4 Distribution-free Methods

Exercises

6.4.1 An approximate .95-confidence interval for μ_3 is given by

$$m_3 \pm z_{\frac{1+\gamma}{2}}\frac{s_3}{\sqrt{n}} = (26.027, 151.373)$$

since $m_3 = 88.7, z_{.975} = 1.96$, and $s_3 = 143.0$.

6.4.2 Recall that, the variance of a random variable can be expressed in terms of the moments as $\sigma_X^2 = \mu_2 - \mu_1^2$. Hence, the method of moments estimator of the population variance is given by $\hat{\sigma}_X^2 = m_2 - m_1^2$. To check if this estimator is unbiased we compute

$$\begin{aligned} E\left(m_2 - m_1^2\right) &= \mu_2 - \left(\operatorname{Var}(m_1) + E^2(m_1)\right) = \mu_2 - \left(\frac{1}{n}\left(\mu_2 - \mu_1^2\right) + \mu_1^2\right) \\ &= \left(1 - \frac{1}{n}\right)\sigma_X^2 \end{aligned}$$

Hence, this estimator is not unbiased.

6.4.3 The method of moments estimator of the coefficient of variation of a random variable X is $\sqrt{m_2 - m_1^2}/m_1$. Now let $Y = cX$. The $E(Y) = cE(X)$ and $\operatorname{Var}(Y) = c^2\operatorname{Var}(X)$. Therefore, the coefficient of variation of Y is

$$c\,\mathrm{Sd}(X)\,/cE(X) = \mathrm{Sd}(X)\,/E(X)$$

which is the coefficient of variation of X.

6.4.4 Let $\psi(\mu) = \exp(\mu)$ then $\psi'(\mu) = \exp(\mu)$. By the delta theorem (6.4.1), an approximate γ-confidence interval for $\psi(\mu)$ is given by

$$\exp\bar{x} \pm \frac{s\exp(\bar{x})}{\sqrt{n}}z_{\frac{1+\gamma}{2}} = \exp(2.9) - \frac{2.997\exp(2.9)}{\sqrt{20}}1.96 = (-5.6975, 42.046).$$

6.4.5 Recall from, Problem 3.4.10 that the moment generating function of a $X \sim N(\mu, \sigma^2)$ is given by $m_X(s) = \exp\left(\mu s + \frac{\sigma^2 s^2}{2}\right)$ Then, by Theorem 3.4.3 the third moment in given by

$$m_X'''(0) = 3\sigma^2\left(\mu + \sigma^2 s\right)e^{\mu s + \frac{1}{2}\sigma^2 s^2} + \left(\mu + \sigma^2 s\right)^3 e^{\mu s + \frac{1}{2}\sigma^2 s^2}\Big|_{s=0} = 3\sigma^2\mu + \mu^3$$

The plug-in estimator of μ_3 is given by $\hat{\mu}_3 = 3\left(m_2 - m_1^2\right)m_1 + m_1^3$, while the method of moments estimator of μ_3 is $m_3 = \frac{1}{n}\sum x_i^3$. So these estimators are different.

6.4.6 The t-statistic for testing $H_0 : \mu = 3$ is 0.47 and the P-value (based on 9 df) is 0.650. Hence, we don't have evidence against H_0 .

To test the hypothesis $H_0 : x_{.5}(\theta) = 3$ using the sign test statistic, we have $S = \sum_{i=1}^{n} I_{(-\infty,3]}(x_i) = 5$. The P-value is given by $P(\{i : |i - 5| \geq 0\}) = 1$. Therefore, we don't have evidence against H_0 .

The following boxplot of the data indicates that the normal assumption is a problem as it is strongly skewed to the right. Under these circumstances we prefer the sign test.

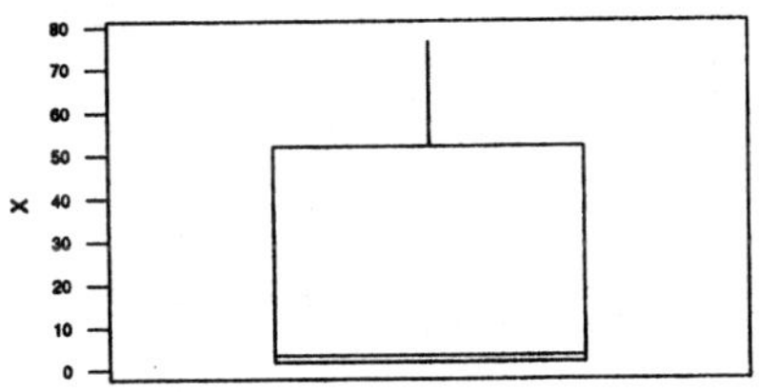

6.4.7 The empirical cdf is given by the following table. The sample median is estimated by $-.03$ and the first quartile is -1.28, while the third quartile is .98. The value $F(2)$ is estimated by $\hat{F}(2) = \hat{F}(1.36) = .90$.

i	$x_{(i)}$	$\hat{F}\left(x_{(i)}\right)$	i	$x_{(i)}$	$\hat{F}\left(x_{(i)}\right)$
1	−1.42	0.06	11	0.00	0.55
2	−1.35	0.10	12	0.38	0.60
3	−1.34	0.15	13	0.40	0.65
4	−1.29	0.20	14	0.44	0.70
5	−1.28	0.25	15	0.98	0.75
6	−1.02	0.30	16	1.06	0.80
7	−0.58	0.35	17	1.06	0.85
8	−0.35	0.40	18	1.36	0.90
9	−0.24	0.45	19	2.05	0.95
10	−0.03	0.50	20	2.13	1.00

Computer Exercises

6.4.8 To test the hypothesis $H_0 : x_{.5}(\theta) = 0$ the sign test statistic is given by $S = \sum_{i=1}^{n} I_{(-\infty,0]}(x_i) = 10$. The P-value is given by $P(\{i : |i - 10| \geq 0\}) = 1$. Hence, we don't have any evidence against H_0.

We have that

$$P(\{i : |i - 10| \geq 10\}) = \binom{20}{0}\left(\frac{1}{2}\right)^{20} = 1.9073 \times 10^{-6}$$

$$P(\{i : |i - 10| \geq 9\}) = 2\binom{20}{0}\left(\frac{1}{2}\right)^{20} + 2\binom{20}{1}\left(\frac{1}{2}\right)^{20} = 4.0054 \times 10^{-5}$$

$$P(\{i : |i - 10| \geq 8\}) = 2\binom{20}{0}\left(\frac{1}{2}\right)^{20} + 2\binom{20}{1}\left(\frac{1}{2}\right)^{20} + 2\binom{20}{2}\left(\frac{1}{2}\right)^{20}$$
$$= 4.0245 \times 10^{-4}$$

$$P(\{i : |i - 10| \geq 7\}) = \sum_{j=0}^{3} 2\binom{20}{j}\left(\frac{1}{2}\right)^{20} = 2.5768 \times 10^{-3}$$

$$P(\{i : |i - 10| \geq 6\}) = \sum_{j=0}^{4} 2\binom{20}{j}\left(\frac{1}{2}\right)^{20} = 1.1818 \times 10^{-2}$$

$$P(\{i : |i - 10| \geq 5\}) = \sum_{j=0}^{5} 2\binom{20}{j}\left(\frac{1}{2}\right)^{20} = 1.1818 \times 10^{-2}$$

$$P(\{i : |i - 10| \geq 4\}) = \sum_{j=0}^{6} 2\binom{20}{j}\left(\frac{1}{2}\right)^{20} = 0.11532.$$

Therefore, $j = 15$ and a .95-confidence interval is given by $[x_{(6)}, x_{(15)}) = [-1.02, 0.98)$. The exact coverage probability of this interval is $1 - 1.1818 \times 10^{-2} = 0.98818$.

6.4.9 To test the hypothesis $H_0 : x_{.25}(\theta) = -1.0$ the sign test statistic is given by $S_0 = \sum_{i=1}^{n} I_{(-\infty,-1.0]}(x_i) = 6$. The P-value, using (6.4.6), is given by $P(\{i : \binom{20}{i}(0.25)^i(0.75)^{20-i} \leq \binom{20}{6}(0.25)^6(0.75)^{14}\})$, and a tabulation of the Binomial$(20, .25)$ probability function reveals that this set is given by all the points except $\{5, 4\}$, so the P-value is given by $1 - \binom{20}{5}(0.25)^5(0.75)^{15} - \binom{20}{4}(0.25)^4(0.75)^{16} = 0.60798$ and we have no evidence against H_0.

6.4.10 The characteristic of the distribution we are interested in is $\psi(\theta) = T(F_\theta) = \mu_3$, which we estimate by $T(\hat{F}) = m_3 = 88.7442$. We want to estimate the MSE of the plug-in MLE of μ_3, which is given by $\hat{\psi} = m_1^3 + 3m_1s^2 = (2.9)^3 + 3(2.9)(2.997)^2 = 102.53$. First, the squared bias in this estimator is given by $(\hat{\psi} - T(\hat{F}))^2 = \left(m_1^3 + 3m_1\hat{\sigma}^2 - m_3\right)^2 = (102.53 - 88.7442)^2 = 190.05$.

Next, based on 10^3 samples, we obtained $\widehat{\text{Var}}_{\hat{F}}(\hat{\psi}) = 956.598$. Hence, $\widehat{\text{MSE}_\theta}(\hat{\psi}) = 102.53 + 956.598 = 1059.1$. Note that, based on 10^4 samples, we obtained $\widehat{\mathrm{Var}}_{\hat{F}}(\hat{\psi}) = 981.057$ and, based on 10^5 samples, we obtained $\widehat{\mathrm{Var}}_{\hat{F}}(\hat{\psi}) = 973.434$. Hence, $m = 1000$ is a large enough sample for accurate results.

The Minitab code for carrying out these simulations is given below.

```
gmacro
bootstrapping
base 34256734
note - original sample is stored in c1
note - bootstrap sample is placed in c2 (each one overwritten)
note - third moments of bootstrap samples are stored in c4 for more
analysis
note - k1 = size of data set (and bootstrap samples)
let k1=15
do k2=1:1000
sample 20 c1 c2;
replace.
let c3=c2**3
let c4(k2)=mean(c3)
enddo
note - k3 equals (6.4.5)
let k3=(stdev(c4))**2
print k3
endmacro
```

6.4.11 The characteristic of the $N(\mu, \sigma^2)$ distribution that we are interested in is $\psi(\theta) = T(F_\theta) = x_{(0.25)}(\theta) = \mu + \sigma z_{0.25}$, which we estimate by $T(\hat{F}) = \hat{x}_{0.25} = 0.15$, i.e., the sample first quartile. We want to estimate the MSE of the plug-in MLE of $x_{(0.25)}(\theta)$, which is given by $\hat{\psi} = m_1 + s z_{0.25} = 2.9 + (2.997)(-0.6745) = 0.87852$. The squared bias in this estimator is given by $(\hat{\psi} - t(\hat{F}))^2 = (0.87852 - 0.15)^2 = 0.53074$.

Based on a 10^3 samples, the variance of this estimator, is estimated as $\widehat{\mathrm{Var}}_{\hat{F}}(\hat{\psi}) = 1.85568$. Hence, $\widehat{\mathrm{MSE}}_\theta(\hat{\psi}) = 0.53074 + 1.85568 = 2.3864$. Based on a 10^4 samples, the variance of this estimator is estimated as $\widehat{\mathrm{Var}}_{\hat{F}}(\hat{\psi}) = 1.89582$. Hence, $\widehat{\mathrm{MSE}}_\theta(\hat{\psi}) = 0.53074 + 1.89582 = 2.4266$.

The Minitab code for this simulation is given below.

```
gmacro
bootstrapping
base 34256734
note - original sample is stored in c1
note - bootstrap sample is placed in c2 (each one overwritten)
note - first quartiles of bootstrap samples are stored in
       c4 for more analysis
note - k1 = size of data set (and bootstrap samples)
let k1=15
do k2=1:20000
sample 20 c1 c2;
```

```
replace.
sort c2 c3
let c4(k2)=c3(5)
enddo
note - k3 equals (6.4.5)
let k3=(stdev(c4))**2
print k3
endmacro
```

6.4.12 The characteristic of the $N(\mu, \sigma^2)$ distribution that we are interested in is $\psi(\mu, \sigma^2) = t(F_{(\mu,\sigma^2)}) = F_{(\mu,\sigma^2)}(3) = \Phi\left(\frac{3-\mu}{\sigma}\right)$, where Φ is the cdf of the $N(0, 1)$ distribution. The plug-in estimator of $\psi(\theta)$ is

$$\hat{\psi}(x_1, ...x_n) = \Phi\left(\frac{3 - 2.9}{2.997}\right) = \Phi\left(3.3367 \times 10^{-2}\right) = 0.5133.$$

The bias squared in this estimator is given by $(\hat{\psi} - t(\hat{F}))^2 = (0.5133 - 0.4)^2 = 0.01284$. Based on a 10^3 samples the variance of this estimator is estimated as $\widehat{\text{Var}}_{\hat{F}}(\hat{\psi}) = 0.0117605$. Hence, $\widehat{\text{MSE}_\theta}(\hat{\psi}) = 0.01284 + 0.0117605 = 2.4601 \times 10^{-2}$. Based on a 10^4 samples, the variance of this estimator is estimated as $\widehat{\text{Var}}_{\hat{F}}(\hat{\psi}) = 0.0118861$. Hence, $\widehat{\text{MSE}_\theta}(\hat{\psi}) = 0.01284 + 0.0118861 = 2.4726 \times 10^{-2}$.

The Minitab code for these simulations is given below.

```
gmacro
bootstrapping
base 34256734
note - original sample is stored in c1
note - bootstrap sample is placed in c2 (each one overwritten)
note - value of the ecdf at 3 of bootstrap samples are stored in
       c5 for more analysis
note - k1 = size of data set (and bootstrap samples)
let k1=15
do k2=1:10000
sample 20 c1 c2;
replace.
sort c2 c3
let c4= c3 le 3
let c5(k2)=mean(c4)
enddo
note - k3 equals (6.4.5)
let k3=(stdev(c5))**2
print k3
endmacro
```

Problems

6.4.13 For a random variable with this distribution, we have that $E_{\hat{F}}\left(X^i\right) = \sum_{j=1}^{n} x_{(j)}^i \left(\hat{F}\left(x_{(j)}\right) - \hat{F}\left(x_{(j-1)}\right)\right)$, where we take $x_{(0)} = -\infty$. Now $\hat{F}\left(x_{(j)}\right) - \hat{F}\left(x_{(j-1)}\right) = 1/n$ since all the $x_{(j)}$ are distinct. This implies the result.

6.4.14 We have that $E_{\hat{F}}\left(X^i\right) = \sum_{j=1}^{n^*}(x_{(j)}^*)^i(\hat{F}(x_{(j)}^*) - \hat{F}(x_{(j)}^*))$, where n^* is the number of distinct values, $x_1^*, \ldots, x_{n^*}^*$ are the distinct values in the sample, $x_{(1)}^*, \ldots, x_{(n^*)}^*$ are the ordered distinct values in the sample, and $\hat{F}(x_{(j)}^*) - \hat{F}(x_{(j)}^*)$ equals the relative frequency of $x_{(j)}^*$ in the original sample.

6.4.15
(a) First, note that for the Poisson distribution we have $\mu_1 = \lambda = \sigma^2$, i.e., the mean and the variance are the same. Now using $\psi(x) = \sqrt{x}$ as a transformation, by the delta theorem, we have $\psi(M_1) = \sqrt{M_1}$ is asymptotically normal with mean $\psi(\mu_1) = \sqrt{\mu_1}$ and variance given by $(\psi'(\mu_1))^2\frac{\sigma^2}{n} = \frac{1}{4\lambda}\frac{\lambda}{n} = \frac{1}{4n}$, which is free of μ_1, and hence this transformation is variance stabilizing.
(b) Using $\psi(x) = \arcsin\sqrt{x}$ as a transformation, by the delta theorem, we have $\psi(M_1) = \arcsin\sqrt{M_1}$ is asymptotically normal with mean $\psi(\mu_1) = \arcsin\sqrt{\mu_1}$ and variance given by

$$\left(\psi'(\mu_1)\right)^2 \frac{\sigma^2}{n} = \left(\frac{1}{2\sqrt{(1-\theta)}\sqrt{\theta}}\right)^2 \frac{\theta(1-\theta)}{n} = \frac{1}{4n},$$

which is free of θ, and hence this transformation is variance stabilizing.
(c) First, we have $\sigma^2 = a\mu_1^2$. Next, the mean of $\psi(M_1) = \ln(M_1)$ is approximately $\psi(\mu_1) = \ln(\mu_1)$ and the variance is approximately

$$\left(\psi'(\mu_1)\right)^2 \frac{\sigma^2}{n} = \frac{1}{\mu_1^2}\frac{a\mu_1^2}{n} = \frac{a}{n},$$

which is free of μ_1, and hence this transformation is variance stabilizing.

Challenges

6.4.16 Let $Y = |X|$, then Y has a distribution on $R^+ = (0, \infty)$ given by

$$\begin{aligned} F_Y(y) &= P(Y \le y) = P(|X| \le y) = P(-y \le X \le y) = F_X(y) - F_X(-y) \\ &= 2F_X(y) - 1 \end{aligned}$$

where the last equality follows by symmetry of the distribution of X. Therefore, the density of Y is $2f$, where f is the density of x.

Next, let $Z = sgn(X)$, then $P(Z = -1) = P(X < 0) = 0.5$, $P(Z = 1) = P(X > 0) = 0.5$, and $P(Z = 0) = P(X = 0) = 0$. Therefore, Z is uniform on $\{-1, 1\}$.

To show that Y and Z are independent we proceed as follows.

$$P(Y \le y, Z = 1) = P(-y \le X \le y, X > 0) = P(0 \le X \le y) = F_X(y) - \frac{1}{2}$$

which is the same as $P(Y \le y)P(Z=1) = (2F_X(y)-1)/2 = F_X(y) - 1/2$. Hence, we have established that Y and Z are independent.

6.4.17

(a) We have that $|x_i - x_0|\, sgn(x_i - x_0) = x_i - x_0$, so $S_o^+ = n(\bar{x} - x_0)$.

(b) Note that, under H_0, $Y = X - x_0$ is distributed from an absolutely continuous distribution that is symmetric about 0. Therefore, by Problem 6.4.17 we have that $|Y|$ and $sgn(Y) = sgn(X - x_0)$ are independent and $sgn(Y)$ is uniform on $\{-1, 1\}$. The conditional distribution of S^+, given the values $|Y_1| = |x_1 - x_0|, \ldots, |Y_n| = |x_n - x_0|$, is therefore determined by $(sgn(Y_1), \ldots, sgn(Y_n))$ and, because of independence, this is uniform on $\{-1,1\}^n$. This implies that the conditional distribution of S^+ is the same no matter which absolutely continuous distribution, symmetric about its median, that we are sampling from. The conditional mean of S^+ is then

$$E\left(S^+ \mid |x_1 - x_0|, \ldots, |x_n - x_0|\right) = \sum_{i=1}^{n} |x_i - x_0|\, E\left(sgn\left(X_i - x_0\right)\right) = 0$$

since $E\left(sgn\left(X_i - x_0\right)\right) = 0$ for each i. Further, it is clear that this conditional distribution is symmetric about 0 since the distribution of each $sgn\left(X_i - x_0\right)$ is symmetric about 0.

(c) We have that

$$S_o^+ = \sum_{i=1}^{n} |x_i - x_0|\, sgn\left(x_i - x_0\right) = .2 - .5 + 1.4 - 1.6 + 3.3 + 2.3 + .1 = 5.2.$$

Now each possible value of $(sgn(X_1 - x_0), \ldots, sgn(X_n - x_0))$ occurs with probability $(1/2)^6 = 1.5625 \times 10^{-2}$ and $4\left(1.5625 \times 10^{-2}\right) = 0.0625$, while $2\left(1.5625 \times 10^{-2}\right) = 0.03125$. So to determine if 5.2 yields a P-value less than .05, we need to evaluate the 4 extreme points (2 on each tail) of the conditional distribution of S^+. Starting from the most extreme values and moving towards the center 0 we have that S^+ takes the values

$$\begin{gathered} .2 + .5 + 1.4 + 1.6 + 3.3 + 2.3 + .1 = 9.4 \\ .2 + .5 + 1.4 + 1.6 + 3.3 + 2.3 - .1 = 9.2 \\ \vdots \\ -.2 - .5 - 1.4 - 1.6 - 3.3 - 2.3 + .1 = -9.2 \\ -.2 - .5 - 1.4 - 1.6 - 3.3 - 2.3 - .1 = -9.4 \end{gathered}$$

so the P-value is greater than .05 and we have no evidence against H_0.

(d) We have that $t = n(\bar{x} - x_0)/s$ and

$$\begin{aligned}(n-1)\,s^2 &= \sum_{i=1}^{n}(x_i - \bar{x})^2 = \sum_{i=1}^{n}(x_i - x_0 + x_0 - \bar{x})^2 \\ &= \sum_{i=1}^{n}|x_i - \bar{x}|^2 + 2\sum_{i=1}^{n}(x_i - x_0)(x_0 - \bar{x}) + n(\bar{x} - \bar{x})^2 \\ &= \sum_{i=1}^{n}|x_i - \bar{x}|^2 - 2n(\bar{x} - x_0)^2 + n(\bar{x} - \bar{x})^2 \\ &= \sum_{i=1}^{n}|x_i - \bar{x}|^2 - n(\bar{x} - \bar{x})^2\end{aligned}$$

and note that $\sum_{i=1}^{n}|x_i - \bar{x}|^2$ is fixed under the conditional distribution. Therefore,

$$t = \frac{n(\bar{x} - x_0)}{s} = \frac{n(\bar{x} - x_0)}{\sqrt{n-1}\sqrt{\sum_{i=1}^{n}|x_i - \bar{x}|^2 - n(\bar{x} - \bar{x})^2}}.$$

Then we see that t is an increasing function of $n(\bar{x} - x_0)$ for $-\sum_{i=1}^{n}|x_i - \bar{x}|^2 \le n(\bar{x}-x_0) \le \sum_{i=1}^{n}|x_i - \bar{x}|^2$ so that t is large whenever S_o^+ is large and conversely.

6.5 Large Sample Behavior of the MLE

Exercises

6.5.1 The score function for the $N(\mu_0, \sigma^2)$ family is given by $S(\sigma^2 \,|\, x_1, ..., x_n) = -\frac{n}{2\sigma^2} + \frac{1}{2\sigma^4}\sum_{i=1}^{n}(x_i - \mu_0)^2$. The Fisher information is then given by

$$nI(\sigma^2) = -E_{\sigma^2}\left(\frac{\partial}{\partial\sigma^2}S(\sigma^2 \,|\, X_1, ..., X_n)\right) = -E_{\sigma^2}\left(\frac{n}{2\sigma^4} - \frac{1}{\sigma^6}\sum_{i=1}^{n}(X_i - \mu_0)^2\right)$$

$$= -\frac{n}{2\sigma^4} + \frac{1}{\sigma^6}E_{\sigma^2}\left(\sum_{i=1}^{n}(X_i - \mu_0)^2\right) = -\frac{n}{2\sigma^4} + \frac{n\sigma^2}{\sigma^6} = \frac{n}{2\sigma^4}.$$

6.5.2 The score function for Gamma(α_0, θ), where α_0 is known, is given by $S(\theta \,|\, x_1, ..., x_n) = n\alpha_0/\theta - n\bar{x}$. The Fisher information is then given by $nI(\theta) = -E_\theta\left(\frac{\partial}{\partial\theta}S(\theta \,|\, X_1, ..., X_n)\right) = -E_\theta\left(-\frac{n\alpha_0}{\theta^2}\right) = \frac{n\alpha_0}{\theta^2}$.

6.5.3 The score function for Pareto(α) is given by $S(\alpha \,|\, x_1, ..., x_n) = n/\alpha - \sum_{i=1}^{n}\ln(1 + x_i)$. The Fisher information is then given by $nI(\alpha) = -E_\alpha\left(\frac{\partial}{\partial\alpha}S(\alpha \,|\, X_1, ..., X_n)\right) = -E_\alpha\left(-\frac{n}{\alpha^2}\right) = \frac{n}{\alpha^2}$.

6.5.4 An approximate .95-confidence for λ in a Poisson model is given by (Example 6.5.5) $\bar{x} \pm (\bar{x}/n)^{1/2} z_{(1+\gamma)/2}$. The average number of calls per day is $\bar{x} = 9.650$. Therefore, the confidence interval is given by $(8.2885, 11.011)$. This contains the value $\lambda_0 = 11$, and therefore we don't have enough evidence against $H_0 : \lambda_0 = 11$ at the 5% level.

An approximate power for this procedure when $\lambda = 10$ is given by

$$\begin{aligned}
&P_{10}\left(2\left\{1-\Phi\left(\sqrt{\frac{n}{\lambda_0}}\left|\bar{X}-\lambda_0\right|\right)\right\} < 0.05\right) \\
&= P_{10}\left(\Phi\left(\sqrt{\frac{20}{11}}\left|\bar{X}-11\right|\right) > 0.975\right) = P_{10}\left(\left|\bar{X}-11\right| > \sqrt{\frac{11}{20}}z_{0.975}\right) \\
&= P_{10}\left((\bar{X}-11) < -\sqrt{\frac{11}{20}}z_{0.975} \text{ or } (\bar{X}-11) > \sqrt{\frac{11}{20}}z_{0.975}\right) \\
&= P_{10}\left((\bar{X}-10)\sqrt{\frac{20}{10}} < \sqrt{\frac{20}{10}} - \sqrt{\frac{11}{10}}z_{0.975}\right) \\
&\quad + P\left((\bar{X}-10)\sqrt{\frac{20}{10}} > \sqrt{\frac{20}{10}} + \sqrt{\frac{11}{10}}z_{0.975}\right) \\
&\approx P(Z < -.64145) + P(Z > 3.4699) = .26062 + .000\,26 = .26088.
\end{aligned}$$

6.5.5 The score function for Gamma$(2, \theta)$ is given by $S(\theta \mid x_1, ..., x_n) = 2n/\theta - n\bar{x}$, so the MLE is $\hat{\theta} = 2/\bar{x} = 2/1627 = 1.2293 \times 10^{-3}$. The Fisher information is then given by,

$$nI(\theta) = -E_\theta\left(\frac{\partial}{\partial\theta}S(\theta \mid X_1, ..., X_n)\right) = -E_\theta\left(-\frac{2n}{\theta^2}\right) = \frac{2n}{\theta^2}.$$

By corollary 6.5.2 we have that

$$\sqrt{\frac{2n}{\hat{\theta}^2}}\left(\hat{\theta} - \theta\right) \xrightarrow{D} N(0,1).$$

Hence, an approximate .90-confidence interval is given by

$$\begin{aligned}
\frac{2}{\bar{x}} \pm \frac{1}{\sqrt{2n}}\left(\frac{2}{\bar{x}}\right)z_{.95} &= \left(1.2293 \times 10^{-3}\right) \pm \frac{1}{\sqrt{54}}\left(1.2293 \times 10^{-3}\right)(1.6449) \\
&= \left(9.5413 \times 10^{-4}, 1.5045 \times 10^{-3}\right).
\end{aligned}$$

6.5.6 The score function for Gamma$(1, \theta)$ is given by $S(\theta \mid x_1, ..., x_n) = n/\theta - n\bar{x}$, so the MLE is $\hat{\theta} = 1/\bar{x} = 1/1627 = 6.1463 \times 10^{-4}$. The Fisher information is then given by

$$nI(\theta) = -E_\theta\left(\frac{\partial}{\partial\theta}S(\theta \mid X_1, ..., X_n)\right) = -E_\theta\left(-\frac{n}{\theta^2}\right) = \frac{n}{\theta^2}.$$

By Corollary 6.5.2 we have that

$$\sqrt{\frac{n}{\hat{\theta}^2}}\left(\hat{\theta} - \theta\right) \xrightarrow{D} N(0,1).$$

Hence, an approximate .90-confidence interval is given by

$$\frac{1}{\bar{x}} \pm \frac{1}{\sqrt{n}}\left(\frac{1}{\bar{x}}\right) z_{.95} = \left(6.1463 \times 10^{-4}\right) \pm \frac{1}{\sqrt{27}}\left(6.1463 \times 10^{-4}\right)(1.6449)$$
$$= \left(4.2006 \times 10^{-4}, 8.0920 \times 10^{-4}\right).$$

Note that this interval is shorter than the one in Exercise 6.5.5 and is shifted to the left.

6.5.7 The score function for Pareto(α) is given by $S(\alpha \mid x_1, ..., x_n) = n/\alpha - \sum_{i=1}^{n} \ln(1 + x_i)$, so the MLE for α is

$$\hat{\alpha} = \frac{n}{\sum_{i=1}^{n} \ln(1 + x_i)}.$$

Using the result of Exercise 6.5.3 the Fisher information is n/α^2. Note this is a continuous function of $\alpha \in (0, \infty)$. Hence, by Corollary 6.5.2 an approximate .95-confidence interval is given by $\hat{\alpha} \pm (\hat{\alpha}/\sqrt{n}) z_{\frac{1+\gamma}{2}}$. Substituting $\hat{\alpha} = 0.322631$, $z_{.975} = 1.96$, we obtain $(0.18123, 0.46403)$ as a .95-confidence interval.

The mean of the Pareto(α) distribution is $1/(\alpha - 1)$. Hence, assessing that the mean income in this population is \$25K is equivalent to assessing $\alpha = 1 + \frac{1}{25} = 1.04$. Since the .95-confidence interval does not contain this value, we have enough evidence against H_0 at the 5% level to conclude that the mean income of this population is not \$25K.

6.5.8 The score function for a sample from Exponential(θ) is given by $S(\theta \mid x_1, ..., x_n) = n/\theta - n\bar{x}$, so the MLE is $\hat{\theta} = 1/\bar{x}$. The Fisher information is given by $nI(\theta) = -E_\theta\left(-\frac{n}{\theta^2}\right) = \frac{n}{\theta^2}$. By Corollary 6.5.2 we have that

$$\sqrt{\frac{n}{\theta^2}}\left(\hat{\theta} - \theta\right) \xrightarrow{D} N(0, 1).$$

A left-sided γ-confidence interval for θ should satisfy $P_\theta\left(\theta \leq c(x_1, ...x_n)\right) \geq \gamma$ for every $\theta > 0$. Using the same method as Exercise 6.3.17 we obtain the interval

$$\left(-\infty, \hat{\theta} + \left(nI\left(\hat{\theta}\right)\right)^{-1/2} z_\gamma\right) = \left(-\infty, \frac{1}{\bar{x}} + \frac{1}{\bar{x}}\frac{z_\gamma}{\sqrt{n}}\right).$$

6.5.9 The likelihood function for the Geometric(θ) is given by $L(\theta \mid x) = \theta(1 - \theta)^x$. The score function is then given by $S(\theta \mid x) = \frac{1}{\theta} - \frac{x}{1-\theta}$. The Fisher information is then

$$I(\theta) = -E_\theta\left(\frac{\partial}{\partial\theta} S(\theta \mid X)\right) = -E_\theta\left(-\frac{1}{\theta^2} - \frac{X}{(1-\theta)^2}\right) = \frac{1}{\theta^2(1-\theta)}$$

since $E_\theta(X) = (1 - \theta)/\theta$.

The score function for a sample is then

$$S(\theta \mid x_1, ..., x_n) = \frac{n}{\theta} - \sum_{i=1}^{n} \frac{x_i}{1-\theta} = \frac{n}{\theta} - \frac{n\bar{x}}{1-\theta},$$

so the MLE for θ in this model is $\hat{\theta} = 1/1 + \bar{x}$ and the Fisher information is given by $n/\left(\theta^2(1-\theta)\right)$. A left-sided γ-confidence interval for θ should satisfy the $P_\theta\left(\theta < c\left(X_1, ..., X_n\right)\right) \geq \gamma$ for every $\theta \in [0,1]$. Using the same method as in Exercise 6.3.17 we obtain the interval

$$\left[0, \min(\hat{\theta} + \left(nI(\hat{\theta})\right)^{-1/2} z_\gamma, 1)\right] = \left[0, \min\left(\frac{1}{1+\bar{x}} + \frac{1}{\sqrt{n}}\frac{1}{1+\bar{x}}\sqrt{\frac{\bar{x}}{1+\bar{x}}} z_\gamma, 1\right)\right].$$

6.5.10 The likelihood function for the Negative-Binomial(r,θ) family is given by (from example 2.3.5) $L(\theta \mid x) = \binom{r-1+x}{x}\theta^r(1-\theta)^x$. The score function is given by $S(\theta \mid x) = \frac{r}{\theta} - \frac{x}{1-\theta}$ and the Fisher information is given by

$$I(\theta) = -E_\theta\left(\frac{\partial}{\partial\theta}S(\theta \mid X)\right) = -E_\theta\left(-\frac{r}{\theta^2} - \frac{X}{(1-\theta)^2}\right) = \frac{r}{\theta^2(1-\theta)}$$

since $E_\theta(X) = r(1-\theta)/\theta$. The score function for a sample is given by

$$S(\theta \mid x_1, ..., x_n) = \frac{rn}{\theta} - \frac{n\bar{x}}{1-\theta},$$

so the MLE for θ in this model is $\hat{\theta} = r/(r+\bar{x})$.

A left-sided γ-confidence interval for θ should satisfy $P_\theta\left(\theta < c\left(X_1, ...X_n\right)\right) \geq \gamma$ for every θ. Using the same method as in Exercise 6.3.17 we obtain the following interval

$$\left(-\infty, \hat{\theta} + \left(nI\left(\hat{\theta}\right)\right)^{-1/2} z_\gamma\right) = \left(-\infty, \frac{r}{r+\bar{x}} + \frac{1}{\sqrt{nr}}\frac{r}{r+\bar{x}}\sqrt{\frac{\bar{x}}{r+\bar{x}}} z_\gamma\right).$$

Problems

6.5.11 (6.5.2) ,(6.5.3), (6.5.4), and (6.5.5) require that

$$\frac{\partial^2 \ln f_\theta(x)}{\partial\theta^2} \text{exists for each } x, E_\theta\left(S(\theta \mid s)\right) = 0,$$
$$E_\theta\left(\frac{\partial^2 \ln f_\theta(X)}{\partial\theta^2} + S^2(\theta \mid X)\right) = 0, \quad E_\theta\left(\left|\frac{\partial^2 \ln f_\theta(X)}{\partial\theta^2}\right|\right) < \infty.$$

We have

$$\frac{\partial^2 \ln f_{\sigma^2}(x)}{\partial\sigma^2} = \frac{1}{2\sigma^4} - \frac{(x-\mu_0)^2}{\sigma^6}$$

and this exists for each x. Also,

$$E_{\sigma^2}\left(S(\theta \mid X)\right) = E_{\sigma^2}\left(-\frac{1}{2\sigma^2} + \frac{(X-\mu_0)^2}{2\sigma^4}\right) = -\frac{1}{2\sigma^2} + \frac{\sigma^2}{2\sigma^4} = 0$$

and

$$E_{\sigma^2}\left(\frac{\partial^2 \ln f_{\sigma^2}(X)}{\partial \sigma^2} + S^2(\sigma^2 \,|\, X)\right)$$
$$= E_{\sigma^2}\left(\frac{1}{2\sigma^4} - \frac{(X-\mu_0)^2}{\sigma^6}\right) + E_{\sigma^2}\left(\left(-\frac{1}{2\sigma^2} + \frac{(X-\mu_0)^2}{2\sigma^4}\right)^2\right)$$
$$= \frac{1}{2\sigma^4} - \frac{1}{\sigma^4} + \frac{1}{4\sigma^4} - \frac{2}{2\sigma^2} E_{\sigma^2}\left(\frac{(X-\mu_0)^2}{2\sigma^4}\right) + \frac{1}{4\sigma^4} E_{\sigma^2}\left(\left(\frac{X-\mu_0}{\sigma}\right)^4\right)$$
$$= -\frac{3}{4\sigma^4} + \frac{1}{4\sigma^4} E\left(Z^4\right) = -\frac{3}{4\sigma^4} + \frac{3}{4\sigma^4} = 0$$

since $E\left(Z^4\right) = 3$ for $Z \sim N(0,1)$. Finally, by the triangular inequality and monotonicity of expected values we have

$$E_{\sigma^2}\left(\left|\frac{\partial^2 \ln f_{\sigma^2}(X)}{\partial \sigma^2}\right|\right) = E_\theta\left(\left|\frac{1}{2\sigma^4} - \frac{(X-\mu_0)^2}{\sigma^6}\right|\right)$$
$$\le E_\theta\left(\left|\frac{1}{2\sigma^4}\right| + \left|\frac{(X-\mu_0)^2}{\sigma^6}\right|\right) = \frac{1}{2\left(\sigma^2\right)^2} + \frac{\sigma^2}{\left(\sigma^2\right)^3} = \frac{3}{2\sigma^4} < \infty.$$

6.5.12 In Exercise 6.5.2 we have $\frac{\partial^2 \ln f_\theta(x)}{\partial \theta^2} = -\frac{\alpha_0}{\theta^2}$ and this exists for each x. Also, $E_\theta\left(S(\theta \,|\, X)\right) = E_\theta\left(\frac{\alpha_0}{\theta} - X\right) = \frac{\alpha}{\theta} - \frac{\alpha}{\theta} = 0$ and

$$E_\theta\left(\frac{\partial^2 \ln f_\theta(X)}{\partial \theta^2} + S^2(\theta \,|\, X)\right) = E_\theta\left(-\frac{\alpha_0}{\theta^2}\right) + E_\theta\left(\left(\frac{\alpha_0}{\theta} - X\right)^2\right)$$
$$= -\frac{\alpha_0}{\theta^2} - \frac{\alpha_0^2}{\theta^2} + E_\theta\left(X^2\right) = -\frac{\alpha_0}{\theta^2} - \frac{\alpha_0^2}{\theta^2} + \frac{\alpha_0\left(\alpha_0+1\right)}{\theta^2} = 0.$$

Finally we have that

$$E_\theta\left(\left|\frac{\partial^2 \ln f_\theta(s)}{\partial \theta^2}\right|\right) = E_\theta\left(\left|-\frac{\alpha_0}{\theta^2}\right|\right) = \frac{\alpha_0}{\theta^2}.$$

6.5.13 In Exercise 6.5.3 we have $\frac{\partial^2 \ln f_\alpha(x)}{\partial \alpha^2} = -\frac{1}{\alpha^2}$ and this exists for each x. Also, since $\ln(1+X) \sim$ Exponential(α) we have

$$E_\alpha\left(S(\theta \,|\, X)\right) = E_\theta\left(\frac{1}{\alpha} - \ln(1+X)\right) = \frac{1}{\alpha} - \frac{1}{\alpha} = 0,$$

and

$$E_\alpha\left(\frac{\partial^2 \ln f_\alpha(X)}{\partial \alpha^2} + S^2(\alpha \,|\, X)\right) = E_\alpha\left(-\frac{1}{\alpha^2} + \left(\frac{1}{\alpha} - \ln(1+X)\right)^2\right)$$
$$= -\frac{2}{\alpha^2} + E_\alpha\left(\left(\ln(1+X)\right)^2\right) = -\frac{2}{\alpha^2} + \frac{2}{\alpha^2} = 0.$$

Finally, we have

$$E_\alpha\left(\left|\frac{\partial^2 \ln f_\theta(s)}{\partial\theta^2}\right|\right) = E_\alpha\left(\left|-\frac{1}{\alpha^2}\right|\right) = \frac{1}{\alpha^2} < \infty.$$

6.5.14 Under i.i.d. sampling from f_θ, where the model $\{f_\theta : \theta \in \Omega\}$ satisfies the appropriate conditions, we have

$$\hat{I}(x_1, ..., x_n) =_{\theta=\hat{\theta}}= -\frac{\partial^2}{\partial\theta^2}\sum_{i=1}^{n} \ln f_\theta(x_i)\bigg|_{\theta=\hat{\theta}} = \sum_{i=1}^{n}\left(-\frac{\partial^2 \ln f_\theta(x_i)}{\partial\theta^2}\right)\bigg|_{\theta=\hat{\theta}}.$$

By the strong law of large numbers (Theorem 4.3.2) we have

$$\frac{\hat{I}(x_1, ..., x_n)}{n} \overset{a.s}{\to} E_\theta\left(-\frac{\partial^2 \ln f_\theta(x_i)}{\partial\theta^2}\right) = I(\theta).$$

6.5.15 Recall that the likelihood function is given by $L(\theta_1, \theta_2 \,|\, x_1, x_2, x_3) = \theta_1^{x_1}\theta_2^{x_2}(1-\theta_1-\theta_2)^{x_3}$. The log-likelihood function is then given by

$$l(\theta_1, \theta_2 \,|\, x_1, x_2, x_3) = x_1 \ln\theta_1 + x_2 \ln\theta_2 + x_3 \ln(1-\theta_1-\theta_2).$$

Using the methods discussed in Section 6.2.1 we obtain the score function as

$$S(\theta_1, \theta_2 \,|\, x) = \begin{pmatrix} \frac{x_1}{\theta_1} - \frac{x_3}{1-\theta_1-\theta_2} \\ \frac{x_2}{\theta_2} - \frac{x_3}{1-\theta_1-\theta_2} \end{pmatrix}.$$

The Fisher information is then given by

$$I(\theta) = -E_\theta\begin{pmatrix} -\frac{X_1}{\theta_1^2} - \frac{X_3}{(1-\theta_1-\theta_2)^2} & -\frac{X_3}{(1-\theta_1-\theta_2)^2} \\ -\frac{X_3}{(1-\theta_1-\theta_2)^2} & -\frac{X_2}{\theta_2^2} - \frac{X_3}{(1-\theta_1-\theta_2)^2} \end{pmatrix}$$

Now $X_i \sim \text{Binomial}(n, \theta_i)$ and so $E_{(\theta_1,\theta_2)}(X_i) = n\theta_i$. Therefore,

$$I(\theta) = n\begin{pmatrix} \frac{\theta_1}{\theta_1^2} + \frac{\theta_3}{(1-\theta_1-\theta_2)^2} & \frac{\theta_3}{(1-\theta_1-\theta_2)^2} \\ \frac{\theta_3}{(1-\theta_1-\theta_2)^2} & \frac{\theta_2}{\theta_2^2} + \frac{\theta_3}{(1-\theta_1-\theta_2)^2} \end{pmatrix} = n\begin{pmatrix} \frac{1}{\theta_1} + \frac{1}{\theta_3} & \frac{1}{\theta_3} \\ \frac{1}{\theta_3} & \frac{1}{\theta_2} + \frac{1}{\theta_3} \end{pmatrix}.$$

6.5.16 The likelihood function is given by $L(\theta_1, \ldots, \theta_{k-1} \,|\, x_1, \ldots, x_k) = \theta_1^{x_1}\theta_2^{x_2}\cdots(1-\theta_1-\cdots-\theta_{k-1})^{x_k}$. The log-likelihood function is then given by $l(\theta_1, \ldots, \theta_{k-1} \,|\, x_1, \ldots, x_k) = x_1 \ln\theta_1 + x_2 \ln\theta_2 + \cdots + x_k \ln(1-\theta_1-\cdots-\theta_{k-1})$. Using the methods discussed in Section 6.2.1 we obtain the score function as

$$S(\theta_1, \ldots, \theta_{k-1} \,|\, x_1, \ldots, x_k) = \begin{pmatrix} \frac{x_1}{\theta_1} - \frac{x_k}{1-\theta_1-\cdots-\theta_{k-1}} \\ \vdots \\ \frac{x_{k-1}}{\theta_{k-1}} - \frac{x_k}{1-\theta_1-\cdots-\theta_{k-1}} \end{pmatrix}.$$

The Fisher information is then given by

$$I_{ij}(\theta) = \frac{E_{(\theta_1,\ldots,\theta_{k-1})}(X_k)}{(1-\theta_1-\cdots-\theta_{k-1})^2} \text{ when } i \neq j,$$
$$I_{ii}(\theta) = \frac{E_{(\theta_1,\ldots,\theta_{k-1})}(X_i)}{\theta_1^2} + \frac{E_{(\theta_1,\ldots,\theta_{k-1})}(X_k)}{(1-\theta_1-\cdots-\theta_{k-1})^2}.$$

Now $X_i \sim \text{Binomial}(n, \theta_i)$ and so $E_{(\theta_1,\ldots,\theta_{k-1})}(X_i) = n\theta_i$. Therefore

$$I_{ij}(\theta) = \frac{n}{\theta_k} \text{ when } i \neq j, \quad I_{ii}(\theta) = \frac{n}{\theta_i} + \frac{n}{\theta_k}.$$

6.5.17 The likelihood function is given by (see Example 2.7.8) $L(\mu_1, \mu_2 \,|\, x_1, x_2) = \frac{1}{2\pi}\exp\left\{-\frac{1}{2}\left((x_1-\mu_1)^2 + (x_2-\mu_2)^2\right)\right\}$. The log-likelihood function is then given by $l(\mu_1, \mu_2 \,|\, x_1, x_2) = -\ln(2\pi) - \frac{1}{2}\left\{(x_1-\mu_1)^2 + (x_2-\mu_2)^2\right\}$. Using the methods discussed in section 6.2.1 we obtain the score function as

$$S(\mu_1, \mu_2 \,|\, x_1, x_2) = \begin{pmatrix} \frac{\partial l(\mu_1,\mu_2 \,|\, x_1,x_2)}{\partial \mu_1} \\ \frac{\partial l(\mu_1,\mu_2 \,|\, x_1,x_2)}{\partial \mu_2} \end{pmatrix} = \begin{pmatrix} x_1 - \mu_1 \\ x_2 - \mu_2 \end{pmatrix}.$$

The Fisher information matrix is then given by

$$I(\theta) = \begin{pmatrix} E_{(\mu_1,\mu_2)}\left(-\frac{\partial^2 l(\mu_1,\mu_2 \,|\, x_1,x_2)}{\partial \mu_1^2}\right) & E_{(\mu_1,\mu_2)}\left(-\frac{\partial^2 l(\mu_1,\mu_2 \,|\, x_1,x_2)}{\partial \mu_1 \partial \mu_2}\right) \\ E_{(\mu_1,\mu_2)}\left(-\frac{\partial^2 l(\mu_1,\mu_2 \,|\, x_1,x_2)}{\partial \mu_1 \partial \mu_2}\right) & E_{(\mu_1,\mu_2)}\left(-\frac{\partial^2 l(\mu_1,\mu_2 \,|\, x_1,x_2)}{\partial \mu_2^2}\right) \end{pmatrix}$$
$$= \begin{pmatrix} 1 & 0 \\ 0 & 1 \end{pmatrix} = I.$$

6.5.18 The likelihood function is given by (see Example 2.7.8) $L(\mu_1, \mu_2, \sigma^2 \,|\, x_1, x_2) = \frac{1}{2\pi\sigma^2}\exp\left\{-\frac{1}{2\sigma^2}\left((x_1-\mu_1)^2 + (x_2-\mu_2)^2\right)\right\}$. The log-likelihood function is then given by $l(\mu_1, \mu_2, \sigma^2 \,|\, x_1, x_2) = -\ln(2\pi) - \ln(\sigma^2) - \frac{1}{2\sigma^2}\left\{(x_1-\mu_1)^2 + (x_2-\mu_2)^2\right\}$. Using the methods discussed in Section 6.2.1 we obtain the score function as

$$S(\mu_1, \mu_2, \sigma^2 \,|\, x_1, x_2) = \begin{pmatrix} \frac{\partial l(\mu_1,\mu_2,\sigma^2 \,|\, x_1,x_2)}{\partial \mu_1} \\ \frac{\partial l(\mu_1,\mu_2,\sigma^2 \,|\, x_1,x_2)}{\partial \mu_2} \\ \frac{\partial l(\mu_1,\mu_2,\sigma^2 \,|\, x_1,x_2)}{\partial \sigma^2} \end{pmatrix}$$
$$= \begin{pmatrix} \frac{x_1-\mu_1}{\sigma^2} \\ \frac{x_2-\mu_2}{\sigma^2} \\ -\frac{1}{\sigma^2} + \frac{1}{2\sigma^4}\left\{(x_1-\mu_1)^2 + (x_2-\mu_2)^2\right\} \end{pmatrix}$$

The Fisher information matrix is then given by

$$(I(\theta))_{11} = E_{(\mu_1,\mu_2,\sigma^2)}\left(-\frac{\partial^2 l(\mu_1,\mu_2,\sigma^2\,|\,x_1,x_2)}{\partial\mu_1^2}\right) = E_{(\mu_1,\mu_2,\sigma^2)}\left(\frac{1}{\sigma^2}\right) = \frac{1}{\sigma^2}$$

$$(I(\theta))_{12} = E_{(\mu_1,\mu_2,\sigma^2)}\left(-\frac{\partial^2 l(\mu_1,\mu_2,\sigma^2\,|\,x_1,x_2)}{\partial\mu_1\partial\mu_2}\right) = 0$$

$$(I(\theta))_{13} = E_{(\mu_1,\mu_2,\sigma^2)}\left(-\frac{\partial^2 l(\mu_1,\mu_2,\sigma^2\,|\,x_1,x_2)}{\partial\mu_1\partial\sigma^2}\right) = E_{(\mu_1,\mu_2,\sigma^2)}\left(\frac{X_1-\mu_1}{\sigma^4}\right) = 0$$

$$(I(\theta))_{22} = E_{(\mu_1,\mu_2,\sigma^2)}\left(-\frac{\partial^2 l(\mu_1,\mu_2,\sigma^2\,|\,x_1,x_2)}{\partial\mu_2^2}\right) = E_{(\mu_1,\mu_2,\sigma^2)}\left(\frac{1}{\sigma^2}\right) = \frac{1}{\sigma^2}$$

$$(I(\theta))_{23} = E_{(\mu_1,\mu_2,\sigma^2)}\left(-\frac{\partial^2 l(\mu_1,\mu_2,\sigma^2\,|\,x_1,x_2)}{\partial\mu_2\partial\sigma^2}\right) = E_{(\mu_1,\mu_2,\sigma^2)}\left(\frac{X_2-\mu_2}{\sigma^4}\right) = 0$$

$$\begin{aligned}(I(\theta))_{33} &= E_{(\mu_1,\mu_2,\sigma^2)}\left(-\frac{\partial^2 l(\mu_1,\mu_2,\sigma^2\,|\,x_1,x_2)}{\partial(\sigma^2)^2}\right)\\ &= E_{(\mu_1,\mu_2,\sigma^2)}\left(-\frac{1}{\sigma^4}+\frac{1}{\sigma^6}\left\{(X_1-\mu_1)^2+(X_2-\mu_2)^2\right\}\right) = \frac{1}{\sigma^4}\end{aligned}$$

and the remaining elements follow by symmetry.

6.5.19 Since Ψ is a 1-1 function of θ, for each $\psi \in \Psi$ there is a unique $\theta \in \Omega$ such that $\Psi(\theta) = \psi$. Therefore, we can write the model as $\{g_\psi : \psi \in \Psi\}$, where $g_\psi = f_{\Psi^{-1}(\psi)}$.

Now, using the chain rule, we have that

$$\frac{\partial \ln g_\psi(X)}{\partial\psi} = \frac{\partial \ln f_{\Psi^{-1}(\psi)}(X)}{\partial\psi} = \frac{\partial \ln f_{\Psi^{-1}(\psi)}(X)}{\partial\theta}\frac{\partial\Psi^{-1}(\psi)}{\partial\psi}$$

$$\frac{\partial^2 \ln g_\psi(X)}{\partial\psi^2} = \frac{\partial^2 \ln f_{\Psi^{-1}(\psi)}(X)}{\partial\theta^2}\left(\frac{\partial\Psi^{-1}(\psi)}{\partial\psi}\right)^2 + \frac{\partial \ln f_{\Psi^{-1}(\psi)}(X)}{\partial\theta}\frac{\partial^2\Psi^{-1}(\psi)}{\partial\psi^2}.$$

Therefore, the Fisher information in the new parameterization is given by

$$\begin{aligned}I^*(\psi) &= E_\psi\left(-\frac{\partial^2 \ln g_\psi(X)}{\partial\psi^2}\right) = E_{\Psi^{-1}(\psi)}\left(\begin{array}{c}-\frac{\partial^2 \ln f_{\Psi^{-1}(\psi)}(X)}{\partial\theta^2}\left(\frac{\partial\Psi^{-1}(\psi)}{\partial\psi}\right)^2\\ -\frac{\partial \ln f_{\Psi^{-1}(\psi)}(X)}{\partial\theta}\frac{\partial^2\Psi^{-1}(\psi)}{\partial\psi^2}\end{array}\right)\\ &= E_{\Psi^{-1}(\psi)}\left(-\frac{\partial^2 \ln f_{\Psi^{-1}(\psi)}(X)}{\partial\theta^2}\right)\left(\frac{\partial\Psi^{-1}(\psi)}{\partial\psi}\right)^2\\ &\quad - E_{\Psi^{-1}(\psi)}\left(\frac{\partial \ln f_{\Psi^{-1}(\psi)}(X)}{\partial\theta}\right)\frac{\partial^2\Psi^{-1}(\psi)}{\partial\psi^2}\\ &= E_\theta\left(-\frac{\partial^2 \ln f_\theta(X)}{\partial\theta^2}\right)\left(\frac{\partial\Psi^{-1}(\psi)}{\partial\psi}\right)^2 - E_\theta\left(\frac{\partial \ln f_\theta(X)}{\partial\theta}\right)\frac{\partial^2\Psi^{-1}(\psi)}{\partial\psi^2}\\ &= I(\theta)\left(\frac{\partial\Psi^{-1}(\psi)}{\partial\psi}\right)^2 = I\left(\Psi^{-1}(\psi)\right)\left(\frac{\partial\Psi^{-1}(\psi)}{\partial\psi}\right)^2\end{aligned}$$

since $E_\theta\left(\frac{\partial \ln f_\theta(X)}{\partial\theta}\right) = 0$.

Chapter 7

Bayesian Inference

7.1 The Prior and Posterior Distributions

Exercises

7.1.1 First, we compute $m(s)$ as follows.

$$m(s) = \sum_{\theta=1}^{3} \pi(\theta) f_\theta(s) = \begin{cases} \frac{1}{5}\frac{1}{2} + \frac{2}{5}\frac{1}{3} + \frac{2}{5}\frac{3}{4} = \frac{8}{15} & s = 1 \\ \frac{1}{5}\frac{1}{2} + \frac{2}{5}\frac{2}{3} + \frac{2}{5}\frac{1}{4} = \frac{7}{15} & s = 2 \end{cases}$$

The posterior distribution of θ is then given by

θ	1	2	3
$\pi(\theta \mid s = 1)$	3/16	1/4	9/16
$\pi(\theta \mid s = 2)$	3/14	4/7	3/14

7.1.2 Since the posterior distribution of θ is Beta$(n\bar{x} + \alpha, n(1 - \bar{x}) + \beta)$ we have that

$$\begin{aligned} &E(\theta \mid x_1, \ldots, x_n) \\ &= \int_0^1 \theta \frac{\Gamma(n + \alpha + \beta)}{\Gamma(n\bar{x} + \alpha)\Gamma(n(1 - \bar{x}) + \beta)} \theta^{n\bar{x}+\alpha-1} (1 - \theta)^{n(1-\bar{x})+\beta-1} \, d\theta \\ &= \frac{\Gamma(n + \alpha + \beta)}{\Gamma(n\bar{x} + \alpha)\Gamma(n(1 - \bar{x}) + \beta)} \int_0^1 \theta^{n\bar{x}+\alpha} (1 - \theta)^{n(1-\bar{x})+\beta-1} \, d\theta \\ &= \frac{\Gamma(n + \alpha + \beta)}{\Gamma(n\bar{x} + \alpha)\Gamma(n(1 - \bar{x}) + \beta)} \frac{\Gamma(n\bar{x} + \alpha + 1)\Gamma(n(1 - \bar{x}) + \beta)}{\Gamma(n + \alpha + \beta + 1)} = \frac{n\bar{x} + \alpha}{n + \alpha + \beta}. \end{aligned}$$

and

$$
\begin{aligned}
&E\left(\theta^2 \mid x_1, \ldots, x_n\right) \\
&= \int_0^1 \theta^2 \frac{\Gamma(n+\alpha+\beta)}{\Gamma(n\bar{x}+\alpha)\Gamma(n(1-\bar{x})+\beta)} \theta^{n\bar{x}+\alpha-1}(1-\theta)^{n(1-\bar{x})+\beta-1}\, d\theta \\
&= \frac{\Gamma(n+\alpha+\beta)}{\Gamma(n\bar{x}+\alpha)\Gamma(n(1-\bar{x})+\beta)} \int_0^1 \theta^{n\bar{x}+\alpha+1}(1-\theta)^{n(1-\bar{x})+\beta-1}\, d\theta \\
&= \frac{\Gamma(n+\alpha+\beta)}{\Gamma(n\bar{x}+\alpha)\Gamma(n(1-\bar{x})+\beta)} \frac{\Gamma(n\bar{x}+\alpha+2)\Gamma(n(1-\bar{x})+\beta)}{\Gamma(n+\alpha+\beta+2)} \\
&= \frac{(n\bar{x}+\alpha)(n\bar{x}+\alpha+1)}{(n+\alpha+\beta)(n+\alpha+\beta+1)},
\end{aligned}
$$

so

$$
\begin{aligned}
\operatorname{Var}(\theta \mid x_1, \ldots, x_n) &= \frac{(n\bar{x}+\alpha)(n\bar{x}+\alpha+1)}{(n+\alpha+\beta)(n+\alpha+\beta+1)} - \left(\frac{n\bar{x}+\alpha}{n+\alpha+\beta}\right)^2 \\
&= \frac{(n\bar{x}+\alpha)(n(1-\bar{x})+\beta)}{(n+\alpha+\beta)^2(n+\alpha+\beta+1)}.
\end{aligned}
$$

7.1.3 First, the prior distribution of θ is $N(0,10)$, therefore, the prior probability that θ is positive is 0.5. Next, the posterior distribution of θ is

$$
N\left(\left(\frac{1}{10}+\frac{10}{1}\right)^{-1}\left(\frac{10}{1}\right), \left(\frac{1}{10}+\frac{10}{1}\right)^{-1}\right) = N\left(0.99010, 9.9010\times 10^{-2}\right).
$$

Therefore, the posterior probability that $\theta > 0$ is $1-\Phi\left((0-0.99010)/\sqrt{9.9010\times 10^{-2}}\right) = 1-\Phi(-3.1466) = 1-0.0008 = 0.9992$.

7.1.4 The likelihood function is given by $L(\lambda \mid x_1, \ldots x_n) = e^{-n\lambda}\lambda^{n\bar{x}}/\prod(x_i!)$. The prior distribution has density given by $\beta^{\alpha}\lambda^{\alpha-1}e^{-\beta\lambda}/\Gamma(\alpha)$. The posterior density of λ is then proportional to $\beta^{\alpha}\lambda^{n\bar{x}+a-1}e^{-\lambda(n+\beta)}/\Gamma(\alpha)\prod(x_i!)$, and we recognize this as being proportional to the density of a Gamma$(n\bar{x}+a, n+\beta)$ distribution.

7.1.5 The likelihood function is given by $L(\theta \mid x_1, \ldots x_n) = \frac{1}{\theta^n} I_{[x_{(n)},\infty)}(\theta)$. The prior distribution is the same as in the previous exercise. The posterior distribution of θ is then given by

$$
\pi(\theta \mid x_1, \ldots x_n) \propto \theta^{\alpha-n-1}e^{-\beta\theta} I_{[x_{(n)},\infty)}(\theta) \Big/ \int_{x_{(n)}}^{\infty} \theta^{\alpha-n-1}e^{-\beta\theta}\, d\theta.
$$

7.1.6 From Problem 3.2.16 the posterior mean of θ_i is

$$
\frac{f_i+\alpha_i}{f_1+\alpha_1+f_2+\alpha_2+f_3+\alpha_3} = \frac{f_i+\alpha_i}{n+\alpha_1+\alpha_2+\alpha_3}
$$

and the posterior variance of θ_i is given by

$$
\frac{(f_i+\alpha_i)(f_1+\alpha_1+f_2+\alpha_2+f_3+\alpha_3-f_i-\alpha_i)}{(n+\alpha_1+\alpha_2+\alpha_3)^2(n+\alpha_1+\alpha_2+\alpha_3+1)}.
$$

7.1.7 From the sample, we have $\bar{x} = 5.567$. Also, $\mu_0 = 3$, $\tau_0^2 = 4$ and $\alpha_0 = \beta_0 = 1$. Hence, the posterior distributions are given by $\mu \,|\, \sigma^2, x_1, ...x_n \sim N(5.5353, \frac{4}{81}\sigma^2)$ and $1/\sigma^2 \,|\, x_1, ..., x_n \sim \text{Gamma}(11, 41.737)$.

Computer Exercises

7.1.8 The posterior distribution is

$$N\left(\left(\frac{1}{\tau_0^2}+\frac{n}{\sigma_0^2}\right)^{-1}\left(\frac{\mu_0}{\tau_0^2}+\frac{n}{\sigma_0^2}\bar{x}\right), \left(\frac{1}{\tau_0^2}+\frac{n}{\sigma_0^2}\right)^{-1}\right)$$
$$= N\left(\left(\frac{1}{1}+\frac{20}{1}\right)^{-1}\left(\frac{2}{1}+\frac{20}{1}8.2\right), \left(\frac{1}{1}+\frac{20}{1}\right)^{-1}\right) = N(7.9048, 4.7619 \times 10^{-2}).$$

Then using Minitab the simulation proceeds as follows.

```
MTB > Random 10000 c1;
SUBC> Normal 7.90480 .218218.
MTB > let c2=1/c1
MTB > let c3=c2>.125
MTB > let k1=mean(c3)
MTB > let k2=sqrt(k1*(1-k1))/sqrt(10000)
MTB > let k3=k1-3*k2
MTB > let k4=k1+3*k2
MTB > print k1 k3 k4
Data Display
K1 0.683900
K3 0.669951
K4 0.697849
```

So the estimate of the posterior probability that the coefficient of variation is greater than .125 is 0.683900, and the true value is in the interval (0.669951, 0.697849) with virtual certainty.

7.1.9 The posterior distribution is

$$N\left(\left(\frac{1}{\tau_0^2}+\frac{n}{\sigma_0^2}\right)^{-1}\left(\frac{\mu_0}{\tau_0^2}+\frac{n}{\sigma_0^2}\bar{x}\right), \left(\frac{1}{\tau_0^2}+\frac{n}{\sigma_0^2}\right)^{-1}\right)$$
$$= N\left(\left(\frac{1}{1}+\frac{20}{1}\right)^{-1}\left(\frac{2}{1}+\frac{20}{1}8.2\right), \left(\frac{1}{1}+\frac{20}{1}\right)^{-1}\right) = N(7.9048, 4.7619 \times 10^{-2}).$$

Then using Minitab the simulation proceeds as follows.

```
MTB > Random 10000 c1;
SUBC> Normal 7.90480 .218218.
MTB > let c2=1/c1
MTB > let k1=mean(c2)
MTB > let k2=stdev(c2)/sqrt(10000)
MTB > let k3=k1-3*k2
```

```
MTB > let k4=k1+3*k2
MTB > print k1 k3 k4
Data Display
K1 0.126677
K3 0.126572
K4 0.126783
```

So the estimate of the posterior expectation of the coefficient of variation is 0.126677, and the true value is in the interval $(0.126572, 0.126783)$ with virtual certainty.

7.1.10 The prior density is given by

$$\frac{\Gamma(\alpha+\beta)}{\Gamma(\alpha)\Gamma(\beta)}\theta^{\alpha-1}(1-\theta)^{\beta-1} = \frac{\Gamma(6)}{\Gamma(3)\Gamma(3)}\theta^2(1-\theta)^2 = \frac{5!}{2^2}\theta^2(1-\theta)^2$$

and is plotted below (• • •). The posterior density is given by

$$\begin{aligned}&\frac{\Gamma(n+\alpha+\beta)}{\Gamma(n\bar{x}+\alpha)\Gamma(n(1-\bar{x})+\beta)}\theta^{n\bar{x}+\alpha-1}(1-\theta)^{n(1-\bar{x})+\beta-1}\\ &= \frac{\Gamma(30+3+3)}{\Gamma(30(.73)+3)\Gamma(30(1-.73)+3)}\theta^{30(.73)+3-1}(1-\theta)^{30(1-.73)+3-1}\\ &= \frac{\Gamma(36)}{\Gamma(24.9)\Gamma(11.1)}\theta^{23.9}(1-\theta)^{10.1}\end{aligned}$$

and is plotted below (—). The posterior density has shifted to the right and is more concentrated.

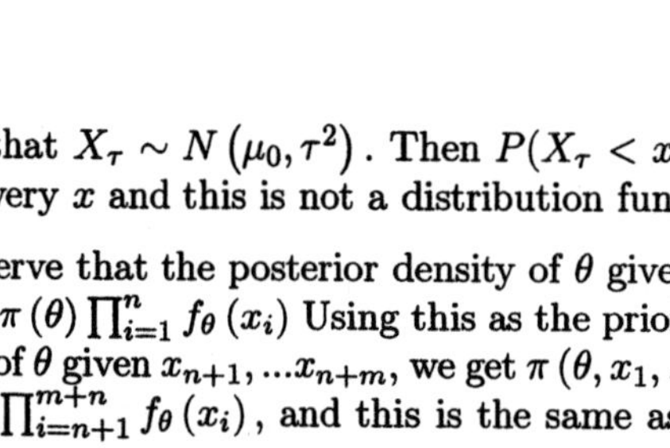

Problems

7.1.11 Suppose that $X_\tau \sim N(\mu_0, \tau^2)$. Then $P(X_\tau < x) = \Phi((x-\mu_0)/\tau) \to \Phi(0) = 1/2$ for every x and this is not a distribution function.

7.1.12 First, observe that the posterior density of θ given $x_1, ...x_n$ is $\pi(\theta \mid x_1, ...x_n) \propto \pi(\theta)\prod_{i=1}^n f_\theta(x_i)$ Using this as the prior density to obtain the posterior density of θ given $x_{n+1}, ...x_{n+m}$, we get $\pi(\theta, x_1, ...x_n \mid x_{n+1}, ...x_{n+m}) \propto \pi(\theta)\prod_{i=1}^n f_\theta(x_i)\prod_{i=n+1}^{m+n} f_\theta(x_i)$, and this is the same as the posterior density of θ given $x_1, ...x_n, x_{n+1}, ...x_{n+m}$.

7.1.13 The joint density of $(\theta, x_1, ...x_n)$ is given by

$$\frac{\Gamma(\alpha+\beta)}{\Gamma(\alpha)\Gamma(\beta)}\theta^{n\bar{x}+\alpha-1}(1-\theta)^{n(1-\bar{x})+\beta-1}$$

and integrating out θ gives the marginal probability function for $(x_1, ...x_n)$ as $m(x_1, ...x_n) = \frac{\Gamma(\alpha+\beta)}{\Gamma(\alpha)\Gamma(\beta)}\frac{\Gamma(n\bar{x}+\alpha)\Gamma(n(1-\bar{x})+\beta)}{\Gamma(\alpha+\beta+n)}$ for $(x_1, ...x_n) \in \{0, 1\}^n$.

To generate from this distribution we can first generate $\theta \sim \text{Beta}(\alpha, \beta)$ and then generate $x_1, ...x_n$ i.i.d. from the Bernoulli(θ) distribution.

7.1.14 First, note that if T is a sufficient statistic, then, by the factorization theorem (Theorem 6.1.1), the density (or probability function) for the model factors as $f_\theta(s) = h(s)\, g_\theta(T(s))$. The posterior density of θ is then given by

$$\pi(\theta \mid s) = \frac{\pi(\theta)\, h(s)\, g_\theta(T(s))}{\int_\Omega \pi(\theta)\, h(s)\, g_\theta(T(s))\, d\theta} = \frac{\pi(\theta)\, g_\theta(T(s))}{\int_\Omega \pi(\theta)\, g_\theta(T(s))\, d\theta}$$

and this depends on the data only through the value of $T(s)$.

Computer Problems

7.1.15 The prior Gamma(1, 1) density of $x = 1/\sigma^2$ is $\frac{1}{\Gamma(1)}x^{1-1}e^{-x} = e^{-x}$ for $x > 0$. Making the transformation $x \to y = 1/x$, the prior density of σ^2 is $x^{-2}e^{-1/x}$ for $x > 0$.

The posterior density of $1/\sigma^2$ is

$$\frac{41.737}{\Gamma(11)}(41.737x)^{11-1}e^{-41.737x} = \frac{41.737}{10!}(41.737x)^{10}e^{-41.737x}$$

for $x > 0$. Making the transformation $x \to y = 1/x$, the posterior density of σ^2 is $\frac{(41.737)^{11}}{10!}x^{-12}e^{-41.737/x}$. Plotting these we see that the posterior of σ^2 (—) is much more diffuse than the prior (• • •).

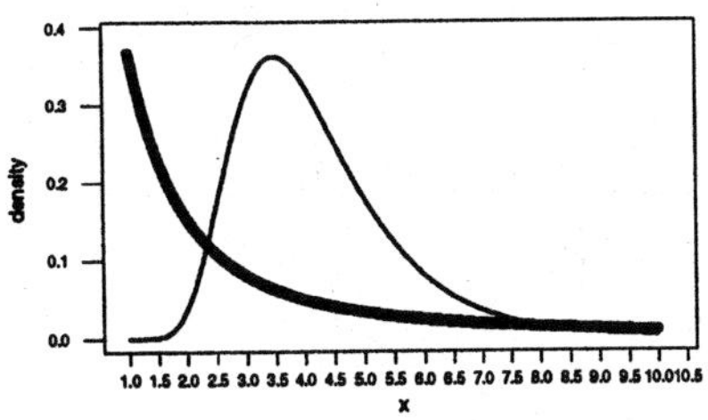

7.1.16 We have that $\mu \,|\, \sigma^2, x_1, \ldots, x_n \sim N\left(7.8095, \left(4.7619 \times 10^{-2}\right)\sigma^2\right)$ and $1/\sigma^2 \,|\, x_1, \ldots, x_n \sim$Gamma(12, 52.969) since

$$\left(n+\frac{1}{\tau_0^2}\right)^{-1}=\left(20+\frac{1}{1}\right)^{-1}=4.7619\times 10^{-2}$$

$$\mu_x=\left(n+\frac{1}{\tau_0^2}\right)^{-1}\left(\frac{\mu_0}{\tau_0^2}+n\bar{x}\right)=(4.7619\times 10^{-2})\left(\frac{0}{1}+20\,(8.2)\right)=7.8095$$

and

$$\beta_x=\beta_0+\frac{n}{2}\bar{x}^2+\frac{\mu_0^2}{2\tau_0^2}+\frac{n-1}{2}s^2-\frac{1}{2}\left(n+\frac{1}{\tau_0^2}\right)^{-1}\left(\frac{\mu_0}{\tau_0^2}+n\bar{x}\right)^2$$

$$=1+\frac{20}{2}(8.2)^2+\frac{0}{2}+\frac{20-1}{2}(2.1)-\frac{1}{2}\left(20+\frac{1}{1}\right)^{-1}\left(\frac{0}{1}+20\,(8.2)\right)^2=52.969.$$

Using Minitab we obtained the following results.

```
MTB > let k1=1/52.969
MTB > print k1
Data Display
K1 0.0188790
MTB > Random 10000 c1;
SUBC> Gamma 12 0.0188790.
MTB > let c2=1/sqrt(c1)
MTB > let c3=c2>2
MTB > let k1=mean(c3)
MTB > let k2=sqrt(k1*(1-k1))/sqrt(10000)
MTB > let k3=k1-3*k2
MTB > let k4=k1+3*k2
MTB > print k1 k3 k4
Data Display
K1 0.671800
K3 0.657713
K4 0.685887
```

So the estimate of the posterior probability that $\sigma > 2$ is 0.671800, and the true value is in the interval $(0.657713, 0.685887)$ with virtual certainty.

7.1.17 We use the distribution determined in 7.1.16. Using Minitab we obtained the following results.

```
MTB > let k1=1/52.969
MTB > print k1
Data Display
K1 0.0188790
MTB > Random 10000 c1;
SUBC> Gamma 12 0.0188790.
MTB > let c2=1/sqrt(c1)
MTB > let k1=mean(c2)
MTB > let k2=stdev(c2)/sqrt(10000)
```

```
MTB > let k3=k1-3*k2
MTB > let k4=k1+3*k2
MTB > print k1 k3 k4
Data Display
K1 2.17083
K3 2.16107
K4 2.18059
```

So the estimate of the posterior expectation of σ is 2.17083 and the true value is in the interval $(2.16107, 2.18059)$ with virtual certainty.

7.2 Inferences Based on the Posterior

Exercises

7.2.1 Recall that for the model discussed in Example 7.1.1, the posterior distribution of θ was Beta$(n\bar{x}+\alpha, n(1-\bar{x})+\beta)$. The posterior density is then given by

$$\pi_{\theta|x_1,\ldots,x_n} = \frac{\Gamma(\alpha+\beta+n)}{\Gamma(n\bar{x}+\alpha)\Gamma(n(1-\bar{x})+\beta)}\theta^{n\bar{x}+\alpha-1}(1-\theta)^{n(1-\bar{x})+\beta-1}$$

The posterior mean is given by

$$\begin{aligned} & E(\theta^m \mid x_1, \ldots, x_n) \\ &= \int_0^1 \frac{\Gamma(\alpha+\beta+n)}{\Gamma(n\bar{x}+\alpha)\Gamma(n(1-\bar{x})+\beta)}\theta^{n\bar{x}+\alpha+m-1}(1-\theta)^{n(1-\bar{x})+\beta-1}\,d\theta \\ &= \frac{\Gamma(\alpha+\beta+n)\Gamma(n\bar{x}+\alpha+m)}{\Gamma(n\bar{x}+\alpha)\Gamma(\alpha+\beta+n+m)}. \end{aligned}$$

7.2.2 Recall that for the model discussed in Example 7.1.2 the posterior distribution of μ is

$$N\left(\left(\frac{1}{\tau_0^2}+\frac{n}{\sigma_0^2}\right)^{-1}\left(\frac{\mu_0}{\tau_0^2}+\frac{n}{\sigma_0^2}\bar{x}\right), \left(\frac{1}{\tau_0^2}+\frac{n}{\sigma_0^2}\right)^{-1}\right)$$

By exercise 2.6.3, the posterior distribution of the third quartile $\Psi = \mu + \sigma_0 z_{0.75}$ is

$$N\left(\left(\frac{1}{\tau_0^2}+\frac{n}{\sigma_0^2}\right)^{-1}\left(\frac{\mu_0}{\tau_0^2}+\frac{n}{\sigma_0^2}\bar{x}\right)+\sigma_0 z_{0.75}, \left(\frac{1}{\tau_0^2}+\frac{n}{\sigma_0^2}\right)^{-1}\right)$$

Since the normal distribution is symmetric about its mode and the mean exists, the posterior mode and mean agree and given by

$$\hat{\psi} = \left(\frac{1}{\tau_0^2}+\frac{n}{\sigma_0^2}\right)^{-1}\left(\frac{\mu_0}{\tau_0^2}+\frac{n}{\sigma_0^2}\bar{x}\right)+\sigma_0 z_{0.75}.$$

7.2.3 Recall that the posterior distribution of σ^2 in Example 7.2.1 is inverse Gamma$(\alpha_0 + n/2, \beta_x)$, where β_x is given by (7.1.8). The posterior mean is then given by $E\left(1/\sigma^2 \,|\, x_1, ..., x_n\right) = (\alpha_0 + n/2)/\beta_x$. To find the posterior mode we need only maximize $\ln\left(y^{\alpha_0+n/2-1}\exp\left(-\beta_x y\right)\right) = (\alpha_0 + n/2 - 1)\ln y - \beta_x y$. This has first derivative given by $(\alpha_0 + n/2 - 1)/y - \beta_x$ and second derivative $-(\alpha_0 + n/2 - 1)/y^2$. Setting the first derivative equal to 0 and solving gives the solution $1/\hat{\sigma}^2 = (\alpha_0 + n/2 - 1)/\beta_x$. The second derivative at this value is negative so this is the unique mode.

7.2.4 Recall that the posterior distribution of σ^2 in Example 7.2.1 is inverse Gamma$(\alpha_0 + n/2, \beta_x)$, where β_x is given by (7.1.8). The posterior mean is then given by

$$
\begin{aligned}
E\left(\sigma^2 \,|\, x_1, ..., x_n\right) &= \int_0^\infty \frac{1}{y}\frac{\beta_x^{\alpha_0+n/2}}{\Gamma(\alpha_0+n/2)} y^{\alpha_0+n/2-1} e^{-\beta_x y}\, dy \\
&= \frac{\beta_x^{\alpha_0+n/2}}{\Gamma(\alpha_0+n/2)} \int_0^\infty y^{\alpha_0+n/2-2} e^{-\beta_x y}\, dy \\
&= \frac{\beta_x^{\alpha_0+n/2}}{\Gamma(\alpha_0+n/2)} \frac{\Gamma(\alpha_0+n/2-1)}{\beta_x^{\alpha_0+n/2-1}} \int_0^\infty \frac{1}{\Gamma(\alpha_0+n/2-1)} y^{\alpha_0+n/2-2} e^{-y}\, dy \\
&= \frac{\beta_x}{\alpha_0+n/2-1}.
\end{aligned}
$$

By Theorem 2.6.2 the posterior density of σ^2 is given by $\pi\left(\sigma^2 \,|\, x_1, .., x_n\right) = \left(\Gamma(\alpha_0 + n/2)\right)^{-1} (\beta_x)^{\alpha_0+n/2} \left(\sigma^2\right)^{-(\alpha_0+n/2+1)} \exp\left(-\beta_x/\sigma^2\right)$. Then to find the posterior mode we need only maximize $\ln\left(y^{-(\alpha_0+n/2+1)}\exp\left(-\beta_x/y\right)\right) = -(\alpha_0 + n/2 + 1)\ln y - \beta_x/y$. This has first derivative given by $-(\alpha_0 + n/2 + 1)/y + \beta_x/y^2$ and second derivative $(\alpha_0 + n/2 + 1)/y^2 - 2\beta_x/y^3$. Setting the first derivative equal to 0 and solving gives the solution $\hat{\sigma}^2 = \beta_x/(\alpha_0 + n/2 + 1)$. The second derivative at this value is $(\alpha_0 + n/2 + 1)^2/\beta_x^2 - 2(\alpha_0 + n/2 + 1)^3/\beta_x^2 = (\alpha_0 + n/2 + 1)^2(-1 - 2\alpha_0 - n)/\beta_x^2 < 0$, so this is the unique mode.

7.2.5 Recall that in Example 7.2.4 the marginal posterior distribution of θ_1 is Beta$(f_1 + \alpha_1, f_2 + ... + f_k + \alpha_2 + ... + \alpha_k)$. The posterior mean is then given by

$$
\begin{aligned}
&E\left(\theta_1 \,|\, x_1, ..., x_n\right) \\
&= \int_0^1 \theta_1 \frac{\Gamma\left(n + \sum_{i=1}^k \alpha_i\right)}{\Gamma(f_1+\alpha_1)\Gamma\left(\sum_{i=2}^k (f_i+\alpha_i)\right)} (\theta_1)^{f_1+\alpha_1-1} (1-\theta_1)^{\sum_{i=2}^k (f_i+\alpha_i)-1}\, d\theta_1 \\
&= \frac{\Gamma\left(n + \sum_{i=1}^k \alpha_i\right)}{\Gamma(f_1+\alpha_1)\Gamma\left(\sum_{i=2}^k (f_i+\alpha_i)\right)} \int_0^1 (\theta_1)^{f_1+\alpha_1} (1-\theta_1)^{\sum_{i=2}^k (f_i+\alpha_i)-1}\, d\theta_1
\end{aligned}
$$

$$= \frac{\Gamma\left(n+\sum_{i=1}^{k}\alpha_i\right)\Gamma\left(f_1+\alpha_1+1\right)}{\Gamma\left(f_1+\alpha_1\right)\Gamma\left(n+\sum_{i=1}^{k}\alpha_i+1\right)} = \frac{f_1+\alpha_1}{n+\sum_{i=1}^{k}\alpha_i}.$$

To find the posterior mode we need to maximize

$$\begin{aligned}
&\ln((\theta_1)^{f_1+\alpha_1-1}(1-\theta_1)^{\sum_{i=2}^{k}(f_i+\alpha_i)-1}) \\
&= (f_1+\alpha_1-1)\ln(\theta_1) + \left(\sum_{i=2}^{k}(f_i+\alpha_i)-1\right)\ln(1-\theta_1).
\end{aligned}$$

This has first derivative given by $(f_1+\alpha_1-1)/\theta_1 - (\sum_{i=2}^{k}(f_i+\alpha_i)) - 1/(1-\theta_1)$ and second derivative $-(f_1+\alpha_1-1)/\theta_1^2 - (\sum_{i=2}^{k}(f_i+\alpha_i)-1)/(1-\theta_1)^2$. Note that this is always negative when $\alpha_i \ge 1$. Setting the first derivative equal to 0 and solving gives the solution $\hat{\theta}_1 = (f_1+\alpha_1-1)/(n+\sum_{i=1}^{k}\alpha_i-2)$. Since the second derivative at this value is negative, $\hat{\theta}_1$ is the unique posterior mode.

7.2.6 Recall that the posterior distribution of θ in Example 7.2.2 is Beta$(n\bar{x}+\alpha, n(1-\bar{x})+\beta)$. To find the posterior variance we need only to find the second moment as follows.

$$\begin{aligned}
&E(\theta^2 \mid x_1, ..., x_n) \\
&= \int_0^1 \theta^2 \frac{\Gamma(n+\alpha+\beta)}{\Gamma(n\bar{x}+\alpha)\Gamma(n(1-\bar{x})+\beta)}\theta^{n\bar{x}+\alpha-1}(1-\theta)^{n(1-\bar{x})+\beta-1}\,d\theta \\
&= \frac{\Gamma(n+\alpha+\beta)}{\Gamma(n\bar{x}+\alpha)\Gamma(n(1-\bar{x})+\beta)}\int_0^1 \theta^{n\bar{x}+\alpha+1}(1-\theta)^{n(1-\bar{x})+\beta-1}\,d\theta \\
&= \frac{\Gamma(n+\alpha+\beta)}{\Gamma(n\bar{x}+\alpha)\Gamma(n(1-\bar{x})+\beta)}\frac{\Gamma(n\bar{x}+\alpha+2)\Gamma(n(1-\bar{x})+\beta)}{\Gamma(n+\alpha+\beta+2)} \\
&= \frac{(n\bar{x}+\alpha+1)(n\bar{x}+\alpha)}{(n+\alpha+\beta+1)(n+\alpha+\beta)}
\end{aligned}$$

The posterior variance is then given by

$$\begin{aligned}
\mathrm{Var}(\theta \mid x_1, ..., x_n) &= E(\theta^2 \mid x_1, ..., x_n) - (E(\theta \mid x_1, ..., x_n))^2 \\
&= \frac{(n\bar{x}+\alpha+1)(n\bar{x}+\alpha)}{(n+\alpha+\beta+1)(n+\alpha+\beta)} - \left(\frac{n\bar{x}+\alpha}{n+\alpha+\beta}\right)^2 \\
&= \frac{(n\bar{x}+\alpha)(n(1-\bar{x})+\beta)}{(n+\alpha+\beta+1)(n+\alpha+\beta)^2}.
\end{aligned}$$

Now $0 \le \bar{x} \le 1$, so

$$\begin{aligned}
\mathrm{Var}(\theta \mid x_1, ..., x_n) &= \frac{(n\bar{x}+\alpha)(n(1-\bar{x})+\beta)}{(n+\alpha+\beta+1)(n+\alpha+\beta)^2} \\
&\le \frac{(1+\alpha/n)(1+\beta/n)}{n(1+\alpha/n+\beta/n+1/n)(1+\alpha/n+\beta/n)^2} \to 0
\end{aligned}$$

as $n \to \infty$.

7.2.7 Recall that the posterior distribution of θ_1 in Example 7.2.2 is Beta$(f_1 + \alpha_1, f_2 + ... + f_k + \alpha_2 + ... + \alpha_k)$. To find the posterior variance we need only find the second moment as follows.

$$
\begin{aligned}
&E\left(\theta_1^2 \,|\, x_1, ..., x_n\right) \\
&= \int_0^1 \theta_1^2 \frac{\Gamma\left(n + \sum_{i=1}^k \alpha_i\right)}{\Gamma(f_1+\alpha_1)\Gamma\left(\sum_{i=2}^k (f_i+\alpha_i)\right)} (\theta_1)^{f_1+\alpha_1-1}(1-\theta_1)^{\sum_{i=2}^k (f_i+\alpha_i)-1}\, d\theta_1 \\
&= \frac{\Gamma\left(n + \sum_{i=1}^k \alpha_i\right)}{\Gamma(f_1+\alpha_1)\Gamma\left(\sum_{i=2}^k (f_i+\alpha_i)\right)} \int_0^1 (\theta_1)^{f_1+\alpha_1+1}(1-\theta_1)^{\sum_{i=2}^k (f_i+\alpha_i)-1}\, d\theta_1 \\
&= \frac{\Gamma\left(n + \sum_{i=1}^k \alpha_i\right)}{\Gamma(f_1+\alpha_1)\Gamma\left(\sum_{i=2}^k (f_i+\alpha_i)\right)} \frac{\Gamma(f_1+\alpha_1+2)\Gamma\left(\sum_{i=2}^k (f_i+\alpha_i)\right)}{\Gamma\left(n + \sum_{i=1}^k \alpha_i + 2\right)} \\
&= \frac{(f_1+\alpha_1+1)(f_1+\alpha_1)}{\left(n+\sum_{i=1}^k \alpha_i + 1\right)\left(n+\sum_{i=1}^k \alpha_i\right)}.
\end{aligned}
$$

The posterior variance is then given by

$$
\begin{aligned}
\mathrm{Var}\left(\theta_1 \,|\, x_1, ..., x_n\right) &= E\left(\theta_1^2 \,|\, x_1, ..., x_n\right) - \left(E\left(\theta_1 \,|\, x_1, ..., x_n\right)\right)^2 \\
&= \frac{(f_1+\alpha_1+1)(f_1+\alpha_1)}{\left(n+\sum_{i=1}^k \alpha_i + 1\right)\left(n+\sum_{i=1}^k \alpha_i\right)} - \left(\frac{f_1+\alpha_1}{n+\sum_{i=1}^k \alpha_i}\right)^2 \\
&= \frac{(f_1+\alpha_1)\left(\sum_{i=2}^k (f_i+\alpha_i)\right)}{\left(n+\sum_{i=1}^k \alpha_i + 1\right)\left(n+\sum_{i=1}^k \alpha_i\right)^2}.
\end{aligned}
$$

Now $0 \le f_1/n \le 1$, so

$$
\mathrm{Var}\left(\theta_1 \,|\, x_1, ..., x_n\right) = \frac{(f_1/n+\alpha_1)\left(\sum_{i=2}^k (f_i/n+\alpha_i)\right)}{n\left(1+\sum_{i=1}^k \alpha_i/n + 1/n\right)\left(1+\sum_{i=1}^k \alpha_i/n\right)^2} \to 0
$$

as $n \to \infty$.

7.2.8 The posterior mode always takes a value in the set $\{0, 1\}$, and the value we are predicting also is in this set. On the other hand, the posterior expectation can take a value anywhere in the interval $(0, 1)$. Accordingly, the mode seems like a more sensible predictor.

7.2.9 We have $x_{n+1} \,|\, \mu, x_1, \ldots, x_n \sim N\left(\bar{x}, \sigma_0^2\right)$ and this is independent of μ. Therefore, since the posterior predictive density of x_{n+1} is obtained by averaging the $N(\bar{x}, \sigma_0^2)$ density with respect to the posterior density of μ, we must have that this is also the posterior predictive distribution.

7.2.10 The likelihood function is given by $L(\lambda \mid x_1, \ldots x_n) = \lambda^n e^{-n\bar{x}\lambda}$. The prior distribution has density given by $\beta_0^{\alpha_0}\lambda^{\alpha_0-1}e^{-\beta_0\lambda}/\Gamma(\alpha_0)$. The posterior density of λ is then given by $\pi(\lambda \mid x_1, \ldots x_n) \propto \lambda^{n+\alpha_0-1}e^{-\lambda(n\bar{x}+\beta_0)}$, and we recognize this as being the density of a Gamma$(n+\alpha_0, n\bar{x}+\beta_0)$ distribution. The posterior mean and variance of λ are then given by $E(\lambda \mid x_1, \ldots x_n) = (n+\alpha_0)/(n\bar{x}+\beta_0)$, $Var(\lambda \mid x_1, \ldots x_n) = (n+\alpha_0)/(n\bar{x}+\beta_0)^2$.

To find the posterior mode we need to maximize $\ln\left(\lambda^{n+\alpha_0-1}e^{-\lambda(n\bar{x}+\beta_0)}\right) = (\alpha_0+n-1)\ln\lambda - \lambda(n\bar{x}+\beta_0)$. This has first derivative given by $(\alpha_0+n-1)/\lambda - (n\bar{x}+\beta_0)$ and second derivative $-(\alpha_0+n-1)/\lambda^2$. Setting the first derivative equal to 0 and solving gives the solution $\hat{\lambda} = (\alpha_0+n-1)/(n\bar{x}+\beta_0)$. The second derivative at this value is $-(n\bar{x}+\beta_0)^2/(\alpha_0+n-1)$, which is clearly negative, so $\hat{\lambda}$ is the unique posterior mode.

7.2.11 First we find the posterior predictive density of $t = x_{n+1}$ as follows.

$$\begin{aligned} q(t \mid x_1, \ldots, x_n) &= \int_0^\infty \lambda e^{-\lambda t}\frac{(\beta_0+n\bar{x})^{n+\alpha_0}}{\Gamma(\alpha_0+n)}\lambda^{n+\alpha_0-1}e^{-\lambda(n\bar{x}+\beta_0)}d\lambda \\ &= \frac{(\beta_0+n\bar{x})^{n+\alpha_0}}{\Gamma(\alpha_0+n)}\int_0^\infty \lambda^{n+\alpha_0}e^{-\lambda(n\bar{x}+\beta_0+t)}d\lambda \\ &= \frac{(\beta_0+n\bar{x})^{n+\alpha_0}}{\Gamma(\alpha_0+n)}\frac{\Gamma(n+\alpha_0+1)}{(n\bar{x}+\beta_0+t)^{n+\alpha_0+1}} \\ &= \frac{(n+\alpha_0)(\beta_0+n\bar{x})^{n+\alpha_0}}{(n\bar{x}+\beta_0+t)^{n+\alpha_0+1}} = \frac{(n+\alpha_0)(\beta_0+n\bar{x})^{-1}}{(1+t/(n\bar{x}+\beta_0))^{n+\alpha_0+1}} \end{aligned}$$

which is a $(n\bar{x}+\beta_0)$Pareto$(n+\alpha_0)$ distribution.

To find the posterior mode we need to maximize $\ln\left((n\bar{x}+\beta_0+t)^{-(n+\alpha_0+1)}\right)$ $= -(\alpha_0+n+1)\ln(n\bar{x}+\beta_0+t)$. This has first derivative (with respect to t) given by $-(\alpha_0+n+1)/(n\bar{x}+\beta_0+t)$. Since the first derivative is negative for all t and $t \geq 0$, the posterior mode is $\hat{t} = 0$.

Now the posterior distribution of $t/(n\bar{x}+\beta_0)$ is Pareto$(n+\alpha_0)$. By Problem 3.2.12 the posterior expectation of t is therefore $(n\bar{x}+\beta_0)/(n+\alpha_0-1)$ and, by Problem 3.3.16, the posterior variance of t is
$(n\bar{x}+\beta_0)^2(n+\alpha_0)/(n+\alpha_0-1)^2(n+\alpha_0-2)$.

7.2.12
(a) As in Example 7.2.1, we have that the posterior distribution of μ is given by the

$$N\left(\left(1+\frac{10}{9}\right)^{-1}\left(65+\left(\frac{10}{9}\right)63.20\right), \left(1+\frac{10}{9}\right)^{-1}\right) = N\left(64.053, \frac{9}{19}\right).$$

The posterior mode is then $\hat{\mu} = 64.053$. A .95-credible interval for μ is given by $64.053 \pm \sqrt{9/19}z_{0.975} = (62.704, 65.402)$. Since this interval has length equal

to 2.698 and the margin of error is less then 1.5 marks)which is quite small) we conclude that the estimate is quite accurate.
(b) Based on the .95-credible interval, we can't reject $H_0 : \mu = 65$, at the 5% level since 65 falls inside the interval.
(c) The posterior probability of the null hypothesis above is given by

$$\Pi(\mu = 65 \,|\, x_1, .., x_n) = \frac{0.5m_1(s)}{0.5m_1(s) + 0.5m_2(s)} \Pi_1(\mu = 65 \,|\, x_1, .., x_n) + \frac{0.5m_1(s)}{0.5m_1(s) + 0.5m_{21}(s)} \Pi_2(\mu = 65 \,|\, x_1, .., x_n)$$

where $\Pi_2(\cdot \,|\, x_1, .., x_n)$ is as given in part (a) and $\Pi_1(\cdot \,|\, x_1, .., x_n)$ is degenerate at $\mu = 65$.

The prior predictive under Π_1 is given by

$$m_1(x_1, .., x_n) = (18\pi)^{-5} \exp\left(-\frac{(10-1)\,252.622}{(2)\,9}\right) \exp\left(-\frac{10}{18}(63.20 - 65)^2\right) = 3.981 \times 10^{-65}$$

while the prior predictive under Π_2 is given by

$$m_2(x_1, .., x_n) = (18\pi)^{-5} \exp\left(-\frac{(10-1)\,252.622}{(2)\,9}\right) \times \exp\left(\frac{1}{2}\frac{9}{19}(135.22)^2\right) \exp\left(-\frac{1}{2}8663.0\right)(.68825) = 6.2662 \times 10^{-65}$$

The posterior probability of the null is then equal to

$$\frac{0.5m_1(s)}{0.5m_1(s) + 0.5m_2(s)} = \frac{3.981 \times 10^{-65}}{3.981 \times 10^{-65} + 6.2662 \times 10^{-65}} = .3885.$$

(d) The Bayes factor in favor of $H_0 : \mu = 65$ is given by

$$BF_{H_0} = \frac{\exp\left(-\frac{10}{18}(63.20 - 65)^2\right)}{\exp\left(\frac{1}{2}\frac{9}{19}(135.22)^2\right)\exp\left(-\frac{1}{2}8663.0\right) \times .68825} = .6353.$$

7.2.13

(a) The likelihood function is given by $L(\sigma^2 \,|\, x_1, ...x_n) = (\sigma^2)^{-n/2} \exp\left(-\frac{n-1}{2\sigma^2}s^2\right) \exp\left(-\frac{n}{2\sigma^2}(\bar{x} - \mu_0)^2\right)$. The prior distribution has density given by $\beta_0^{\alpha_0}(\sigma^2)^{-(\alpha_0 - 1)} e^{-\beta_0/\sigma^2}/\Gamma(\alpha_0)$. The posterior density of $1/\sigma^2$ is then proportional to

$$(\sigma^2)^{-n/2} \exp\left(-\frac{1}{2\sigma^2}\left((n-1)s^2 + n(\bar{x} - \mu_0)^2\right)\right)(\sigma^2)^{-(\alpha_0 - 1)} \exp\left(-\frac{\beta_0}{\sigma^2}\right)$$
$$= (\sigma^2)^{-(n/2 + \alpha_0 - 1)} \exp\left(-\frac{1}{2\sigma^2}\left((n-1)s^2 + n(\bar{x} - \mu_0)^2 + 2\beta_0\right)\right)$$

which we recognize as being proportional to the Gamma$(n/2+\alpha_0, \beta_x)$ density, where $\beta_x = (n-1)\, s^2/2 + n\,(\bar{x}-\mu_0)^2/2 + \beta_0$. Therefore, the posterior distribution of σ^2 is inverse Gamma$(n/2+\alpha_0, \beta_x)$.
(b) The posterior mean of σ^2 is given by $E\left(\sigma^2 \,|\, x_1, ...x_n\right) = \beta_x / \left(n/2+\alpha_0-1\right)$.
(c) To assess the hypothesis $H_0 : \sigma^2 \le \sigma_0^2$, which is equivalent to assessing $H_0 : 1/\sigma^2 \ge 1/\sigma_0^2$, we compute

$$\begin{aligned}\Pi\left(1/\sigma^2 \ge 1/\sigma_0^2 \,|\, x_1, .., x_n\right) &= \Pi\left(2\beta_x/\sigma^2 \ge 2\beta_x/\sigma_0^2 \,|\, x_1, .., x_n\right) \\ &= 1 - G\left(2\beta_x/\sigma_0^2; 2\alpha_0 + n\right)\end{aligned}$$

where $G(\cdot\,; 2\alpha_0+n)$ is the $\chi^2(2\alpha_0+n)$ distribution function.

Problems

7.2.14 The likelihood function is given by $L(\theta\,|\,x_1, ...x_n) = \theta^{-n} I_{[x_{(n)},\infty)}(\theta)$. The posterior distribution of θ is then given by $\pi(\theta\,|\,x_1, ...x_n) \propto \theta^{\alpha-n-1}e^{-\beta\theta} I_{[x_{(n)},\infty)}(\theta)$. Note that this is not differentiable at $x_{(n)}$. The maximum of $\theta^{\alpha-n-1}e^{-\beta\theta}$ occurs at the same point as the maximum of $\ln\left(\theta^{\alpha-n-1}e^{-\beta\theta}\right) = (\alpha-n-1)\ln\theta - \beta\theta$, which has first derivative $(\alpha-n-1)/\theta - \beta$ and second derivative $-(\alpha-n-1)/\theta^2$. Setting the first derivative equal to 0 and solving we have that the maximum occurs at $\hat{\theta} = (\alpha-n-1)/\beta$ whenever $\alpha-n-1>0$. Therefore, the posterior mode is given by $\max\left\{(\alpha-n-1)/\beta, x_{(n)}\right\}$.

7.2.15 The likelihood function is given by $L(\theta\,|\,x_1, .., x_n) = \theta^{-n} I_{(x_{(n)},\infty)}(\theta)$ and the prior is $I_{(0,1)}(\theta)$, so the posterior is

$$\frac{\theta^{-n} I_{(x_{(n)},1)}(\theta)}{\int_{x_{(n)}}^1 \theta^{-n}\, d\theta} = \frac{\theta^{-n} I_{(x_{(n)},1)}(\theta)}{(n-1)\left(x_{(n)}^{1-n}-1\right)}.$$

Since this density strictly increases in $\left(x_{(n)}, 1\right)$ and HPD interval is of the form $(c,1)$, c is determined by

$$\gamma = \int_c^1 \frac{\theta^{-n} I_{(x_{(n)},1)}(\theta)}{(n-1)\left(x_{(n)}^{1-n}-1\right)}\, d\theta = \frac{c^{1-n}-1}{x_{(n)}^{1-n}-1},$$

so $c = \left\{1+\gamma\left(x_{(n)}^{1-n}-1\right)\right\}^{1/(1-n)}$.

7.2.16 The posterior distribution of μ given σ^2 is the $N(\mu_x, \left(n+1/\tau_0^2\right)^{-1}\sigma^2)$ distribution where μ_x is given by (7.1.7). The posterior distribution of σ^2 is the Gamma$(\alpha_0+n/2, \beta_x)$ distribution, where β_x is given by (7.1.8). Therefore, the integral (7.2.2) is given by

$$\psi_0^{-2}\int_0^\infty \frac{1}{\sqrt{2\pi}}\left(n+\frac{1}{\tau_0^2}\right)^{1/2} \exp\left(-\frac{\lambda}{2}\left(n+\frac{1}{\tau_0^2}\right)\left(\psi_0^{-1}\lambda^{-\frac{1}{2}}-\mu_x\right)^2\right)\times$$
$$\frac{(\beta_x)^{\alpha_0+n/2}}{\Gamma(\alpha_0+n/2)}\lambda^{\alpha_0+n/2-1}\exp(-\beta_x\lambda)\, d\lambda.$$

7.2.17 Let $\psi\left(\mu,\sigma^2\right) = \mu + \sigma z_{0.75} = \mu + \left(1/\sigma^2\right)^{-1/2} z_{0.75}$ and $\lambda = \lambda\left(\mu,\sigma^2\right) = 1/\sigma^2$, so

$$J\left(\theta\left(\psi,\lambda\right)\right) = \left|\det\begin{pmatrix} \frac{\partial\psi}{\partial\mu} & \frac{\partial\psi}{\partial\left(\frac{1}{\sigma^2}\right)} \\ \frac{\partial\lambda}{\partial\mu} & \frac{\partial\lambda}{\partial\left(\frac{1}{\sigma^2}\right)} \end{pmatrix}\right| = \left|\det\begin{pmatrix} 1 & -\frac{1}{2}z_{0.75}\left(\frac{1}{\sigma^2}\right)^{-\frac{3}{2}} \\ 0 & 1 \end{pmatrix}\right| = 1.$$

Therefore, the posterior density of ψ is given by

$$\int_0^\infty \frac{1}{\sqrt{2\pi}}\left(n+\frac{1}{\tau_0^2}\right)^{1/2}\lambda^{1/2}\exp\left(-\frac{\lambda}{2}\left(n+\frac{1}{\tau_0^2}\right)\left(\left(\psi_0-\lambda^{-1/2}z_{0.75}\right)-\mu_x\right)^2\right)$$
$$\times\frac{\left(\beta_x\right)^{\alpha_0+n/2}}{\Gamma\left(\alpha_0+n/2\right)}\lambda^{\alpha_0+n/2-1}\exp\left(-\beta_x\lambda\right)d\lambda.$$

which is a difficult integral to evaluate.

7.2.18

(a) We can write

$$\psi = \sigma_0/\mu = \sigma_0\left(\left(\frac{1}{\tau_0^2}+\frac{n}{\sigma_0^2}\right)^{-1}\left(\frac{\mu_0}{\tau_0^2}+\frac{n}{\sigma_0^2}\bar{x}\right)+\left(\frac{1}{\tau_0^2}+\frac{n}{\sigma_0^2}\right)^{-1/2}Z\right)^{-1}$$
$$= \sigma_0\left(a+bZ\right)^{-1}$$

where $Z \sim N(0,1)$,

$$a = \left(\frac{1}{\tau_0^2}+\frac{n}{\sigma_0^2}\right)^{-1}\left(\frac{\mu_0}{\tau_0^2}+\frac{n}{\sigma_0^2}\bar{x}\right), b = \left(\frac{1}{\tau_0^2}+\frac{n}{\sigma_0^2}\right)^{-1/2}.$$

The posterior mean of ψ is $E\left(\psi\,|\,x_1,\ldots,x_n\right) = \int_{-\infty}^{\infty}\frac{\sigma_0}{a+bz}\frac{1}{\sqrt{2\pi}}e^{-z^2/2}\,dz$, and this integral doesn't exist because, noting that the integrand becomes infinite at $z=-a/b$, for $\epsilon>0$

$$\int_{-a/b}^{\infty}\frac{1}{a+bz}e^{-z^2/2}\,dz \geq \int_{-a/b}^{-a/b+\epsilon}\frac{1}{a+bz}e^{-z^2/2}\,dz$$
$$\geq \min\left\{e^{-z^2/2} : -a/b \leq z \leq -a/b+\epsilon\right\}\int_{-a/b}^{-a/b+\epsilon}\frac{1}{a+bz}\,dz$$
$$= \min\left\{e^{-z^2/2} : -a/b \leq z \leq -a/b+\epsilon\right\}\left.\frac{\ln\left(a+bz\right)}{b}\right|_{-a/b}^{-a/b+\epsilon} = \infty$$

while

$$\int_{-\infty}^{-a/b} \frac{1}{a+bz} e^{-z^2/2} dz \leq \int_{-a/b-\epsilon}^{-a/b} \frac{1}{a+bz} e^{-z^2/2} dz$$

$$\leq \min\left\{ e^{-z^2/2} : -a/b \leq z \leq -a/b+\epsilon \right\} \int_{-a/b-\epsilon}^{-a/b} \frac{1}{a+bz} dz$$

$$= \min\left\{ e^{-z^2/2} : -a/b \leq z \leq -a/b+\epsilon \right\} \left. \frac{-\ln(a+bz)}{b} \right|_{-a/b}^{-a/b+\epsilon} = -\infty.$$

Therefore, $E(\psi \mid x_1, ..., x_n) = \infty - \infty$, which is not defined.
(b) The posterior density of μ is given by

$$\pi(\mu \mid x_1, .., x_n) = \frac{1}{\sqrt{2\pi} b^{1/2}} \exp\left(-\frac{1}{2b} (\mu - a)^2 \right).$$

Using Theorem 2.6.2 we can find the posterior density of $\psi = \sigma_0/\mu$ (since this is a differentiable and strictly decreasing function of μ and excluding the 0 line from the parameter space) as

$$\pi\left(\psi^{-1}(\varphi) \mid x_1, .., x_n\right) / \left|\psi'\left(\psi^{-1}(\varphi)\right)\right| = \frac{1}{\sqrt{2\pi} b} \exp\left(-\frac{1}{2b^2} \left(\frac{\sigma_0}{\varphi} - a \right)^2 \right) \frac{\sigma_0}{\varphi^2}.$$

(c) To find the posterior mode we need to maximize

$$\ln\left(\exp\left(-\frac{1}{2b^2} \left(\frac{\sigma_0}{\varphi} - a \right)^2 \right) \frac{1}{\varphi^2} \right) = -\frac{1}{2b^2} \left(\frac{\sigma_0}{\varphi} - a \right)^2 - 2 \ln \varphi.$$

This has first derivative given by

$$\frac{1}{b^2} \left(\frac{\sigma_0}{\varphi} - a \right) \frac{\sigma_0}{\varphi^2} - \frac{2}{\varphi} = \frac{\sigma_0}{b^2 \varphi^3} - \frac{a\sigma_0}{b^2 \varphi^2} - \frac{2}{\varphi}.$$

Setting the first derivative equal to 0 gives the quadratic equation

$$\varphi^2 + \frac{a\sigma_0}{2b^2} \varphi - \frac{\sigma_0}{2b^2} = 0.$$

Solving this gives the solutions

$$\hat{\varphi} = -\frac{a\sigma_0}{4b^2} \pm \frac{1}{2} \sqrt{\frac{a^2\sigma_0^2}{4b^4} + \frac{4a\sigma_0}{2b^2}} = -\frac{a\sigma_0}{4b^2} \pm \frac{1}{2b} \sqrt{\frac{a^2\sigma_0^2}{4b^2} + 2a\sigma_0}$$

and these are real numbers since $b > 0$. Since the posterior density is finite everywhere, goes to 0 at $\pm\infty$, and is 0 at $\varphi = 0$, we know that these must both correspond to peaks. Therefore, we can determine the mode by evaluating the posterior density at these values, and the mode is the one that gives the largest value.

7.2.19
(a) The marginal density of $(\theta_1, ..., \theta_{k-2})$ is given by

$$f_{(\theta_1,...,\theta_{k-2})}(z_1, ..., z_{k-2})$$
$$= \int_0^{1-z_1-\cdots-z_{k-2}} \frac{\Gamma(\alpha_1+\cdots+\alpha_k)}{\Gamma(\alpha_1)\cdots\Gamma(\alpha_k)} z_1^{\alpha_1-1} z_2^{\alpha_2-1} \cdots z_{k-2}^{\alpha_{k-2}-1} z_{k-1}^{\alpha_{k-1}-1}$$
$$\times (1-z_1-\cdots-z_{k-1})^{\alpha_k-1}\, dz_{k-1}$$
$$= \frac{\Gamma(\alpha_1+\cdots+\alpha_k)}{\Gamma(\alpha_1)\cdots\Gamma(\alpha_k)} z_1^{\alpha_1-1} z_2^{\alpha_2-1} \cdots z_{k-2}^{\alpha_{k-2}-1} \int_0^{1-z_1-\cdots-z_{k-2}} z_{k-1}^{\alpha_{k-1}-1}$$
$$\times (1-z_1-\cdots-z_{k-1})^{\alpha_k-1}\, dz_{k-1}$$
$$= \frac{\Gamma(\alpha_1+\cdots+\alpha_k)}{\Gamma(\alpha_1)\cdots\Gamma(\alpha_k)} z_1^{\alpha_1-1} z_2^{\alpha_2-1} \cdots z_{k-2}^{\alpha_{k-2}-1} (1-z_1-\cdots-z_{k-2})^{\alpha_{k-1}+\alpha_k-2}$$
$$\times \int_0^{1-z_1-\cdots-z_{k-2}} \left(\frac{z_{k-1}}{1-z_1-\cdots-z_{k-2}}\right)^{\alpha_{k-1}-1}$$
$$\times \left(1-\frac{z_{k-1}}{1-z_1-\cdots-z_{k-2}}\right)^{\alpha_k-1} dz_{k-1}$$
$$= \frac{\Gamma(\alpha_1+\cdots+\alpha_k)}{\Gamma(\alpha_1)\cdots\Gamma(\alpha_k)} z_1^{\alpha_1-1} z_2^{\alpha_2-1} \cdots z_{k-2}^{\alpha_{k-2}-1} (1-z_1-\cdots-z_{k-2})^{\alpha_{k-1}+\alpha_k-1}$$
$$\times \int_0^1 u^{\alpha_{k-1}-1} (1-u)^{\alpha_k-1}\, dz_{k-1}$$
$$= \frac{\Gamma(\alpha_1+\cdots+\alpha_k)}{\Gamma(\alpha_1)\cdots\Gamma(\alpha_k)} \frac{\Gamma(\alpha_{k-1})\Gamma(\alpha_k)}{\Gamma(\alpha_{k-1}+\alpha_k)} z_1^{\alpha_1-1} z_2^{\alpha_2-1} \cdots z_{k-2}^{\alpha_{k-2}-1}$$
$$\times (1-z_1-\cdots-z_{k-2})^{\alpha_{k-1}+\alpha_k-1}$$
$$= \frac{\Gamma(\alpha_1+\cdots+\alpha_k)}{\Gamma(\alpha_1)\cdots\Gamma(\alpha_{k-2})\Gamma(\alpha_{k-1}+\alpha_k)} z_1^{\alpha_1-1} z_2^{\alpha_2-1} \cdots z_{k-2}^{\alpha_{k-2}-1}$$
$$\times (1-z_1-\cdots-z_{k-2})^{\alpha_{k-1}+\alpha_k-1}$$

and this establishes the result.
(b) Iterating the part (a) gives the result.
(c) The Jacobian matrix of this transformation has a 1 in the i_1th position of the first row, a 1 in the of the i_2th position of the second row, etc. The absolute value of the determinant of this transformation is therefore equal to 1. By the change of variable theorem this implies that $(\theta_{i_1}, \ldots, \theta_{i_{k-1}}) \sim$ Dirichlet$(\alpha_{i_1}, \alpha_{i_2}, \ldots, \alpha_{i_k})$.
(d) This is immediate from parts (c) and (b) as we just choose a permutation that puts θ_i as the first coordinate.

7.2.20 The likelihood function is given by $\theta_1^{f_1}\theta_2^{f_2}\cdots(1-\theta_1-\cdots-\theta_{k-1})^{f_k}$, so the log-likelihood is given by $f_1 \ln\theta_1 + f_2 \ln\theta_2 + \cdots + f_k \ln(1-\theta_1-\cdots-\theta_{k-1})$. Then vector of partial derivatives has ith element equal to
$f_i/\theta_i - f_k/(1-\theta_1-\cdots-\theta_{k-1})$. Setting these equal to 0 we get the system of

equations

$$f_k\theta_1 = f_1(1-\theta_1-\cdots-\theta_{k-1})$$

$$\vdots$$

$$f_k\theta_{k-1} = f_{k-1}(1-\theta_1-\cdots-\theta_{k-1})$$

and summing both sides we obtain $f_k(\theta_1+\cdots+\theta_{k-1}) = (n-f_k)(1-\theta_1-\cdots-\theta_{k-1})$ or $n(\theta_1+\cdots+\theta_{k-1}) = (n-f_k)$, which implies that $(1-\theta_1-\cdots-\theta_{k-1}) = f_k/n$. From this we deduce that the unique solution is $\left(\hat{\theta}_1,\ldots,\hat{\theta}_{k-1}\right) = (f_1/n,\ldots,f_{k-1}/n)$. Now since the log-likelihood is bounded above, continuously differentiable, and goes to $-\infty$ whenever $\theta_i \to 0$, this establishes that $(f_1/n,\ldots,f_{k-1}/n)$ is the MLE, so f_1/n is the plug-in MLE.

7.2.21 In Exercise 7.2.3 we showed that $E\left(1/\sigma^2 \,|\, x_1,\ldots,x_n\right) = (\alpha_0+n/2)/\beta_x$, while in Exercise 7.2.4 we showed that $E\left(\sigma^2 \,|\, x_1,\ldots,x_n\right) = \beta_x/(\alpha_0+n/2-1)$. So the estimate of σ^2 is not equal to one over the estimate of $1/\sigma^2$.

In Exercise 7.2.3 we showed that the posterior mode of $1/\sigma^2$ is $1/\hat{\sigma}^2 = (\alpha_0+n/2-1)/\beta_x$, while in Exercise 7.2.4 we showed that the posterior mode of σ^2 is $\hat{\sigma}^2 = \beta_x/(\alpha_0+n/2+1)$. So the estimate of σ^2 is not equal to one over the estimate of $1/\sigma^2$.

These differences indicate that these estimation procedures do not have the invariance property possessed by the MLE.

7.2.22 Since the variance of a $t(\lambda)$ distribution is $\lambda/(\lambda-2)$, the posterior variance of μ is given by

$$Var\left(\mu_x + \sqrt{\frac{1}{n+2\alpha_0}}\sqrt{\frac{2\beta_x}{n+1/\tau_0^2}}t(n+2\alpha_0)\right)$$

$$= \left(\sqrt{\frac{1}{n+2\alpha_0}}\sqrt{\frac{2\beta_x}{n+1/\tau_0^2}}\right)^2 \frac{n+2\alpha_0}{n+2\alpha_0-2} = \left(\frac{2\beta_x}{n+1/\tau_0^2}\right)\left(\frac{1}{n+2\alpha_0-2}\right).$$

7.2.23 The joint density of (θ,s,t) is given by $q_\theta(t\,|\,s)\,f_\theta(s)\,\pi(\theta)$. The prior predictive density for t is then the marginal density of t and is given by $q(t) = \int_\Omega\int_{-\infty}^{\infty} q_\theta(t\,|\,s)\,f_\theta(s)\,\pi(\theta)\,ds\,d\theta$.

7.2.24 The posterior predictive distribution for $t = (x_{n+1}, x_{n+2})$ is given by

$$q(t\,|\,x_1,\ldots,x_n)$$

$$= \int_0^1 \theta^{t_1+t_2}(1-\theta)^{2-t_1-t_2}\frac{\Gamma(n+\alpha+\beta)}{\Gamma(n\bar{x}+\alpha)\Gamma(n(1-\bar{x})+\beta)}\theta^{n\bar{x}+\alpha-1}$$

$$\times(1-\theta)^{n(1-\bar{x})+\beta-1}\,d\theta$$

$$= \frac{\Gamma(n+\alpha+\beta)}{\Gamma(n\bar{x}+\alpha)\Gamma(n(1-\bar{x})+\beta)}\int_0^1 \theta^{t_1+t_2+n\bar{x}+\alpha-1}(1-\theta)^{2-t_1-t_2+n(1-\bar{x})+\beta-1}\,d\theta$$

$$= \frac{\Gamma(n+\alpha+\beta)}{\Gamma(n\bar{x}+\alpha)\Gamma(n(1-\bar{x})+\beta)} \times \frac{\Gamma(t_1+t_2+n\bar{x}+\alpha)\Gamma(2-t_1-t_2+n(1-\bar{x})+\beta)}{\Gamma(n+\alpha+\beta+2)}$$

$$= \begin{cases} \frac{(n(1-\bar{x})+\beta+1)(n(1-\bar{x})+\beta)}{(n+\alpha+\beta+1)(n+\alpha+\beta)} & t_1 = t_2 = 0 \\ \frac{(n\bar{x}+\alpha)(n(1-\bar{x})+\beta)}{(n+\alpha+\beta+1)(n+\alpha+\beta)} & t_1 = 0, t_2 = 1 \\ \frac{(n\bar{x}+\alpha)(n(1-\bar{x})+\beta)}{(n+\alpha+\beta+1)(n+\alpha+\beta)} & t_1 = 1, t_2 = 0 \\ \frac{(n\bar{x}+\alpha+1)(n\bar{x}+\alpha)}{(n+\alpha+\beta+1)(n+\alpha+\beta)} & t_1 = t_2 = 0. \end{cases}$$

7.2.25 Put

$$a = \left(\frac{1}{\tau_0^2} + \frac{n}{\sigma_0^2}\right)^{-1}\left(\frac{\mu_0}{\tau_0^2} + \frac{n}{\sigma_0^2}\bar{x}\right), \quad b = \left(\frac{1}{\tau_0^2} + \frac{n}{\sigma_0^2}\right)^{-1/2}.$$

We can write $X_{n+1} = \mu + \sigma_0 Z$, where $\mu \sim N(a, b^2)$ independent of $Z \sim N(0,1)$. Therefore, the posterior predictive of X_{n+1} is given by $X_{n+1} \sim N(a, b^2 + \sigma_0^2)$.

7.2.26 We can write $X_{n+1} = \mu + \sigma U$, where $U \sim N(0,1)$ independent of $X_1, \ldots, X_n, \mu, \sigma$. We also have that $\mu = \mu_x + (n + 1/\tau_0^2)^{-1/2}\sigma Z$, where $Z \sim N(0,1)$ is independent of $X_1, \ldots, X_n, \sigma$. Therefore, we can write

$$X_{n+1} = \mu_x + (n+1/\tau_0^2)^{-1/2}\sigma Z + \sigma U$$
$$= \mu_x + \sigma\left\{(n+1/\tau_0^2)^{-1/2} Z + U\right\} = \mu_x + \left\{(n+1/\tau_0^2)^{-1} + 1\right\}^{1/2}\sigma W$$

where

$$W = \left\{(n+1/\tau_0^2)^{-1} + 1\right\}^{-1/2}\left\{(n+1/\tau_0^2)^{-1/2} Z + U\right\}$$
$$= \frac{X_{n+1} - \mu_x}{\left\{(n+1/\tau_0^2)^{-1} + 1\right\}^{1/2}\sigma} \sim N(0,1)$$

independent of $X_1, \ldots, X_n, \sigma$. Therefore, just as in Example 7.2.1,

$$T = \frac{W}{\sqrt{\left(2\frac{\beta_x}{\sigma^2}\right)/(2\alpha_0+n)}}$$
$$= \frac{X_{n+1} - \mu_x}{\left\{(n+1/\tau_0^2)^{-1} + 1\right\}^{1/2}((2\beta_x)/(2\alpha_0+n))^{1/2}} \sim t(2\alpha_0+n).$$

7.2.27 Using the result in Problem 7.2.26 and the fact that the $t(2\alpha_0+n)$ distribution is unimodal with mode at 0 and is symmetric about this mode, we have that a γ-prediction interval for X_{n+1} is given by (following Example 7.2.8)

$$\mu_x \pm \sqrt{\frac{2\beta_x\left\{\left(n+1/\tau_0^2\right)^{-1}+1\right\}}{(2\alpha_0+n)}}\, t_{\frac{1+\gamma}{2}}\left(2\alpha_0+n\right).$$

7.2.28 The prior predictive probability measure for the data s with a mixture of Π_1 and Π_2 prior distributions is given by

$$\begin{aligned}
m(s) &= E_\Pi\left(f_\theta(s)\right) = \sum_\theta f_\theta(s)\,\Pi(\{\theta\}) \\
&= \sum_\theta f_\theta(s)\left(p\Pi_1(\{\theta\}) + (1-p)\,\Pi_2(\{\theta\})\right) \\
&= p\sum_\theta f_\theta(s)\,\Pi_1(\{\theta\})\,(1-p)\sum_\theta f_\theta(s)\,\Pi_2(\{\theta\}) \\
&= pf_{\theta_0}(s) + (1-p)\sum_\theta f_\theta(s)\,\Pi_2(\{\theta\}) = pm_1(s) + (1-p)\,m_2(s).
\end{aligned}$$

The posterior probability measure is given by

$$\begin{aligned}
\Pi(A\,|\,s) &= \sum_{\theta\in A}\frac{f_\theta(s)\,\Pi(\{\theta\})}{m(s)} = \sum_{\theta\in A}\frac{f_\theta(s)\left(p\Pi_1(\{\theta\}) + (1-p)\,\Pi_2(\{\theta\})\right)}{pm_1(s)+(1-p)\,m_2(s)} \\
&= \frac{pm_1(s)}{pm_1(s)+(1-p)\,m_2(s)}\sum_{\theta\in A}\frac{f_\theta(s)\,\Pi_1(\{\theta\})}{m_1(s)} \\
&\quad + \frac{(1-p)\,m_2(s)}{pm_1(s)+(1-p)\,m_2(s)}\sum_{\theta\in A}\frac{f_\theta(s)\,\Pi_2(\{\theta\})}{m_2(s)} \\
&= \frac{pm_1(s)}{pm_1(s)+(1-p)\,m_2(s)}\Pi_1(A\,|\,s) + \frac{(1-p)\,m_2(s)}{pm_1(s)+(1-p)\,m_2(s)}\Pi_2(A\,|\,s).
\end{aligned}$$

7.2.29 The posterior density of θ is $\pi(\theta\,|\,s)$. Now make the transformation $\theta \to h(\theta) = (\psi(\theta), \lambda(\theta))$. Then following Section 2.9.2, we have that putting

$$J(\theta_1,\theta_2) = \begin{pmatrix} \frac{\partial\psi(\theta_1,\theta_2)}{\partial\theta_1} & \frac{\partial\psi(\theta_1,\theta_2)}{\partial\theta_2} \\ \frac{\partial\lambda(\theta_1,\theta_2)}{\partial\theta_1} & \frac{\partial\lambda(\theta_1,\theta_2)}{\partial\theta_2} \end{pmatrix}$$

and an application of Theorem 2.9.2 establishes that the joint density of (ψ,λ) is given by $\pi\left(h^{-1}(\psi,\lambda)\,|\,s\right)\left|J\left(h^{-1}(\psi,\lambda)\right)\right|^{-1}$. Then the marginal density of ψ is given by $\omega(\psi\,|\,s) = \int_{-\infty}^{\infty}\pi\left(h^{-1}(\psi,\lambda)\,|\,s\right)|J(\psi,\lambda)|^{-1}\,d\lambda$.

Challenge

7.2.30 First, let $t = h(\psi)$ be a 1-1 continuously differentiable transformation ψ. The null hypothesis that we want to test is $H_0 : h(\psi) = h(\psi_0) = t_0$. By Theorem 2.6.2 the prior density of t is given by $q(t) = w\left(h^{-1}(t)\right)/\left|h'\left(h^{-1}(t)\right)\right|$. Similarly, the posterior density of t is given by $q(t\,|\,x) = w\left(h^{-1}(t)\,|\,x\right)/\left|h'\left(h^{-1}(t)\right)\right|$.

Hence, since $h^{-1}(t) = \psi$, the ratio of the two is given by $w(t \mid x)/w(t) = w\left(h^{-1}(t) \mid x\right)/w\left(h^{-1}(t)\right) = w(\psi \mid x)/w(\psi)$, which is the ratio given in (7.2.9). The observed ratio is given by $q(t_0 \mid x)/q(t_0) = w\left(h^{-1}(t_0) \mid x\right)/w\left(h^{-1}(t_0)\right) = w(\psi_0 \mid x)/w(\psi_0)$. Therefore, the P-value computed by (7.2.9) would give the same result, and therefore it is invariant.

7.3 Bayesian Computations

Exercises

7.3.1 The likelihood function is given by

$$L(\mu \mid x_1, ..., x_n) = (4\pi)^{-10} \exp\left(-5(\bar{x}-\mu)^2\right) \exp\left(-\frac{19}{4}s^2\right).$$

The prior distribution has density given by $\pi(\mu) = \frac{1}{4}I_{[2,6]}(\mu)$. The posterior density is then proportional to $(4\pi)^{-10} \exp\left(-5(\bar{x}-\mu)^2\right) \exp\left(-\frac{19}{4}s^2\right) \frac{1}{4}I_{[2,6]}(\mu)$. To find the posterior mode we need only maximize $\exp\left(-5(\bar{x}-\mu)^2\right) I_{[2,6]}(\mu)$, which is clearly maximized at $\hat{\mu} = \bar{x}$ when $\bar{x} \in [2,6]$, at $\hat{\mu} = 2$ when $\bar{x} < 2$, and at $\hat{\mu} = 6$ when $\bar{x} > 6$. In this case the posterior mode is then $\hat{\mu} = \bar{x} = 3.825$. It has variance, estimated by

$$\hat{\sigma}^2(x_1, ..., x_n) = \left(- \left. \frac{\partial^2 \ln \left(\begin{array}{c} (4\pi)^{-10} \exp\left(-5(\bar{x}-\mu)^2\right) \\ \times \exp\left(-\frac{19}{4}s^2\right) \frac{1}{4}I_{[2,6]}(\mu) \end{array} \right)}{\partial \mu^2} \right|_{\mu=3.825} \right)^{-1} = \frac{1}{10}$$

A .95 credible interval for μ base on the large sample result is then given by

$$\bar{x} \pm \frac{\hat{\sigma}(x_1, ..., x_n)}{\sqrt{n}} z_{0.975} = 3.825 \pm \frac{1/\sqrt{10}}{\sqrt{20}} 1.96 = (3.6864, 3.9636).$$

Computer Exercises

7.3.2 Below is the Minitab program (modifying the one in Appendix B for Example 7.3.1) used to generate the sample of size $N = 10^4$ from the posterior distribution of $\psi = \mu + \sigma z_{0.25}$.

```
gmacro
normalpost
note - the base command sets the seed for the random number
generator (so you can repeat a simulation)
base 34256734
note - the parameters of the posterior
note - k1 = first parameter of the gamma distribution = (alpha_0 +
n/2)
```

```
let k1=9.5
note - k2 = 1 / beta
let k2=1/77.578
note - k3 = posterior mean
let k3=5.161
note - k4 = (n + 1/(tau_0 squared) )^(-1)
let k4=1/15.5
note - z_.25 = -0.6745
note - main loop
note - c3 contains generated value of sigma**2
note - c4 contains generated value of mu
note - c5 contains generated value of first quartile
do k5=1:10000
random 1 c1;
gamma k1 k2.
let c3(k5)=1/c1(1)
let k6=sqrt(k4/c1(1))
random 1 c2;
normal k3 k6.
let c4(k5)=c2(1)
let c5(k5)=c4(k5)-(0.6745)*sqrt(c3(k5))
enddo
endmacro
```

Below are the density histogram based on samples of $N = 5 \times 10^3$ and $N = 10^4$, respectively.

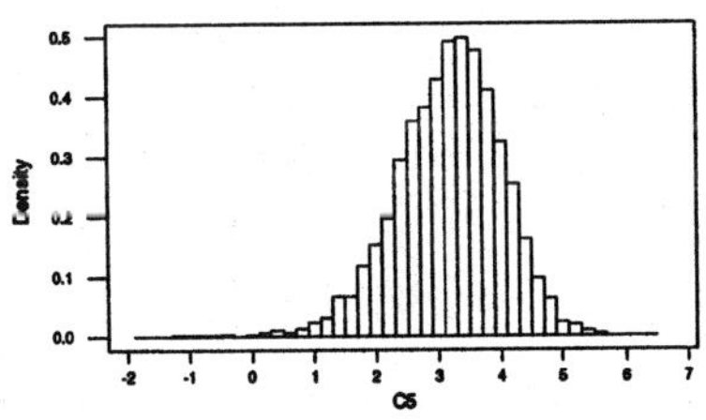

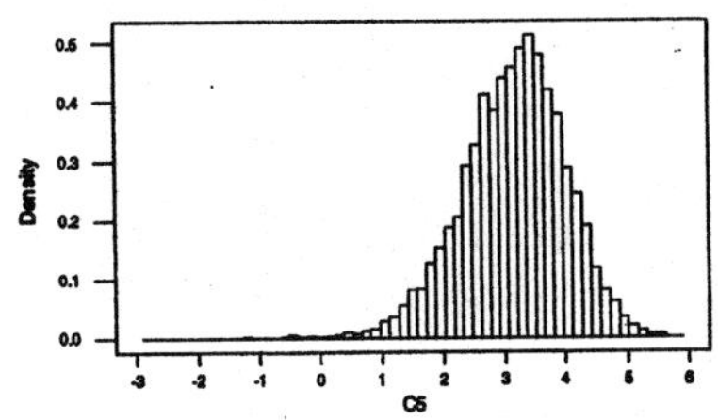

For $N = 5 \times 10^3$ we obtained the following estimates.

```
MTB > let k1=mean(c5)
MTB > let k2=stdev(c5)/sqrt(5000)
MTB > let k3=k1-3*k2
MTB > let k4=k1+3*k2
MTB > print k1 k3 k4
Data Display
K1 3.17641
K3 3.14068
K4 3.21214
```

So the estimate of the posterior mean of the first quartile is 3.17641, and the exact value lies in the interval $(3.14068, 3.21214)$ with virtual certainty.

For $N = 10^4$ we obtained the following estimates.

```
MTB > let k1=mean(c5)
MTB > let k2=stdev(c5)/sqrt(10000)
MTB > let k3=k1-3*k2
MTB > let k4=k1+3*k2
MTB > print k1 k3 k4
Data Display
K1 3.15800
K3 3.13253
K4 3.18346
```

So the estimate of the posterior mean of the first quartile is 3.15800 and the exact value lies in the interval $(3.13253, 3.18346)$ with virtual certainty.

7.3.3 Recall that from Example 7.1.1 we have that the posterior distribution of θ is Beta$(5.5, 105)$. Below is the Minitab code for doing this problem.

```
MTB > Random 1000 c1;
SUBC> Beta 5.5 105.
MTB > let c2=c1 < .1
MTB > let k1=mean(c2)
MTB > let k2=sqrt(k1*(1-k1))/sqrt(1000)
MTB > let k3=k1-3*k2
MTB > let k4=k1+3*k2
MTB > print k1 k3 k4
Data Display
K1 0.980000
K3 0.966718
K4 0.993282
```

The estimate of the posterior probability that $\theta < 0.1$ based on a sample of 1000 from the posterior is 0.980000, and the exact value lies in the interval $(0.966718, 0.993282)$ with virtual certainty.

7.3.4 Recall that from Exercise 7.2.10 we have that the posterior distribution of λ, with $n = 7$, $\bar{x} = 5.9$, is Gamma$(17, 43.3)$.

(a) The estimate of the posterior probability that $1/\lambda \in [3, 6]$ based on a sample of $N = 1000$ from the posterior of $1/\lambda$ is obtained via the following Minitab program.

```
MTB > let k1=17
MTB > let k2=1/43.3
MTB > Random 1000 c1;
SUBC> Gamma k1 k2.
MTB > let c2=1/c1
MTB > let c3=c2 ge 3 and c2 le 6
MTB > let k3=mean(c3)
MTB > let k4=sqrt(k3*(1-k3)/1000)
MTB > let k5=k3-3*k4
MTB > let k6=k3+3*k4
MTB > print k3 k5 k6
Data Display
K3 0.296000
K5 0.252693
K6 0.339307
```

The estimate of the posterior probability that $1/\lambda \in [3, 6]$ is 0.296000, and the exact value of the posterior probability lies in the interval (0.252693, 0.339307) with virtual certainty.

(b) The probability function of $\lfloor 1/\lambda \rfloor$ is estimated as follows.

```
MTB > let c4=floor(c2)
MTB > Tally C4;
SUBC> Counts;
SUBC> Percents.
Tally for Discrete Variables:  C4
 C4 Count Percent
 1 130 13.00
 2 572 57.20
 3 247 24.70
 4 42 4.20
 5 7 0.70
 6 1 0.10
 7 1 0.10
 N= 1000
```

So, for example, we estimate the posterior probability that $\lfloor 1/\lambda \rfloor$ equals 0 by 0 and the posterior probability that $\lfloor 1/\lambda \rfloor$ equals 1 by .13, etc.

(c) The estimate of the posterior expectation of $\lfloor \lambda \rfloor$ base on a Monte Carlo sample of size $N = 10^3$ is given below.

```
MTB > let k1=mean(c2)
MTB > let k2=stdev(c2)/sqrt(1000)
MTB > let k3=k1-3*k2
MTB > let k4=k1+3*k2
MTB > print k1 k3 k4
```

```
Data Display
K1 2.72523
K3 2.65789
K4 2.79257
```

The estimate of the posterior mean of $\lfloor 1/\lambda \rfloor$ is 2.725230 and the true value of the posterior expectation lies in the interval $(2.65789, 2.79257)$ with virtual certainty.

7.3.5 The inverse cdf of a Pareto(α) distribution is given by $x = F^{-1}(u) = (1-u)^{-1/\alpha} - 1$. Therefore, the following Minitab code generates a sample of 100 from the Pareto(2) distribution.

```
MTB > Random 10 c1;
SUBC> Uniform 0.0 1.0.
MTB > let c2=(1-c1)**(-1/2) - 1
MTB > Random 100 c1;
SUBC> Uniform 0.0 1.0.
MTB > let c2=(1-c1)**(-1/2) - 1
```

The likelihood function is given by $L(\alpha \mid x_1, ..., x_n) = \alpha^n \prod (1+x_i)^{-\alpha-1}$. The prior distribution has density given by $\pi(\alpha) = \alpha e^{-\alpha}$. The posterior density is proportional to $\alpha^{n+1} e^{-\alpha} \prod (1+x_i)^{-\alpha} = \alpha^{n+1} e^{-\alpha} \exp(-\alpha \ln(\prod(1+x_i)))$ $= \alpha^{n+1} \exp[-\alpha(\ln(\prod(1+x_i)) + 1)]$, and we recognize this as being proportional to the Gamma($n+1, \ln(\prod(1+x_i)) + 1$) density. The following Minitab code estimates the posterior expectation of $1/(\alpha+1)$.

```
MTB > let c3=loge(1+c2)
MTB > let k1=101
MTB > let k2=1/(sum(c3)+1)
MTB > print k1 k2
Data Display
K1 101.000
K2 0.0190936
MTB > Random 10000 c4;
SUBC> Gamma k1 k2.
MTB > let c4=1/(c4+1)
MTB > let k5=mean(c4)
MTB > let k6=stdev(c4)/sqrt(10000)
MTB > let k7=k5-3*k6
MTB > let k8=k5+3*k6
MTB > print k5 k7 k8
Data Display
K5 0.343067
K7 0.342392
K8 0.343741
```

The estimate of the posterior mean of $1/(\alpha+1)$ is 0.343067, and the exact value of the posterior expectation lies in the interval $(0.342392, 0.343741)$ with virtual certainty.

The true value of $1/(\alpha+1)$, however, is .33333, so note that it is not contained in the above interval. Note that the above interval is in essence a confidence interval for the exact value of the posterior expectation and not the true value of $1/(\alpha+1)$.

Problems

7.3.6
(a) We have that $\frac{1}{n}\ln\left(L\left(\hat{\theta}\,|\,x_1,\ldots,x_n\right)\pi\left(\hat{\theta}\right)\right)=\frac{1}{n}\sum_{i=1}^{n}\ln L\left(\hat{\theta}\,|\,x_i\right)+\frac{1}{n}\ln\pi\left(\hat{\theta}\right)\overset{a.s}{\to}E_{\theta}\left(\ln L\left(\theta\,|\,X\right)\right)=I(\theta)$ by the strong law of large numbers.
(b) Then from the results of part (a) we have that, denoting the true value of θ by θ_0,

$$\frac{\theta-\hat{\theta}\left(X_1,\ldots,X_n\right)}{\hat{\sigma}\left(X_1,\ldots,X_n\right)/\sqrt{n}}\overset{a.s}{\to}\sqrt{nI(\theta_0)}\left(\theta-\theta_0\right)$$

when $\theta\sim\Pi\left(\cdot\,|\,X_1,\ldots,X_n\right)$. This implies that when the sample size is large then inferences will be independent of the prior.

7.3.7 As we increase the Monte Carlo sample size N, the interval that contains the exact value of the posterior expectation with virtual certainty becomes shorter and shorter. But for a given sample size n for the data, the posterior expectation will not be equal to the true value of $1/(\alpha+1)$, so this interval will inevitably exclude the true value.

7.3.8 From Problem 2.8.23 we have that Y given $X=x$ is distributed $N(\mu_2+\rho\sigma_2\left(x-\mu_1\right)/\sigma_1,\left(1-\rho^2\right)\sigma_2^2)$ and similarly X given $Y=y$ is distributed $N(\mu_1+\rho\sigma_1\left(y-\mu_2\right)/\sigma_2,\left(1-\rho^2\right)\sigma_1^2)$. Therefore, a Gibbs sampling algorithm for this problem is given by the following. Select x_0, then generate $Y_1\sim N(\mu_2+\rho\sigma_2\left(x_0-\mu_1\right)/\sigma_1,\left(1-\rho^2\right)\sigma_2^2)$ obtaining y_1, then generate $X_1\sim N(\mu_1+\rho\sigma_1\left(y-\mu_2\right)/\sigma_2,\left(1-\rho^2\right)\sigma_1^2)$ obtaining x_1, then generate $Y_2\sim N(\mu_2+\rho\sigma_2\left(x_0-\mu_1\right)/\sigma_1,\left(1-\rho^2\right)\sigma_2^2)$ obtaining y_2, etc. The sample is $\left(x_1,y_1\right),\left(x_2,y_2\right),\ldots.$ Since this method is not exact, a better method for generating from the bivariate normal is to use (2.7.1), which is exact.

7.3.9 The marginal density of X is given by $\int_x^1 8xy\,dy=4x\left(1-x^2\right)$ and the marginal density of Y is given by $\int_0^y 8xy\,dx=4y^3$. Therefore, $f_{X|Y}\left(x\,|\,y\right)=8xy/4y^3=2x/y^2$ for $0<x<y$ and $f_{Y|X}\left(y\,|\,x\right)=8xy/\left(4x\left(1-x^2\right)\right)=2y/\left(1-x^2\right)$ for $x<y<1$.

The distribution function associated with $f_{Y|X}$ is given by $F_{Y|X}(y)=y^2/\left(1-x^2\right)$ for $x<y<1$. Therefore, the inverse cdf is given by $F_{Y|X}^{-1}(u)=\left(\left(1-x^2\right)u\right)^{1/2}$ for $0<u<1$. Therefore, we can generate Y given $X=x$ by generating $U\sim$ Uniform[0, 1] and putting $Y=\left(\left(1-x^2\right)U\right)^{1/2}$.

The distribution function associated with $f_{X|Y}$ is given by $F_{X|Y}(x)=x^2/y^2$ for $0<x<y$. Therefore the inverse cdf is given by $F_{X|Y}^{-1}(u)=\left(y^2u\right)^{1/2}=yu^{1/2}$ for $0<u<1$. Therefore we can generate X given $Y=y$ by generating $U\sim$ Uniform[0, 1] and putting $X=yU^{1/2}$.

So we select x_0. Then we generate $Y \sim f_{Y|X}(\cdot \,|\, x_0)$, using the above algorithm, obtaining y_1. Next we generate $X \sim f_{X|Y}(\cdot|\, y_1)$, using the above algorithm, obtaining x_1. Then we generate $Y \sim f_{Y|X}(\cdot \,|\, x_1)$, using the above algorithm, obtaining y_2, etc.

We can generate exactly from this distribution as follows. The marginal cdf of Y is $F_Y(y) = y^4$ for $0 < y < 1$. Then the inverse cdf is given by $F_Y^{-1}(u) = u^{1/4}$ for $0 < u < 1$. So we can generate $Y \sim F_Y$ by generating $U \sim$ Uniform[0, 1] and putting $y = U^{1/4}$. Then we use the above algorithm to generate $X \sim f_{X|Y}(\cdot \,|\, y)$. Then we have that $(X, Y) \sim F_{X,Y}$ by the theorem of total probability.

7.3.10 Suppose that the posterior expectation of ψ exists. Then by the theorem of total expectation we have that

$$
\begin{aligned}
E(\psi \,|\, x_1, \ldots, x_n) &= E\left(\frac{\sigma}{\mu} \,|\, x_1, \ldots, x_n\right) \\
&= E\left(\frac{\sigma}{\mu}\left(I_{(-\infty,0)}(\mu) + I_{(0,\infty)}(\mu)\right) \,|\, x_1, \ldots, x_n\right) \\
&= E\left(\frac{\sigma}{\mu} I_{(-\infty,0)}(\mu) \,|\, x_1, \ldots, x_n\right) + E\left(\frac{\sigma}{\mu} I_{(0,\infty)}(\mu) \,|\, x_1, \ldots, x_n\right) \\
&= E\left(E\left(\frac{\sigma}{\mu} I_{(-\infty,0)}(\mu) \,|\, \sigma, x_1, \ldots, x_n\right) \,|\, x_1, \ldots, x_n\right) \\
&\quad + E\left(E\left(\frac{\sigma}{\mu} I_{(0,\infty)}(\mu) \,|\, \sigma, x_1, \ldots, x_n\right) \,|\, x_1, \ldots, x_n\right)
\end{aligned}
$$

and reasoning as in Problem 7.2.18, we have that $E\left(\frac{\sigma}{\mu} I_{(-\infty,0)}(\mu) \,|\, \sigma, x_1, \ldots, x_n\right)$ $= -\infty$ and $E\left(\frac{\sigma}{\mu} I_{(-\infty,0)}(\mu) \,|\, \sigma, x_1, \ldots, x_n\right) = \infty$, so $E(\psi \,|\, x_1, \ldots, x_n) = \infty - \infty$ which is undefined.

Computer Problem

7.3.11 We use the program in Appendix B for Example 7.3.2 to generate a sample of 10^4 from the joint posterior distribution of (μ, σ^2).

The values of μ are stored in C21 and the values of σ^2 are stored in C20. The values of $\mu + \sigma z_{.25}$ are stored in C22.

```
MTB > invcdf .25;
SUBC> normal 0 1.
Inverse Cumulative Distribution Function
Normal with mean = 0 and standard deviation = 1.00000
P( X <= x ) x
 0.2500 -0.6745
MTB > let c22= c21 - (sqrt(c20))*(0.6745)
MTB > let k1=mean(c22)
MTB > print k1
Data Display
K1 3.30113
```

The estimate of the posterior mean of $\mu + \sigma z_{.25}$ is then 3.30113.

To estimate the error in this approximation we use the batching method and for this we used the Minitab code given in Appendix B. For a batch size of $m = 10$, we obtained the following standard error.

```
MTB > let k1=stdev(c2)/sqrt(1000)
MTB > print k1
Data Display
K1 0.0147612
```

For a batch size of $m = 20$ we obtained the following standard error.

```
MTB > let k1=stdev(c2)/sqrt(500)
MTB > print k1
Data Display
K1 0.0150728
```

For a batch size of $m = 40$ we obtained the following standard error.

```
MTB > let k1=stdev(c2)/sqrt(250)
MTB > print k1
Data Display
K1 0.0151834
```

This leads to the interval $3.30113 - 3\,(0.0151834) = (3.2556, 3.3467)$ that contains the true value of the posterior mean with virtual certainty.

7.4 Choosing Priors

Exercises

7.4.1 The likelihood function is given by $L(\lambda\,|\,x_1, ..., x_n) = \lambda^n \prod (1 + x_i)^{-\lambda-1}$. The prior distribution has density given by $\pi(\lambda) = \beta^\alpha \lambda^{\alpha-1} e^{-\beta\lambda} / \Gamma(\alpha)$. The posterior density is then proportional to $\lambda^{n+\alpha-1} \prod (1 + x_i)^{-\lambda} e^{-\beta\lambda} =$
$\lambda^{n+\alpha-1} \exp(-\lambda \ln(\prod (1 + x_i)))\, e^{-\beta\lambda} = \lambda^{n+\alpha-1} \exp[-\lambda(\ln(\prod(1 + x_i)) + \beta)]$, and we recognize this as being proportional to the Gamma$(n + \alpha, \ln(\prod(1 + x_i)) + \beta)$ density. Hence, this is a conjugate family.

7.4.2 The likelihood function is given by $L(\theta\,|\,x_1, ..., x_n) = \theta^{-n} I_{[x_{(n)},\infty)}(\theta)$. The prior distribution has density given by $\pi(\theta) = \theta^{-\alpha} I_{[\beta,\infty)}(\theta) / (\alpha - 1)\beta^{\alpha-1}$, where $\alpha \geq 1$ and $\beta > 0$. The posterior density is then proportional to
$\theta^{-n-\alpha} I_{[x_{(n)},\infty)}(\theta)\, I_{[\beta,\infty)}(\theta) = \theta^{-n-\alpha} I_{[\max\{x_{(n)},\beta\},\infty)}$, which is of the same form as the family of priors and so this is a conjugate family for this problem.

7.4.3
(a) First, we compute the prior predictive for the data as follows.

$$m_\tau(1,1,3) = \sum_{\theta=1}^{2} \pi(\theta) f_\theta(1,1,3) = \begin{cases} \frac{1}{2}\left(\frac{1}{3}\right)^3 + \frac{1}{2}\left(\frac{1}{2}\right)^2_2 \frac{1}{8} = \frac{59}{1728} & \tau = 1 \\ \frac{1}{3}\left(\frac{1}{3}\right)^3 + \frac{2}{3}\left(\frac{1}{2}\right)^2 \frac{1}{8} = \frac{43}{1296} & \tau = 2 \end{cases}$$

The maximum value of the prior predictive is obtained when $\tau = 1$, therefore we choose the first prior.

(b) The posterior of θ given $\tau = 1$ is

$$\pi_1(\theta \mid 1,1,3) = \begin{cases} \frac{\frac{1}{2}\left(\frac{1}{3}\right)^3}{\frac{59}{1728}} = \frac{32}{59} & \theta = a \\ \frac{\frac{1}{2}\left(\frac{1}{2}\right)^2\frac{1}{8}}{\frac{59}{1728}} = \frac{27}{59} & \theta = b. \end{cases}$$

7.4.4 The posterior of θ given $\tau = 2$ is

$$\pi_2(\theta \mid 1,1,3) = \begin{cases} \frac{\frac{1}{3}\left(\frac{1}{3}\right)^3}{\frac{43}{1296}} = \frac{16}{43} & \theta = a \\ \frac{\frac{2}{3}\left(\frac{1}{2}\right)^2\frac{1}{8}}{\frac{43}{1296}} = \frac{27}{43} & \theta = b. \end{cases}$$

Therefore, by the theorem of total probability the unconditional posterior of θ is given by

$$\begin{aligned} \pi(\theta \mid 1,1,3) &= \frac{1}{2}\pi_1(\theta \mid 1,1,3) + \frac{1}{2}\pi_2(\theta \mid 1,1,3) \\ &= \begin{cases} \frac{1}{2}\frac{32}{59} + \frac{1}{2}\frac{16}{43} = \frac{1160}{2537} & \theta = a \\ \frac{1}{2}\frac{27}{59} + \frac{1}{2}\frac{27}{43} = \frac{1377}{2537} & \theta = b. \end{cases} \end{aligned}$$

7.4.5 The prior predictive for the model described in Example 7.1.1 is given by

$$\begin{aligned} m_{\alpha,\beta}(x_1,\ldots,x_n) &= \int_0^1 \frac{\Gamma(\alpha+\beta)}{\Gamma(\alpha)\Gamma(\beta)}\theta^{n\bar{x}+\alpha-1}(1-\theta)^{n(1-\bar{x})+\beta-1}\,d\theta \\ &= \frac{\Gamma(\alpha+\beta)}{\Gamma(\alpha)\Gamma(\beta)}\frac{\Gamma(\alpha+n\bar{x})\Gamma(\beta+n(1-\bar{x}))}{\Gamma(\alpha+\beta+n)}. \end{aligned}$$

When $n = 10, n\bar{x} = 7, \alpha = 1, \beta = 1$ then

$$m_{1,1}(x_1,\ldots,x_n) = \frac{\Gamma(2)}{\Gamma(1)\Gamma(1)}\frac{\Gamma(8)\Gamma(4)}{\Gamma(12)} = \frac{7!3!}{11!} = \frac{1}{1320}.$$

When $\alpha = 5, \beta = 5$ then

$$m_{5,5}(x_1,\ldots,x_n) = \frac{\Gamma(10)}{\Gamma(5)\Gamma(5)}\frac{\Gamma(12)\Gamma(8)}{\Gamma(22)} = \frac{9!}{4!4!}\frac{11!7!}{21!} = \frac{1}{403104}.$$

Therefore, using the prior predictive we would select the prior given by $\alpha = 1, \beta = 1$ for further inferences about θ.

7.4.6 First, for $c > 0$, $\int f_\theta(s)\,\pi(\theta)\,d\theta < \infty$ if and only if $\int f_\theta(s)\,c\pi(\theta)\,d\theta < \infty$. Then assuming this, the posterior density under π is given by $\pi(\theta \mid s) = f_\theta(s)\,\pi(\theta) / \int f_\theta(s)\,\pi(\theta)\,d\theta = f_\theta(s)\,c\pi(\theta) / \int f_\theta(s)\,c\pi(\theta)\,d\theta$, and the result is established.

7.4.7 The likelihood function is given by $L(\theta \mid x_1,\ldots x_n) = \theta^{n\bar{x}}(1-\theta)^{n(1-\bar{x})}$. By Example 6.5.4, the Fisher information function for this model is given by $n/\{\theta(1-\theta)\}$. Therefore, Jeffreys' prior for this model is $\sqrt{n}\theta^{-1/2}(1-\theta)^{-1/2}$.

The posterior density of θ is then proportional to $\theta^{n\bar{x}-1/2}(1-\theta)^{n(1-\bar{x})-1/2}$ which we recognize as the unnormalized density of a Beta$(n\bar{x}+1/2, n(1-\bar{x})+1/2)$ distribution.

7.4.8
(a) The likelihood function is given by $\theta^{-n}I_{[x_{(n)},\infty)}(\theta)$. The posterior exists whenever $\int_{-\infty}^{\infty}\theta^{-n}I_{[x_{(n)},\infty)}(\theta)\,d\theta = \int_{x_{(n)}}^{\infty}\theta^{-n}\,d\theta = -\frac{\theta^{-n+1}}{n-1}\Big|_{x_{(n)}}^{\infty} = \frac{x_{(n)}^{-n+1}}{n-1} < \infty$, and this is the case whenever $n > 1$. Therefore, the posterior exists except when $n = 1$.
(b) From Example 6.5.1 we have that the Fisher information does not exist and so Jeffrey's prior can't exist.

Computer Exercises

7.4.9
(a) The prior predictive, as a function of α is given by $m_\alpha(x_1, ..., x_n) = \frac{\Gamma(2\alpha)}{(\Gamma(\alpha))^2}\frac{\Gamma(\alpha+7)\Gamma(\alpha+3)}{\Gamma(2\alpha+10)}$. Then $\ln m_\alpha(x_1, ..., x_n) = \ln\Gamma(2\alpha) - 2\ln\Gamma(\alpha) + \ln\Gamma(\alpha+7) + \ln\Gamma(\alpha+3) - \ln\Gamma(2\alpha+10)$. The plot of this function is given below.

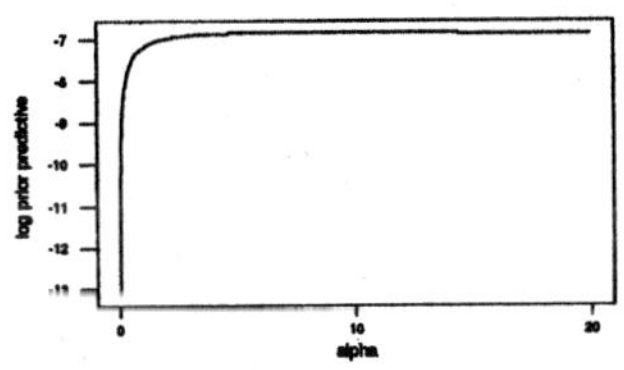

Note that this graph does not discriminate amongst large values of α. Actually from the numbers used to compute the plot the maximum occurs around 7.4, but it is difficult to detect this on the graph.
(b) When $n\bar{x} = 9$ the log prior predictive is plotted below. This is much more discriminating.

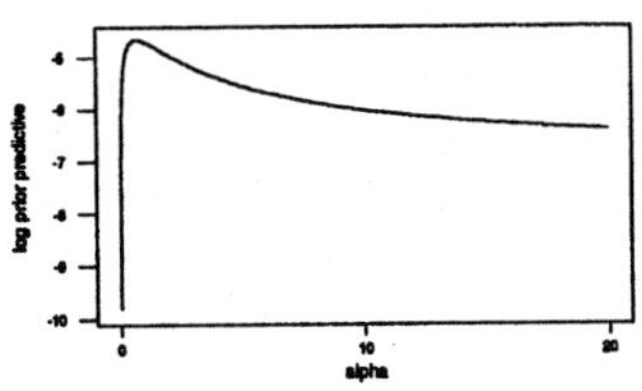

Problems

7.4.10 First, if $X \sim N(\mu_0, \tau_0^2)$, then the p quantile of this distribution satisfies $x_p = \mu_0 + \tau_0 z_p$, where z_p is the pth quantile of the $N(0,1)$ distribution. Therefore, once we specify two quantiles of the distribution, say x_{p_1} and x_{p_2}, we can solve $x_{p_1} = \mu_0 + \tau_0 z_{p_1}$, $x_{p_2} = \mu_0 + \tau_0 z_{p_2}$ to obtain $\tau_0 = (x_{p_1} - x_{p_2}) / (z_{p_1} - z_{p_2})$ and $\mu_0 = x_{p_1} - ((x_{p_1} - x_{p_2}) / (z_{p_1} - z_{p_2})) z_{p_1}$.

7.4.11 From Exercise 6.5.1 the Fisher information is $n/2\sigma^4$. Therefore, Jeffreys' prior is given by $1/\sigma^2$.

7.4.12 We use the prior $1/\sigma^2$. The posterior distribution is proportional to

$$\left(\frac{1}{\sigma^2}\right)^{\frac{n}{2}} \exp\left(-\frac{n}{2\sigma^2}(\bar{x}-\mu)^2\right) \exp\left(-\frac{(n-1)s^2}{2\sigma^2}\right) \frac{1}{\sigma^2}$$

$$= \left(\frac{1}{\sigma^2}\right)^{\frac{1}{2}} \exp\left(-\frac{n}{2\sigma^2}(\bar{x}-\mu)^2\right) \exp\left(-\frac{(n-1)s^2}{2\sigma^2}\right) \left(\frac{1}{\sigma^2}\right)^{\frac{n+1}{2}}.$$

So the posterior distribution of (μ, σ^2) is given by $\mu \mid \sigma^2, x_1, ..., x_n \sim N(\bar{x}, \sigma^2/n)$ and $1/\sigma^2 \mid x_1, ...x_n \sim$Gamma$\left(\frac{n+3}{2}, \frac{n-1}{2}s^2\right)$.

7.4.13
(a) The joint density of $(\mu, X_1, \ldots, X_n)$ is given by

$$\left(2\pi\tau_0^2\right)^{-1/2} \exp\left(-\frac{1}{2\tau_0^2}(\mu-\mu_0)^2\right) \times$$
$$\left(2\pi\sigma_0^2\right)^{-n/2} \exp\left(-\frac{n-1}{2\sigma_0^2}s^2\right) \exp\left(-\frac{n}{2\sigma_0^2}(\bar{x}-\mu)^2\right).$$

To calculate $m(x_1, \ldots, x_n)$ we need to integrate out μ. Using (7.1.2) we see that this integral equals

$$\left(2\pi\tau_0^2\right)^{-1/2} \left(2\pi\sigma_0^2\right)^{-n/2} \left(\frac{1}{\tau_0^2} + \frac{n}{\sigma_0^2}\right)^{-1/2} \exp\left(-\frac{n-1}{2\sigma_0^2}s^2\right) \times$$
$$\exp\left(\frac{1}{2}\left(\frac{1}{\tau_0^2} + \frac{n}{\sigma_0^2}\right)^{-1} \left(\frac{\mu_0}{\tau_0^2} + \frac{n}{\sigma_0^2}\bar{x}\right)^2\right) \exp\left(-\frac{1}{2}\left(\frac{\mu_0^2}{\tau_0^2} + \frac{n\bar{x}^2}{\sigma_0^2}\right)\right)$$

(b) We have that $(X_1, \ldots, X_n)$ given μ is a sample from the $N(\mu, \sigma_0^2)$ and $\mu \sim N(\mu, \tau_0^2)$. So we can generate a value $(X_1, \ldots, X_n)$ from m by generating $\mu \sim N(\mu_0, \tau_0^2)$ and then generating $X_1, \ldots, X_n$ i.i.d. $N(\mu, \sigma_0^2)$.
(c) No, they are not mutually independent, for we can write $X_i = \mu + \sigma_0 Z_i$, where $Z_1, \ldots, Z_n$ are i.i.d. $N(0,1)$ and $\mu = \mu_0 + \tau_0 Z$ where $Z \sim N(0,1)$ independent of $Z_1, \ldots, Z_n$. Therefore, $E(X_i) = E(\mu_0 + \tau_0 Z + \sigma_0 Z_i) = \mu_0 + \tau_0 E(Z) + \sigma_0 E(Z_i) = \mu_0$, and when $i \neq j$,

$$\text{Cov}(X_i, X_j) = E((\mu_0 + \tau_0 Z + \sigma_0 Z_i - \mu_0)(\mu_0 + \tau_0 Z + \sigma_0 Z_j - \mu_0))$$
$$= E((\tau_0 Z + \sigma_0 Z_i)(\tau_0 Z + \sigma_0 Z_j)) = E\left(\tau_0^2 Z^2 + \sigma_0 \tau_0 Z Z_j + \sigma_0 \tau_0 Z Z_i + \sigma_0^2 Z_i Z_j\right)$$
$$= \tau_0^2 E\left(Z^2\right) + \sigma_0 \tau_0 E(Z) E(Z_j) + \sigma_0 \tau_0 E(Z) E(Z_i) + \sigma_0^2 E(Z_i) E(Z_j) = \tau_0^2 \neq 0$$

and so they are not independent.

7.4.14 The joint posterior distribution of $\left(X_1,\ldots,X_n,\mu,1/\sigma^2\right)$ is proportional to $\left(\frac{1}{\sigma^2}\right)^{n/2}\prod_{i=1}^n\left[1+\frac{1}{\lambda}\left(\frac{x_i-\mu}{\sigma}\right)^2\right]^{-\frac{\lambda+1}{2}}\frac{1}{\sigma^2}$. Following Example 7.3.2, we introduce the n latent or hidden variables $(V_1,\ldots,V_n)$, which are i.i.d. $\chi^2(\lambda)$ and suppose $X_i\,|\,v_i\sim N\left(\mu,\sigma^2\lambda/v_i\right)$. With the same prior structure as before, we have that the joint density of $(X_1,V_1),\ldots,(X_n,V_n),\mu,1/\sigma^2$ is proportional to $\left(\frac{1}{\sigma^2}\right)^{\frac{n}{2}}\prod_{i=1}^n\exp\left(-\frac{v_i}{2\sigma^2\lambda}(x_i-\mu)^2\right)v_i^{\frac{\lambda}{2}-\frac{1}{2}}\exp\left(-\frac{v_i}{2}\right)\left(\frac{1}{\sigma^2}\right)$ From this we have that the conditional density of μ is proportional to $\exp\{-\frac{1}{2\sigma^2}\sum_{i=1}^n\frac{v_i}{\lambda}(x_i-\mu)^2\}$, which is proportional to $\exp\left\{-\frac{1}{2\sigma^2}\left(\sum_{i=1}^n\frac{v_i}{\lambda}\right)\mu^2+\frac{2}{2\sigma^2}\left(\sum_{i=1}^n\frac{v_i}{\lambda}x_i\right)\mu\right\}$. From this we immediately deduce that

$$\mu\,|\,x_1,\ldots,x_n,v_1,\ldots,v_n,\sigma^2\sim N\left(\left(\sum_{i=1}^n\frac{v_i}{\lambda}\right)^{-1}\left(\sum_{i=1}^n\frac{v_i}{\lambda}x_i\right),\left(\sum_{i=1}^n\frac{v_i}{\lambda}\right)^{-1}\sigma^2\right).$$

The conditional density of $1/\sigma^2$ is proportional to $\left(\frac{1}{\sigma^2}\right)^{\frac{n}{2}+1}\exp\{-\frac{1}{2}\left(\sum_{i=1}^n\frac{v_i}{\lambda}(x_i-\mu)^2\right)\frac{1}{\sigma^2}\}$ and we immediately deduce that $1/\sigma^2\,|\,x_1,\ldots,x_n,\,v_1,\ldots,v_n,\mu\sim\text{Gamma}(\frac{n}{2}+2,\,\frac{1}{2}\sum_{i=1}^n\frac{v_i}{\lambda}(x_i-\mu)^2)$. The conditional density of V_i is proportional to $v_i^{\frac{\lambda}{2}-\frac{1}{2}}\exp\{-\left[(x_i-\mu)^2/2\sigma^2\lambda+1/2\right]v_i\}$, so $V_i\,|x_1,\ldots,x_n,\,v_1,\ldots,v_{i-1},v_{i+1},\ldots v_n,\mu,\sigma^2\sim\text{Gamma}(\frac{\lambda}{2}+\frac{1}{2},\frac{1}{2}((x_i-\mu)^2/\sigma^2\lambda+1))$.

Computer Problems

7.4.15 First, note that the posterior distribution of $\left(\mu,\sigma^2\right)$ is $\mu\mid\sigma^2,x_1,\ldots,x_n\sim N\left(3.825,\frac{\sigma^2}{20}\right)\mu\mid\sigma^2,x_1,\ldots,x_n\sim N\left(3.825,\frac{\sigma^2}{20}\right)$ and $1/\sigma^2\mid x_1,\ldots x_n\sim$ Gamma$(11.5,10.75)$. We modified the Minitab program in Appendix B for Example 7.3.1 as follows.

```
gmacro
normalpost
note - the base command sets the seed for the random number generator

        (so you can repeat a simulation)
base 34256734
note - the parameters of the posterior
note - k1 = first parameter of the gamma distribution = (n+3)/2
let k1=11.5
note - k2 = 2/(n-1)s**2
let k2=1/10.75
note - k3 = posterior mean of mu
let k3=3.825
note - k4 = 1/n
let k4=1/20
note - main loop
```

```
note - c3 contains generated value of sigma**2
note - c4 contains generated value of mu
note - c5 contains generated value of coefficient of variation
do k5=1:10000
random 1 c1;
gamma k1 k2.
let c3(k5)=1/c1(1)
let k6=sqrt(k4/c1(1))
random 1 c2;
normal k3 k6.
let c4(k5)=c2(1)
let c5(k5)=sqrt(c3(k5))/c4(k5)
enddo
endmacro
```

A density histogram of the sample of 10^4 from the posterior distribution of $\psi = \sigma/\mu$ is given below.

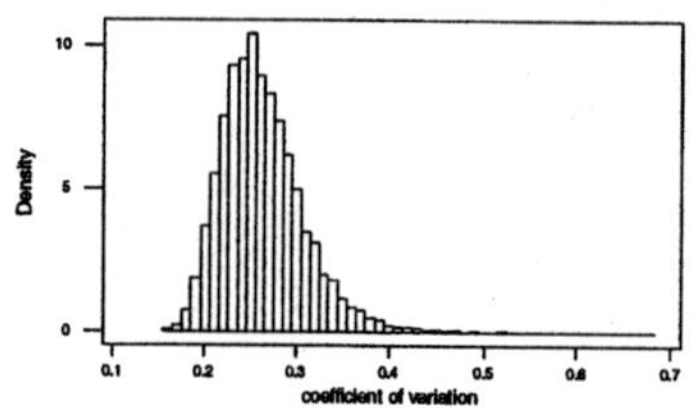

Challenge

7.4.16 The likelihood function is given by

$$L(\mu, \sigma^2 \,|\, x_1, ...x_n) = \left(2\pi\sigma^2\right)^{-n/2} \exp\left(-\frac{n}{2\sigma^2}(\bar{x}-\mu)^2\right) \exp\left(-\frac{n-1}{2\sigma^2}s^2\right).$$

The log-likelihood function is then given by

$$l(\mu, \sigma^2 \,|\, x_1, ...x_n) = -\frac{n}{2}\ln(2\pi) - \frac{n}{2}\ln\sigma^2 - \frac{n(\bar{x}-\mu)^2}{2\sigma^2} - \frac{n-1}{2\sigma^2}s^2.$$

Using the methods discussed in Section 6.2.1 we obtain the score function as follow

$$S(\mu, \sigma^2 \,|\, x) = \begin{pmatrix} \frac{n}{\sigma^2}(\bar{x}-\mu) \\ -\frac{n}{2\sigma^2} + \frac{n}{2\sigma^4}(\bar{x}-\mu)^2 + \frac{n-1}{2\sigma^4}s^2 \end{pmatrix}.$$

The Fisher information matrix is then given by

$$\begin{aligned} I\left(\mu, \sigma^2\right) &= -E_{(\mu,\sigma^2)} \begin{pmatrix} -\frac{n}{\sigma^2} & -\frac{n}{\sigma^4}(\bar{x}-\mu) \\ -\frac{n}{\sigma^4}(\bar{x}-\mu) & \frac{n}{2\sigma^4} - \frac{n}{\sigma^6}(\bar{x}-\mu)^2 - \frac{(n-1)}{\sigma^6}s^2 \end{pmatrix} \\ &= \begin{pmatrix} \frac{n}{\sigma^2} & 0 \\ 0 & -\frac{n}{2\sigma^4} + \frac{1}{\sigma^4} + \frac{n-1}{\sigma^4} \end{pmatrix} = \begin{pmatrix} \frac{n}{\sigma^2} & 0 \\ 0 & \frac{n}{2\sigma^4} \end{pmatrix}. \end{aligned}$$

Jeffreys' prior is then given by

$$\left(\det I(\mu,\sigma^2)\right)^{1/2} = \left(\frac{n}{\sigma^2}\frac{n}{2\sigma^4}\right)^{1/2} = \frac{n}{\sqrt{2}\sigma^3}.$$

Note that this is different than the prior used in 7.4.12.

Chapter 8

Optimal Inferences

8.1 Optimal Unbiased Estimation

Exercises

8.1.1 We have that $L(1\,|\,\cdot) = (3/2)\,L(2\,|\,\cdot)$ and so by Section 6.1.1 T is a sufficient statistic. Given $T = 1$, then the conditional distributions of s are given by the following table.

	$s=1$	$s=2$	$s=3$	$s=4$
$f_a(s\,|\,T=1)$	$\frac{1/3}{1/3+1/6}=\frac{2}{3}$	$\frac{1/6}{1/3+1/6}=\frac{1}{3}$	0	0
$f_b(s\,|\,T=1)$	$\frac{1/2}{1/2+1/4}=\frac{2}{3}$	$\frac{1/4}{1/2+1/4}=\frac{1}{3}$	0	0

We see that these are the same (i.e., independent of θ). When $T = 3$ the conditional distributions of s are given by

	$s=1$	$s=2$	$s=3$	$s=4$
$f_a(s\,|\,T=3)$	0	0	1	0
$f_b(s\,|\,T=3)$	0	0	1	0

and when $T = 4$ the conditional distributions are given by

	$s=1$	$s=2$	$s=3$	$s=4$
$f_a(s\,|\,T=4)$	0	0	0	1
$f_b(s\,|\,T=4)$	0	0	0	1

and these are also independent of θ.

8.1.2 Using $\text{Var}(x) = E\left(x^2\right) - (E(x))^2$, $\text{Var}(\bar{x}) = \text{Var}(x)/n$ we have that

$$E\left(\sum_{i=1}^{n}(x_i-\bar{x})^2\right) = E\left(\sum_{i=1}^{n}x_i^2 - n\bar{x}^2\right) = \sum_{i=1}^{n}E\left(x_i^2\right) - nE\left(\bar{x}^2\right)$$
$$= nE\left(x_1^2\right) - nE\left(\bar{x}^2\right) = n\left(\sigma^2+\mu^2\right) - n\left(\frac{\sigma^2}{n}+\mu^2\right) = (n-1)\sigma^2$$

and the result follows. This estimator will be UMVU whenever $(\bar{x}, s^2)$ is a complete sufficient statistic for the class of possible distributions that we are sampling from.

8.1.3 From Example 8.1.3 we know that $\bar{x}$ is a complete sufficient statistic. Therefore, any function of $\bar{x}$ is a UMVU estimator of its mean. We have that $E(\bar{x}^2) = \mu^2 + \sigma_0^2/n$ and so $\bar{x}^2 - \sigma_0^2/n + \sigma_0^2 = \bar{x}^2 + (1 - 1/n)\,\sigma_0^2$ is UMVU for $\mu^2 + \sigma_0^2$.

8.1.4 We have that $E(\bar{x} + \sigma_0 z_{.25}) = E(\bar{x}) + \sigma_0 z_{.25} = \mu + \sigma_0 z_{.25}$. Since $\bar{x}$ is complete this implies that $\bar{x} + \sigma_0 z_{.25}$ is UMVU.

8.1.5 This is a UMVU estimator of $5 + 2\mu$.

8.1.6 We have that $E(\bar{x}) = \theta$ and Challenge 8.1.22 says that $\bar{x}$ is complete. Therefore, $\bar{x}$ is UMVU for θ.

8.1.7 We have that the mean of a Gamma(α_0, β) random variable is given by $E(X) = \int_0^\infty x \frac{(\beta x)^{\alpha_0 - 1}}{\Gamma(\alpha_0)} e^{-\beta x} \beta\, dx = \frac{1}{\beta\Gamma(\alpha_0)} \int_0^\infty (\beta x)^{\alpha_0} e^{-\beta x} \beta\, dx = \frac{\Gamma(\alpha_0 + 1)}{\beta\Gamma(\alpha_0)} = \frac{\alpha_0}{\beta}$. Therefore, $\bar{x}/\alpha_0$ is an unbiased estimator of β^{-1}, and since $\bar{x}$ is complete, this implies that $\bar{x}/\alpha_0$ is UMVU.

8.1.8 The likelihood function is given by $L(x_1, \ldots, x_n \,|\, \sigma^2) =$ $\sigma^{-2n} \exp\left\{-\frac{1}{2\sigma^2} \sum_{i=1}^n (x_i - \mu_0)^2\right\}$. By factorization (Theorem 6.1.1), $\sum_{i=1}^n (x_i - \mu_0)^2$ is sufficient. Further, $E_{\sigma^2}\left(\sum_{i=1}^n (x_i - \mu_0)^2\right) = n\sigma^2$, so $n^{-1} \sum_{i=1}^n (x_i - \mu_0)^2$ is unbiased for σ^2. Since this sufficient statistic is complete, we have that $n^{-1} \sum_{i=1}^n (x_i - \mu_0)^2$ is UMVU for σ^2.

Problems

8.1.9 The likelihood function is given by $L(x_1, \ldots, x_n \,|\, \theta) = \theta^{-n}$ whenever $\theta > x_{(n)}$ and 0 otherwise. Therefore, when we know $x_{(n)}$ we know the likelihood function and so $x_{(n)}$ is sufficient. Then $x_{(n)}$ has density given by $\theta^{-n} n x^{n-1}$ for $0 < x < \theta$ and $E(x_{(n)}) = \int_0^\theta \theta^{-n} n x^n\, dx = \frac{\theta^{-n} n}{n+1} x^{n+1}\Big|_0^\theta = \frac{n}{n+1}\theta$. So $\frac{n+1}{n} x_{(n)}$ is UMVU for θ.

8.1.10 We have that the joint conditional probability function of $(x_1, \ldots, x_n)$ given $n\bar{x}$ is

$$\prod_{i=1}^n \theta^{x_i} (1 - \theta)^{1 - x_i} \Big/ \left\{ \binom{n}{n\bar{x}} \theta^{n\bar{x}} (1 - \theta)^{n - n\bar{x}} \right\} = 1 \Big/ \binom{n}{n\bar{x}}.$$

This is the uniform distribution on the set of all sequences of 0's and 1's of length n that have $n\bar{x}$ 1's.

8.1.11 We have that

$$\begin{aligned}
&(\theta - \alpha a_1 - (1-\alpha)\, a_2)^2 = \theta^2 - 2\theta\,(\alpha a_1 + (1-\alpha)\, a_2) + (\alpha a_1 + (1-\alpha)\, a_2)^2 \\
&= \alpha\,(\theta - a_1)^2 + (1-\alpha)\,(\theta - a_2)^2 - \alpha a_1^2 - (1-\alpha)\, a_1^2 + (\alpha a_1 + (1-\alpha)\, a_2)^2 \\
&= \alpha\,(\theta - a_1)^2 + (1-\alpha)\,(\theta - a_2)^2 - \alpha\,(1-\alpha)\, a_1^2 - \alpha\,(1-\alpha)\, a_1^2 + \alpha\,(1-\alpha)\, 2a_1a_2 \\
&= \alpha\,(\theta - a_1)^2 + (1-\alpha)\,(\theta - a_2)^2 - \alpha\,(1-\alpha)\,(a_1 - a_2)^2 \\
&\le \alpha\,(\theta - a_1)^2 + (1-\alpha)\,(\theta - a_2)^2 .
\end{aligned}$$

Then by Jensen's inequality we have that $\text{MSE}_\theta\,(T) = E_\theta((T - \psi\,(\theta))^2) = E_\theta(E_{P(\cdot\,|\,U)}\,(T - \psi\,(\theta))^2) \ge E_\theta((E_{P(\cdot\,|\,U)}\,(T) - \psi\,(\theta))^2) = \text{MSE}_\theta\,(T_U)$.

8.1.12 We have that

$$|x+y| = \begin{cases} x+y & x+y \ge 0 \\ -(x+y) & x+y \le 0 \end{cases} \le |x| + |y|$$

for any x, y. Therefore,

$$\begin{aligned}
&|\theta - \alpha a_1 - (1-\alpha)\, a_2| = |\alpha\,(\theta - a_1) + (1-\alpha)\,(\theta - a_2)| \\
&\le |\alpha\,(\theta - a_1)| + |(1-\alpha)\,(\theta - a_2)| = \alpha\,|\theta - a_1| + (1-\alpha)\,|\theta - a_2| .
\end{aligned}$$

Then by Jensen's inequality $E_\theta\,(|T - \psi\,(\theta)|) = E_\theta\,(E_{P(\cdot\,|\,U)}\,(|T - \psi\,(\theta)|)) \ge E_\theta\,(|E_{P(\cdot\,|\,U)}\,(T) - \psi\,(\theta)|) = E_\theta\,(|T_U - \psi\,(\theta)|)$.

8.1.13 We have that

$$\begin{aligned}
\text{MSE}_{(\mu,\sigma^2)}\,(cs^2) &= E_{(\mu,\sigma^2)}\left((cs^2 - \sigma^2)^2\right) = \sigma^4 E_{(\mu,\sigma^2)}\left(\left(\frac{c}{(n-1)}X - 1\right)^2\right) \\
&= \sigma^4 \left\{\left(\frac{c}{n-1}\right)^2 E_{(\mu,\sigma^2)}\,(X^2) - \frac{2c}{n-1} E_{(\mu,\sigma^2)}\,(X) + 1\right\}
\end{aligned}$$

where $X = (n-1)\, s^2/\sigma^2 \sim \chi^2(n-1)$. So $E_{(\mu,\sigma^2)}\,(X) = n-1$ and $\text{Var}_{(\mu,\sigma^2)}\,(X) = 2(n-1)$, which implies $E_{(\mu,\sigma^2)}\,(X^2) = 2(n-1) + (n-1)^2$. Differentiating the above expression with respect to c and setting this equal to 0, gives the optimal value satisfies

$$\frac{c}{(n-1)^2} E_{(\mu,\sigma^2)}\,(X^2) - \frac{1}{n-1} E_{(\mu,\sigma^2)}\,(X) = 0$$

or

$$c = (n-1)\,\frac{E_{(\mu,\sigma^2)}\,(X)}{E_{(\mu,\sigma^2)}\,(X^2)} = (n-1)\,\frac{(n-1)}{2(n-1) + (n-1)^2} = \frac{n-1}{n+1}.$$

We have that the bias equals

$$E_{(\mu,\sigma^2)}\,(cs^2) - \sigma^2 = (c-1)\,\sigma^2 = \left(\frac{n-1}{n+1} - 1\right)\sigma^2 = \frac{-2\sigma^2}{n+1}\sigma^2.$$

8.1.14 Suppose that c is a function such that $E_\theta(c(U)) = 0$ for every θ. Then $E_\theta(c(h(T))) = 0$ for every θ, and the completeness of T implies that $P_\theta(\{s : c(h(T(s))) = 0\}) = 1$ for every θ. Now suppose u is such that $c(u) \neq 0$. Then $P_\theta(U = u) = P_\theta(h(T) = u) = P_\theta(T = h^{-1}(u)) =$ $P_\theta(\{s : T(s) = h^{-1}(u)\}) = 0$ since $c(h(T(s))) = c(u)$ for s in $\{s : T(s) = h^{-1}(u)\}$. This implies that U is complete.

8.1.15 We have that $X = (n-1)s^2/\sigma^2 \sim \chi^2(n-1) = \text{Gamma}((n-1)/2, 1/2)$ and so

$$E\left(X^{1/2}\right) = \frac{1}{\Gamma\left(\frac{n-1}{2}\right)} \int_0^\infty x^{1/2} \left(\frac{x}{2}\right)^{\frac{n-1}{2}-1} e^{-\frac{x}{2}} \frac{1}{2}\, dx$$
$$= \frac{2^{1/2}}{\Gamma\left(\frac{n-1}{2}\right)} \int_0^\infty \left(\frac{x}{2}\right)^{\frac{n}{2}-1} e^{-\frac{x}{2}} \frac{1}{2}\, dx = \frac{2^{1/2}\Gamma\left(\frac{n}{2}\right)}{\Gamma\left(\frac{n-1}{2}\right)}.$$

Therefore, $(n-1)^{1/2}\Gamma\left(\frac{n-1}{2}\right)s/2^{1/2}\Gamma\left(\frac{n}{2}\right)$ is an unbiased estimator of σ. Since it is a function of the complete sufficient statistic, it is UMVU.

8.1.16 From Problem 8.1.13 we have that $\bar{x} + (n-1)^{1/2}\Gamma\left(\frac{n-1}{2}\right) s\, z_{.25}/2^{1/2}\Gamma\left(\frac{n}{2}\right)$ is an unbiased estimator of the first quartile. Since it is a function of a complete sufficient statistic, it is UMVU.

8.1.17 The likelihood function is given by $L(x_1, \ldots, x_n \mid \mu) =$ $\exp\left\{-n(\bar{x}-\mu)^2/2\sigma_0^2\right\}$ for $\mu \in \{\mu_1, \mu_2\}$. Clearly, given $\bar{x}$ we can determine the likelihood function so $\bar{x}$ is sufficient. Now the log-likelihood function takes the values $-n(\bar{x}-\mu_1)^2/2\sigma_0^2$ and $-n(\bar{x}-\mu_2)^2/2\sigma_0^2$, which give

$$\frac{-n(\bar{x}-\mu_1)^2}{2\sigma_0^2} + \frac{n(\bar{x}-\mu_2)^2}{2\sigma_0^2} = \frac{n\bar{x}(\mu_1-\mu_2)}{\sigma_0^2} - \frac{n\left(\mu_1^2-\mu_2^2\right)}{2\sigma_0^2},$$

so we can determine $\bar{x}$ from the likelihood, and $\bar{x}$ is a minimal sufficient statistic.

Now supposing $\mu_1 < \mu_2$, we have that

$$P_\mu(\mu_1 < \bar{x} < \mu_2) = P_\mu\left(\frac{\mu_1-\mu}{\sigma_0/\sqrt{n}} < \frac{\bar{x}-\mu}{\sigma_0/\sqrt{n}} < \frac{\mu_2-\mu}{\sigma_0/\sqrt{n}}\right)$$
$$= \Phi\left(\frac{\mu_2-\mu}{\sigma_0/\sqrt{n}}\right) - \Phi\left(\frac{\mu_1-\mu}{\sigma_0/\sqrt{n}}\right) = \begin{cases} \Phi\left(\frac{\mu_2-\mu_1}{\sigma_0/\sqrt{n}}\right) - \Phi(0) & \mu = \mu_1 \\ \Phi(0) - \Phi\left(\frac{\mu_1-\mu_2}{\sigma_0/\sqrt{n}}\right) & \mu = \mu_2 \end{cases}$$

and, since $\Phi\left(\frac{\mu_2-\mu_1}{\sigma_0/\sqrt{n}}\right) = 1 - \Phi\left(\frac{\mu_1-\mu_2}{\sigma_0/\sqrt{n}}\right)$, we see that this probability is independent of $\mu \in \{\mu_1, \mu_2\}$. Now $I_{(\mu_1,\mu_2)}(\bar{x})$ is unbiased for $P_\mu(\mu_1 < \bar{x} < \mu_2)$ and therefore $I_{(\mu_1,\mu_2)}(\bar{x}) - P_\mu(\mu_1 < \bar{x} < \mu_2)$ is an unbiased estimator of 0 that is not 0 with probability 1. Therefore, $\bar{x}$ is not complete.

8.1.18 The log-likelihood function is given by $-n(\bar{x}-\mu)^2/2\sigma_0^2$, so $S(\mu \mid x_1, \ldots, x_n) = n(\bar{x}-\mu)/\sigma_0^2$. Then $S'(\mu \mid x_1, \ldots, x_n) = -n/\sigma_0^2$, so $I(\mu) = n/\sigma_0^2$. Since $\psi(\mu) = \mu^2 + \sigma_0^2$, then $\psi'(\mu) = 2\mu$ and the information lower bound for an unbiased estimator is given by $4\mu^2\sigma_0^2/n$. The estimator that obtains this lower bound is given by $\mu^2 + \sigma_0^2 - 2\mu\frac{\sigma_0^2}{n}\frac{n(\bar{x}-\mu)}{\sigma_0^2} = -2\mu\bar{x} + 3\mu^2 + \sigma_0^2$, which is not equal to the UMVU estimator obtained in Exercise 8.1.3. Therefore, the UMVU estimator cannot obtain the lower bound.

8.1.19 The log-likelihood function is given by $l(\beta \mid x_1, \ldots, x_n) = n\alpha_0 \ln\beta - \beta n\bar{x}$, so $S(\beta \mid x_1, \ldots, x_n) = n\alpha_0/\beta - n\bar{x}$, $S'(\beta \mid x_1, \ldots, x_n) = -n\alpha_0/\beta^2$, which implies $I(\beta) = n\alpha_0/\beta^2$. Since $\psi(\beta) = \beta^{-1}, \psi'(\beta) = -\beta^{-2}$ the information lower bound for unbiased estimators is given by $(1/\beta^4)(\beta^2/n\alpha_0) = 1/n\alpha_0\beta^2$. Note that by Exercise 8.1.7 $\bar{x}/\alpha_0$ is UMVU for β^{-1} and this has variance $\alpha_0/n\alpha_0^2\beta^2 = 1/n\alpha_0\beta^2$, which is the Cramer-Rao lower bound.

8.1.20 The log-likelihood function is given by $l(\theta \mid x_1, \ldots, x_n) = n\ln\theta + (\theta - 1)\sum_{i=1}^n \ln x_i$, so $S(\theta \mid x_1, \ldots, x_n) = n/\theta + \sum_{i=1}^n \ln x_i$, $S'(\beta \mid x_1, \ldots, x_n) = -n/\theta^2$, which implies $I(\beta) = n/\theta^2$. Since $\psi(\theta) = \theta, \psi'(\beta) = 1$ the information lower bound for unbiased estimators is given by θ^2/n. This is attained by the estimator $\theta + \frac{\theta^2}{n}\left(\frac{n}{\theta} + \sum_{i=1}^n \ln x_i\right) = 2\theta + \frac{\theta^2}{n}\sum_{i=1}^n \ln x_i$. Since this depends on θ, this implies that any UMVU estimator, if it exists, cannot have variance at the lower bound.

Challenges

8.1.21 We can assume that $c = 0$ (or else replace X by $Y = X - c$). We have that $X(s) = X(s)I_{(-\infty,-1)}(X(s)) + X(s)I_{[-1,1]}(X(s)) + X(s)I_{(1,\infty)}(X(s))$. Then $E(X) = E\left(XI_{(-\infty,-1)}(X)\right) + E\left(XI_{[-1,1]}(X)\right) + E\left(XI_{(1,\infty)}(X)\right)$ and $-1 \le E\left(XI_{[-1,1]}(X)\right) \le 1$. Also, $E(X^2) = E\left(X^2 I_{(-\infty,-1)}(X)\right) + E\left(X^2 I_{[-1,1]}(X)\right) + E\left(X^2 I_{(1,\infty)}(X)\right)$, and each term in this sum is nonnegative (possibly infinite).

Now suppose $E\left(XI_{(1,\infty)}(X)\right) = \infty$. Then, when $X > 1$, we have $X^2 > X$ and so $E\left(X^2 I_{(1,\infty)}(X)\right) \ge E\left(XI_{(1,\infty)}(X)\right) = \infty$. If $E\left(XI_{(-\infty,-1)}(X)\right) = -\infty$ then, when $X < -1$, we have $-X^2 < X$, so $E\left(-X^2 I_{(1,\infty)}(X)\right) \le E\left(XI_{(1,\infty)}(X)\right) = -\infty$, which implies $E\left(X^2 I_{(1,\infty)}(X)\right) = \infty$. In both cases, we have that $E(X^2) = \infty$.

8.1.22 The log-likelihood function is given by $l(\beta \mid x_1, \ldots, x_n) = n\alpha_0 \ln\beta - \beta n\bar{x}$, so $\bar{x}$ determines the likelihood and as such is sufficient. Now $S(\beta \mid x_1, \ldots, x_n) = n\alpha_0/\beta - n\bar{x}$, and, setting the score function equal to 0, we obtain the MLE of β as $\alpha_0/\bar{x}$, so we can also obtain $\bar{x}$ from the likelihood. This implies that $\bar{x}$ is minimal sufficient.

We know that $\bar{x} \sim \text{Gamma}(n\alpha_0, n\beta)$. We will present the argument when $n = 1$ and the proof is the same for $n > 1$.

Now suppose h is such that $E_\beta(h(X)) = 0$ for every $\beta > 0$.

Suppose that $h \ge 0$. Then this implies that $\int_0^\infty h(x)x^{\alpha-1}e^{-\beta x}\,dx = 0$ for any β, but since the integrand is nonnegative, this can only happen when $h(x) = 0$. Similarly if $h \le 0$.

Now write $h = h^+ - h^-$ where $h^+(x) = \max\{0, h(x)\}, h^-(x) = \max\{0, -h(x)\}$, and we must have $\int_0^\infty h^+(x)x^{\alpha-1}e^{-\beta x}\,dx = \int_0^\infty h^-(x)x^{\alpha-1}e^{-\beta x}\,dx$. This implies $\int_0^\infty h^+(x)x^{\alpha-1}e^{-x}\,dx = \int_0^\infty h^-(x)x^{\alpha-1}e^{-x}\,dx$. Now suppose $h^+ > 0$ and

$h^- > 0$ on subintervals of $(0,\infty)$. Note they can't both be nonzero on the same subinterval. Then we have that

$$\frac{\int_0^\infty e^{-(\beta-1)x}h^+(x)x^{\alpha-1}e^{-x}\,dx}{\int_0^\infty h^+(x)x^{\alpha-1}e^{-x}\,dx} = \frac{\int_0^\infty e^{-(\beta-1)x}h^-(x)x^{\alpha-1}e^{-x}\,dx}{\int_0^\infty h^-(x)x^{\alpha-1}e^{-x}\,dx}$$

or $m_+(\beta-1) = m_-(\beta-1)$ for every $\beta > 1$, where m_+ is the mgf of the distribution on $(0,\infty)$with density given by $h^+(x)x^{\alpha-1}e^{-x}/\int_0^\infty h^+(x)x^{\alpha-1}e^{-x}\,dx$, and m_- is the mgf of the distribution on $(0,\infty)$with density given by $h^-(x)x^{\alpha-1}e^{-x}/\int_0^\infty h^-(x)x^{\alpha-1}e^{-x}\,dx$. But the equality of the mgf's implies the equality of the distributions (Theorem 3.4.6), and these distributions are concentrated on disjoint subsets, so we have a contradiction to the supposition that $h^+ > 0$ and $h^- > 0$ on subintervals of $(0,\infty)$. This implies that $h^+ = h^- = 0$ and so $h = 0$.

8.2 Optimal Hypothesis Testing

Exercises

8.2.1 The ratio $f_b(s)/f_a(s)$ has the following distribution when $\theta = a$: $P_a(f_b(s)/f_a(s) = 3/2) = P_a(\{1,2\}) = 1/2, P_a(f_b(s)/f_a(s) = 2) = P_a(\{3\}) = 1/12$, and $P_a(f_b(s)/f_a(s) = 1/5) = P_a(\{4\}) = 5/12$.When $\alpha = .1$, using (8.2.4) and (8.2.5), we have that $c_0 = 3/2$ and $\gamma = ((1/10) - (1/12))/(1/2) = 1/30$. The power of the test is $P_b(\{3\}) + P_b(\{1,2\})/30 = 1/6 + (3/4)/30 = 23/120$.

When $\alpha = .05$ we have that $c_0 = 2$ and $\gamma = ((1/20) - 0)/(1/12) = 3/5$.The power of the test is $P_b(\{3\})(3/5) = (1/6)(3/5) = 1/10$.

8.2.2 Such a test completely ignores the data and so makes no use of any information that this provides about the true value of θ. The power of such a test is clearly 1/20, and this is smaller than the power 1/10 of the optimal size .05 test derived in Exercise 8.2.1.

8.2.3 By (8.2.6) the optimal .01 test is of the form (using $z_{.99} = 2.3263$)

$$\varphi_0(\bar{x}) = \begin{cases} 1 & \bar{x} \ge 1 + \frac{\sqrt{2}}{\sqrt{10}}2.3263 \\ 0 & \bar{x} < 1 + \frac{\sqrt{2}}{\sqrt{10}}2.3263 \end{cases} = \begin{cases} 1 & \bar{x} \ge 2.0404 \\ 0 & \bar{x} < 2.0404. \end{cases}$$

Problems

8.2.4 From the argument in the text we have that (8.2.7) is UMP size α for $H_0 : \mu = \mu_0$ versus $H_a : \mu = \mu_1$ for some $\mu_1 < \mu_0$. Since the test doesn't depend on μ_1, this says that (8.2.7) is UMP size α for $H_0 : \mu = \mu_0$ versus $H_a : \mu < \mu_0$. Now the power function is given by $\beta_{\varphi_0}(\mu) = P_\mu\left(\bar{x} \le \mu_0 + \frac{\sigma_0}{\sqrt{n}}z_\alpha\right) = P_\mu\left(\frac{\bar{x}-\mu}{\sigma_0/\sqrt{n}} \le \frac{\mu_0-\mu}{\sigma_0/\sqrt{n}} + z_\alpha\right) = \Phi\left(\frac{\mu_0-\mu}{\sigma_0/\sqrt{n}} + z_\alpha\right)$. Note that this is decreasing in μ. This implies that φ_0 is a size α test function for $H_0 : \mu \ge \mu_0$ versus $H_a : \mu < \mu_0$.

Observe that, if φ is a size α test function for $H_0 : \mu \geq \mu_0$ versus $H_a : \mu < \mu_0$, then it is also a size α test for $H_0 : \mu = \mu_0$ versus $H_a : \mu < \mu_0$. From this we conclude that φ_0 is UMP size α for $H_0 : \mu \leq \mu_0$ versus $H_a : \mu > \mu_0$.

8.2.5 We have that $E_\theta(\varphi) = \alpha$ for every θ, so it is of exact size α. For this test, no matter what data is obtained, we randomly decide to reject H_0 with probability α.

8.2.6 Suppose that φ_0 is a size α UMP test for a specific problem and let φ be the test function of Problem 8.2.5. Then for θ such that the alternative is true we have that $E_\theta(\varphi_0) \geq E_\theta(\varphi) = \alpha$, so φ_0 is unbiased.

8.2.7 The likelihood function is given by $L(\beta \mid x_1, \ldots, x_n) = \beta^{n\alpha_0} \exp\{-\beta n\bar{x}\}$. Therefore, we reject $H_0 : \beta = \beta_0$ versus $H_a : \beta = \beta_1$ whenever $\beta_1^{n\alpha_0} \exp\{-\beta_1 n\bar{x}\} / \beta_0^{n\alpha_0} \exp\{-\beta_0 n\bar{x}\} > c_0$ or, equivalently, whenever $(\beta_0 - \beta_1) n\bar{x} > n\alpha_0 (\beta_0 - \beta_1) + \ln c_0$ or, since $\beta_0 < \beta_1$, whenever $n\bar{x} < (n\alpha_0 (\beta_0 - \beta_1) + \ln c_0) / (\beta_0 - \beta_1)$. When H_0 is true we have that $n\bar{X} \sim$ Gamma$(n\alpha_0, \beta_0)$, so with $x_{1-\alpha}(\beta_0)$ denoting the $(1-\alpha)$th quantile of this distribution, the UMP size α test is to reject whenever $n\bar{x} \leq x_\alpha(\beta_0)$.

Since this test does not depend on β_1, it is also UMP size α for $H_0 : \beta = \beta_0$ versus $H_a : \beta > \beta_0$. Now observe that when $X \sim$ Gamma(α, β), $Z = \beta X \sim$ Gamma$(\alpha, 1)$. Therefore, $P_\beta(n\bar{x} \leq x_{1-\alpha}(\beta_0)) = P_\beta(\beta n\bar{x} \leq \beta x_\alpha(\beta_0)) = P_1(Z \leq \beta x_\alpha(\beta_0))$, where $Z \sim$ Gamma$(\alpha, 1)$. The above implies that the power function is increasing in β. Therefore, the above test is size α for $H_0 : \beta \leq \beta_0$ versus $H_a : \beta > \beta_0$. Now suppose φ is also size α for $H_0 : \beta \leq \beta_0$ versus $H_a : \beta > \beta_0$. Then φ is also size α for $H_0 : \beta = \beta_0$ versus $H_a : \beta > \beta_0$ and so must have its power function uniformly less than or equal to the power function for the above test when $\beta > \beta_0$. This implies that the above test is UMP size α for $H_0 : \beta \leq \beta_0$ versus $H_a : \beta > \beta_0$.

8.2.8 Without loss of generality, assume $\mu_0 = 0$. Then for $H_0 : \sigma^2 = \sigma_0^2$ versus $H_a : \sigma^2 = \sigma_1^2$, the UMP size α test rejects H_0 whenever

$$\frac{L(\sigma_1^2 \mid x_1, \ldots, x_n)}{L(\sigma_0^2 \mid x_1, \ldots, x_n)} = \frac{\sigma_1^{-2n} \exp\left\{-\frac{1}{2\sigma_1^2} \sum_{i=1}^n x_i^2\right\}}{\sigma_0^{-2n} \exp\left\{-\frac{1}{2\sigma_0^2} \sum_{i=1}^n x_i^2\right\}} > c_0$$

or, equivalently, whenever $n(\sigma_0^2 - \sigma_1^2) + \frac{1}{2}\left(\frac{1}{\sigma_0^2} - \frac{1}{\sigma_1^2}\right) \sum_{i=1}^n x_i^2 > \ln c_0$ or, using $\sigma_0^2 < \sigma_1^2$, whenever

$$\frac{1}{\sigma_0^2} \sum_{i=1}^n x_i^2 > \frac{2}{\sigma_0^2}\left(\frac{1}{\sigma_0^2} - \frac{1}{\sigma_1^2}\right)^{-1} (\ln c_0 - n(\sigma_0^2 - \sigma_1^2)).$$

Under H_0 we have that $\frac{1}{\sigma_0^2} \sum_{i=1}^n x_i^2 \sim \chi^2(n)$, so the test is to reject whenever $\frac{1}{\sigma_0^2} \sum_{i=1}^n x_i^2 > x_{1-\alpha}$, where $x_{1-\alpha}$ is the $(1-\alpha)$th quantile of the $\chi^2(n)$ distribution. Since the test does not involve σ_1^2, it is UMP size α for $H_0 : \sigma^2 = \sigma_0^2$ versus $H_a : \sigma^2 > \sigma_0^2$. The power function of this test is given by

$P_{\sigma^2}\left(\frac{1}{\sigma_0^2}\sum_{i=1}^n x_i^2 \geq x_{1-\alpha}\right) = P_{\sigma^2}\left(\frac{1}{\sigma^2}\sum_{i=1}^n x_i^2 \geq \frac{\sigma_0^2}{\sigma^2}x_{1-\alpha}\right) = P\left(Z \geq \frac{\sigma_0^2}{\sigma^2}x_{1-\alpha}\right)$ where $Z = \left(\sum_{i=1}^n x_i^2\right)/\sigma^2 \sim \chi^2(n)$, so the power function is increasing in σ^2. This implies that the above test is of size α for $H_0 : \sigma^2 \leq \sigma_0^2$ versus $H_a : \sigma^2 > \sigma_0^2$. Now suppose φ is also size α for $H_0 : \sigma^2 \leq \sigma_0^2$ versus $H_a : \sigma^2 > \sigma_0^2$. Then φ is also size α for $H_0 : \sigma^2 = \sigma_0^2$ versus $H_a : \sigma^2 > \sigma_0^2$ and so must have its power function uniformly less than or equal to the power function for the above test when $\sigma^2 > \sigma_0^2$. This implies that the above test is UMP size α for $H_0 : \sigma^2 \leq \sigma_0^2$ versus $H_a : \sigma^2 > \sigma_0^2$.

8.2.9 For $H_0 : \theta = \theta_0$ versus $H_a : \theta = \theta_1$, the UMP size α test rejects H_0 whenever $L(\theta_1 \,|\, x_1, \ldots, x_n)/L(\theta_0 \,|\, x_1, \ldots, x_n) = \theta_1^{-n} I_{(0,\theta_1)}(x_{(n)})/\theta_0^{-n} I_{(0,\theta_0)}(x_{(n)}) > c_0$ or, equivalently, whenever $I_{(0,\theta_1)}(x_{(n)})/I_{(0,\theta_0)}(x_{(n)}) > \theta_0^{-n} c_0/\theta_1^{-n}$. So we reject categorically whenever $x_{(n)} > \theta_0$ because the likelihood ratio equals ∞. When $0 \leq x_{(n)} \leq 1$ the likelihood ratio equals $\theta_1^{-n}/\theta_0^{-n}$. This implies that the UMP size α rejects whenever the likelihood ratio is greater than $\theta_1^{-n}/\theta_0^{-n}$ (i.e., equals ∞) and otherwise we randomly reject with probability α (i.e., when the likelihood ratio doesn't equal ∞).

Note that the above test doesn't depend on θ_1 and so is UMP size α for $H_0 : \theta = \theta_0$ versus $H_a : \theta > \theta_0$. Further, if $\theta < \theta_0$ we have that $P_\theta\left(\frac{L(\theta_1 \,|\, x_1,\ldots,x_n)}{L(\theta \,|\, x_1,\ldots,x_n)} < \infty\right) = 1$, so the above test has size α for $H_0 : \theta \leq \theta_0$ versus $H_a : \theta > \theta_0$. Now suppose φ is size α for $H_0 : \theta \leq \theta_0$ versus $H_a : \theta > \theta_0$. Then φ is also size α for $H_0 : \theta = \theta_0$ versus $H_a : \theta > \theta_0$ and so must have its power function uniformly less than or equal to the power function for the above test when $\theta > \theta_0$. This implies that the above test is UMP size α for $H_0 : \theta \leq \theta_0$ versus $H_a : \theta > \theta_0$.

8.2.10 Suppose $X \sim$ Binomial(n, θ) and put

$$F(x) = \frac{\Gamma(n+1)}{\Gamma(x+1)\Gamma(n-x)} \int_\theta^1 y^x (1-y)^{n-x-1}\, dy.$$

So $F(0) = \frac{\Gamma(n+1)}{\Gamma(0+1)\Gamma(n-0)} \int_\theta^1 y^0 (1-y)^{n-0-1}\, dy = n \int_\theta^1 (1-y)^{n-1}\, dy = (1-\theta)^n = P(X = 0)$. Using integration by parts we put $u = y^x$, giving $du = xy^{x-1}$ and $dv = (1-y)^{n-x-1}$, giving $v = -(1-y)^{n-x}/(n-x)$, so that

$$\begin{aligned}
F(x) &= \frac{\Gamma(n+1)}{\Gamma(x+1)\Gamma(n-x)} \int_\theta^1 y^x (1-y)^{n-x-1}\, dy \\
&= \frac{\Gamma(n+1)}{\Gamma(x+1)\Gamma(n-x)} \left\{ -y^x \frac{(1-y)^{n-x}}{n-x}\Bigg|_\theta^1 + \frac{x}{n-x} \int_\theta^1 y^{x-1} (1-y)^{n-x}\, dy \right\} \\
&= \binom{n}{x} \theta^x (1-\theta)^{n-x} + \frac{\Gamma(n+1)}{\Gamma(x)\Gamma(n-x+1)} \int_\theta^1 y^{x-1} (1-y)^{n-x}\, dy \\
&= P(X = x) + F(x-1).
\end{aligned}$$

Continuing this recursively establishes the result.

8.2.11 Let $X \sim$ Poisson(λ) and put $F(x) = \frac{1}{x!}\int_\lambda^\infty y^x e^{-y}\,dy$. Then $F(0) = \frac{1}{0!}\int_\lambda^\infty y^0 e^{-y}\,dy = e^{-y} = P(X = 0)$. Using integration by parts with $u = y^x$, giving $du = xy^{x-1}$, $dv = e^{-y}$, giving $v = -e^{-y}$, we have that

$$\begin{aligned} F(x) &= \frac{1}{x!}\left\{-y^x e^{-y}\Big|_\lambda^\infty + x\int_\lambda^\infty y^{x-1}e^{-y}\,dy\right\} = \frac{\lambda^x e^{-\lambda}}{x!} + F(x-1) \\ &= P(X = x) + F(x-1). \end{aligned}$$

Continuing this recursively establishes the result.

8.2.12 The UMP size α test for $H_0 : \lambda = \lambda_0$ versus $H_0 : \lambda = \lambda_1$ is of the form

$$\frac{L(\lambda_1 \,|\, x_1, \ldots, x_n)}{L(\lambda_0 \,|\, x_1, \ldots, x_n)} = \frac{(\lambda_1)^{n\bar{x}}\, e^{-\lambda_1}}{(\lambda_0)^{n\bar{x}}\, e^{-\lambda_0}} > c_0$$

or, equivalently, whenever $n\bar{x}(\ln\lambda_1 - \ln\lambda_0) > (\lambda_1 - \lambda_0)\ln c_0$, and since $\lambda_1 > \lambda_0$, this is equivalent to rejecting whenever $n\bar{x} > (\lambda_1 - \lambda_0)\ln c_0/(\ln\lambda_1 - \ln\lambda_0)$. Now recall that $n\bar{x} \sim$ Poisson($n\lambda_0$) under H_0 so we must determine the smallest k such that $P_{\lambda_0}(n\bar{x} > k) \le \alpha$ and then put $\gamma = (\alpha - P_{\lambda_0}(n\bar{x} > k))/P_{\lambda_0}(n\bar{x} = k)$. Since this test does not involve λ_1, it is UMP size α for $H_0 : \lambda = \lambda_0$ versus $H_0 : \lambda > \lambda_0$. From Problem 8.2.11 we have that $P_\lambda(n\bar{x} > k) \le 1 - \frac{1}{x!}\int_\lambda^\infty y^x e^{-y}\,dy$, and we see that this is increasing in λ. Therefore, this test is UMP size α for $H_0 : \lambda \le \lambda_0$ versus $H_0 : \lambda > \lambda_0$.

8.2.13 When $H_0 : \mu = \mu_0$ holds, the log-likelihood and score functions are given by $l(x_1, \ldots, x_n \,|\, \sigma^2) = -n\ln\sigma^2 - \frac{1}{2\sigma^2}\sum_{i=1}^n (x - \mu_0)^2$, $S(x_1, \ldots, x_n \,|\, \sigma^2) = -\frac{n}{\sigma^2} + \frac{1}{2\sigma^4}\sum_{i=1}^n (x - \mu_0)^2$. Then $S(x_1, \ldots, x_n \,|\, \sigma^2) = 0$ leads to the MLE $\hat{\mu}_{H_0} = \mu_0$, $\hat{\sigma}^2_{H_0} = \frac{1}{n}\sum_{i=1}^n (x - \mu_0)^2$, and the maximized log-likelihood equals $l(x_1, \ldots, x_n \,|\, \hat{\sigma}^2_{H_0}) = n\ln n - n\ln\sum_{i=1}^n (x - \mu_0)^2 - \frac{n}{2}$.

When $H_a : \mu \ne \mu_0$ holds, the log-likelihood function is given by $l(x_1, \ldots, x_n \,|\, \mu, \sigma^2) = -n\ln\sigma^2 - \frac{1}{2\sigma^2}\sum_{i=1}^n (x - \mu)^2$.

By Example 6.2.6 the MLE is given by $\hat{\mu} = \bar{x}$, $\hat{\sigma}^2 = \frac{1}{n}\sum_{i=1}^n (x - \bar{x})^2$ and the maximized log-likelihood is given by $l(x_1, \ldots, x_n \,|\, \hat{\mu}, \hat{\sigma}^2) = n\ln n - n\ln\sum_{i=1}^n (x - \bar{x})^2 - \frac{n}{2}$. Then the likelihood ratio test rejects whenever

$$\begin{aligned} &2\left(l(x_1, \ldots, x_n \,|\, \hat{\mu}, \hat{\sigma}^2) - l(x_1, \ldots, x_n \,|\, \hat{\sigma}^2_{H_0})\right) \\ &= 2\left(-n\ln\sum_{i=1}^n (x - \bar{x})^2 + n\ln\sum_{i=1}^n (x - \mu_0)^2\right) = 2n\ln\frac{\sum_{i=1}^n (x - \mu_0)^2}{\sum_{i=1}^n (x - \bar{x})^2} > x_{1-\alpha} \end{aligned}$$

where $x_{1-\alpha}$ is the $(1 - \alpha)$th quantile of the $\chi^2(2 - 1) = \chi^2(1)$ distribution.

8.2.14

(a) We have that

$$\begin{aligned} P_\theta(\psi(\theta) \in C(s)) &= P_\theta(\{s : \psi(\theta) \in C(s)\}) = P_\theta\left(\{s : \varphi_{\psi(\theta)}(s) = 0\}\right) \\ &= 1 - P_\theta\left(\{s : \varphi_{\psi(\theta)}(s) = 1\}\right) = 1 - E_\theta\left(\varphi_{\psi(\theta)}\right) \ge 1 - \alpha \end{aligned}$$

since $E_\theta\left(\varphi_{\psi(\theta)}\right) \le \alpha$.
(b) We have that $E_\theta\left(\varphi^*_{\psi(\theta)}\right) = P_\theta\left(\psi(\theta) \notin C^*(s)\right) = 1 - P_\theta\left(\psi(\theta) \in C^*(s)\right) \le 1 - (1-\alpha) = \alpha$ since $P_\theta\left(\psi(\theta) \in C^*(s)\right) \ge 1 - \alpha$.
(c) Suppose now that, for each value of ψ_0, the test function φ_{ψ_0} is UMP size α for $H_0 : \psi(\theta) = \psi_0$ versus $H_a : \psi(\theta) \ne \psi_0$. Then, if C is the confidence set corresponding to this family of tests, we have that $P_\theta\left(\psi(\theta^*) \in C(s)\right) = P_\theta\left(\{s : \psi(\theta^*) \in C(s)\}\right) = 1 - P_\theta\left(\psi(\theta^*) \notin C^*(s)\right) = 1 - E_\theta\left(\varphi_{\psi(\theta^*)}\right)$, and since $E_\theta\left(\varphi_{\psi(\theta^*)}\right)$ is maximized when $\psi(\theta) \ne \psi(\theta^*)$, part (b) implies that the probability of covering a false value is uniformly minimized by C.

Challenge

8.2.15 Suppose that a test φ is size α and UMP for $H_0 : \theta = \theta_0$ versus $H_a : \theta = \theta_1$. Let φ_0 be as in Theorem 8.2.1. Following the proof of Theorem 8.2.1, let $S^* = \{s : \varphi_0(s) \ne \varphi(s)\} \cap \{s : f_{\theta_1}(s) \ne c_0 f_{\theta_0}(s)\}$. Then $E_{\theta_1}(\varphi) = E_{\theta_1}(\varphi_0)$ and, since φ is size α, we have that $0 \ge E_{\theta_1}(\varphi_0) - E_{\theta_1}(\varphi) - c_0\left(\alpha - E_{\theta_0}(\varphi)\right) = \sum_{s \in S}\left(\varphi_0(s) - \varphi(s)\right)\left(f_{\theta_1}(s) - c_0 f_{\theta_0}(s)\right) = \sum_{s \in S^*}\left(\varphi_0(s) - \varphi(s)\right)$
$\times \left(f_{\theta_1}(s) - c_0 f_{\theta_0}(s)\right) \ge 0$ since $\left(\varphi_0(s) - \varphi(s)\right)\left(f_{\theta_1}(s) - c_0 f_{\theta_0}(s)\right) > 0$ on S^*. But this implies that $S^* = \phi$ and we have that $\varphi_0(s) = \varphi(s)$ whenever $f_{\theta_1}(s) \ne c_0 f_{\theta_0}(s)$. The values φ_0 and φ may differ on $B = \{s : f_{\theta_1}(s) = c_0 f_{\theta_0}(s)\}$.

Since $E_{\theta_1}(\varphi_0) - E_{\theta_1}(\varphi) = 0$ the inequalities above establish that $c_0\left(\alpha - E_{\theta_0}(\varphi)\right) = 0$. If $c_0 = 0$, then $\varphi_0(s) = 1$ whenever $f_{\theta_1}(s) > 0$, so the power of the UMP test is 1. If $c_0 \ne 1$, then $E_{\theta_0}(\varphi) = \alpha$ and φ has exact size α.

8.3 Optimal Bayesian Inferences

Exercises

8.3.1 The posterior distribution of θ is given by

$$\Pi(\theta = 1 \,|\, 2) = \frac{\frac{1}{2}\frac{1}{6}}{\frac{1}{2}\frac{1}{6} + \frac{1}{2}\frac{1}{4}} = \frac{2}{5}, \quad \Pi(\theta = 2 \,|\, 2) = \frac{\frac{1}{2}\frac{1}{4}}{\frac{1}{2}\frac{1}{6} + \frac{1}{2}\frac{1}{4}} = \frac{3}{5},$$

so $\Pi(\theta = 2 \,|\, 2) > \Pi(\theta = 1 \,|\, 2)$ and we accept $H_0 : \theta = 2$.

8.3.2 The Bayes rule is given by the posterior mean and this is given by $\frac{2}{5} + \frac{3}{5}2 = \frac{8}{5}$.

8.3.3 In Example 7.1.2 we determined that the posterior distribution of μ is given by the

$$N\left(\left(\frac{1}{\tau_0^2} + \frac{n}{\sigma_0^2}\right)^{-1}\left(\frac{\mu_0}{\tau_0^2} + \frac{n}{\sigma_0^2}\bar{x}\right), \left(\frac{1}{\tau_0^2} + \frac{n}{\sigma_0^2}\right)^{-1}\right)$$

distribution. Then the Bayes rule is given by the posterior mean

$$\left(\frac{1}{\tau_0^2} + \frac{n}{\sigma_0^2}\right)^{-1}\left(\frac{\mu_0}{\tau_0^2} + \frac{n}{\sigma_0^2}\bar{x}\right) \to \left(\frac{n}{\sigma_0^2}\right)^{-1}\left(\frac{n}{\sigma_0^2}\bar{x}\right) = \bar{x}$$

as $\tau_0 \to \infty$.

8.3.4 From Example 7.1.1 we have that the posterior distribution of θ is Beta$(n\bar{x} + \alpha, n(1 - \bar{x}) + \beta)$. The Bayes rule is given by the posterior mean and this is evaluated in Example 7.2.2 to be $(n\bar{x} + \alpha) / (n + \alpha + \beta)$.

8.3.5 The likelihood is given by $L(\beta \mid x_1, \ldots, x_n) = \beta^{n\alpha_0} \exp\{-\beta n\bar{x}\}$ and the prior density is $\pi(\beta) \propto \beta^{\tau_0 - 1} e^{-v_0 \beta}$. Therefore, the posterior density is proportional to $\beta^{n\alpha_0 + \tau_0 - 1} \exp\{-(n\bar{x} + v_0)\beta\}$, and from this we deduce that the posterior distribution of β is Gamma$(n\alpha_0 + \tau_0, n\bar{x} + v_0)$. The Bayes rule is given by the posterior mean and this equals $(n\alpha_0 + \tau_0) / (n\bar{x} + v_0)$. By the weak law of large numbers this converges in probability to (since $E_\beta(\bar{x}) = \alpha_0/\beta$) $\alpha_0 / (\alpha_0/\beta) = \beta$ as $n \to \infty$.

8.3.6 The Bayes rule is given by the posterior mean of $1/\beta$ and this equals

$$\begin{aligned}
&\frac{(n\bar{x} + v_0)^{n\alpha_0 + \tau_0}}{\Gamma(n\alpha_0 + \tau_0)} \int_0^\infty \left(\frac{1}{\beta}\right) \beta^{n\alpha_0 + \tau_0 - 1} \exp\{-(n\bar{x} + v_0)\beta\}\, d\beta \\
&= \frac{(n\bar{x} + v_0)^{n\alpha_0 + \tau_0}}{\Gamma(n\alpha_0 + \tau_0)} \int_0^\infty \beta^{n\alpha_0 + \tau_0 - 2} \exp\{-(n\bar{x} + v_0)\beta\}\, d\beta \\
&= \frac{(n\bar{x} + v_0)^{n\alpha_0 + \tau_0}}{\Gamma(n\alpha_0 + \tau_0)} \frac{\Gamma(n\alpha_0 + \tau_0 - 1)}{(n\bar{x} + v_0)^{n\alpha_0 + \tau_0 - 1}} = \frac{n\bar{x} + v_0}{n\alpha_0 + \tau_0 - 1}
\end{aligned}$$

and this converges to$(\alpha_0/\beta) / \alpha_0 = \beta^{-1}$ as $n \to \infty$.

Problems

8.3.7 Suppose $T(s) \in \{\theta_1, \theta_2\}$ for each s. The Bayes rule will minimize

$$\begin{aligned}
&E_\Pi(P_\theta(T(s) \neq_\theta)) = E_\Pi\left(E_\theta\left(1 - I_{\{\theta\}}(T(s))\right)\right) \\
&= 1 - E_\Pi\left(E_\theta\left(I_{\{\theta\}}(T(s))\right)\right) = 1 - E_M\left(E_{\Pi(\cdot\,|\,s)}\left(I_{\{\theta\}}(T(s))\right)\right).
\end{aligned}$$

Therefore, the Bayes rule at s is given by $T(s)$ which maximizes

$$E_{\Pi(\cdot\,|\,s)}\left(I_{\{\theta\}}(T(s))\right) = \Pi(\{\theta_1\} \mid s)\, I_{\{\theta_1\}}(T(s)) + \Pi(\{\theta_2\} \mid s)\, I_{\{\theta_2\}}(T(s))$$

and this is clearly given by

$$T(s) = \begin{cases} \theta_1 & \Pi(\{\theta_1\} \mid s) > \Pi(\{\theta_2\} \mid s) \\ \theta_2 & \Pi(\{\theta_2\} \mid s) > \Pi(\{\theta_1\} \mid s) \end{cases}$$

and when $\Pi(\{\theta_1\} \mid s) = \Pi(\{\theta_2\} \mid s)$ we can take $T(s)$ to be either θ_1 or θ_2. So the Bayes rule is given by the posterior mode.

8.3.8 Since $\Pi(\theta = 2 \mid 2) > \Pi(\theta = 1 \mid 2)$, we have that the Bayes rule takes the value $T(s) = 2$ for this data. An advantage for this estimator over the posterior mean is that the posterior mode is always an element of the parameter space, while the posterior mean may not be, as in Exercise 8.3.1. So if we use the posterior mean, we may estimate the parameter by a value that it could never possibly take.

8.3.9 In Example 7.1.4 we derived the posterior distribution of $(\mu, 1/\sigma^2)$, and in Example 7.2.1 we derived the marginal posterior distribution of μ to be

$$\mu_x + \sqrt{\frac{2\beta_x}{(2\alpha_0+n)(n+1/\tau_0^2)}}Z.$$

where $Z \sim t(n+2\alpha_0)$ where $\mu_x = (n+1/\tau_0^2)^{-1}(\mu_0/\tau_0^2 + n\bar{x})$ and

$$\beta_x = \beta_0 + \frac{n}{2}\bar{x}^2 + \frac{\mu_0^2}{2\tau_0^2} + \frac{n-1}{2}s^2 - \frac{1}{2}\left(n+\frac{1}{\tau_0^2}\right)^{-1}\left(\frac{\mu_0}{\tau_0^2}+n\bar{x}\right)^2.$$

We know the Bayes rule is given by the posterior mean of μ and this equals

$$\mu_x + \sqrt{\frac{2\beta_x}{(2\alpha_0+n)(n+1/\tau_0^2)}}E_{\Pi(\cdot\,|\,x_1,\ldots,x_n)}(Z) = \mu_x$$

since the mean of a Student(λ) random variable is 0 (provided $\lambda > 1$). So the Bayes rule is given by μ_x.

8.3.10 Suppose $T(s) \in \{\theta_1, \ldots, \theta_k\}$ for each s. The Bayes rule will minimize

$$\begin{aligned} &E_\Pi(P_\theta(T(s) \neq_\theta)) = E_\Pi(E_\theta(1 - I_{\{\theta\}}(T(s)))) \\ &= 1 - E_\Pi(E_\theta(I_{\{\theta\}}(T(s)))) = 1 - E_M(E_{\Pi(\cdot\,|\,s)}(I_{\{\theta\}}(T(s)))). \end{aligned}$$

Therefore, the Bayes rule at s is given by $T(s)$, which maximizes

$$E_{\Pi(\cdot\,|\,s)}(I_{\{\theta\}}(T(s))) = \sum_{i=1}^{k}\Pi(\{\theta_i\}\,|\,s)\,I_{\{\theta_i\}}(T(s))$$

and this is clearly given by $T(s) = \theta_i$ whenever $\Pi(\{\theta_i\}\,|\,s) > \Pi(\{\theta_j\}\,|\,s)$ for every $j \neq i$. When more than one value of θ maximizes $\Pi(\{\theta\}\,|\,s)$ we can take $T(s)$ to be any of these values.

Challenge

8.3.11 We have that

$$\begin{aligned} E\left(\left(\tilde{t}(s)-t\right)^2\right) &= E_\Pi\left(E_{P_\theta}\left(E_{Q_\theta(\cdot\,|\,s)}\left(\left(\tilde{T}(s)-t\right)^2\right)\right)\right) \\ &= E_M\left(E_{P_\theta(\cdot\,|\,s)}\left(E_{Q_\theta(\cdot\,|\,s)}\left(\left(\tilde{T}(s)-t\right)^2\right)\right)\right) \\ &= E_M\left(\left(E_{Q(\cdot\,|\,s)}\left(\left(\tilde{T}(s)-t\right)^2\right)\right)\right) \end{aligned}$$

where $Q(\cdot \mid s)$ is the posterior predictive measure for t given s (has density or probability function $q(t \mid s)$ as specified in Section 7.2.4). This is minimized if we can find $\tilde{T}(s)$ that minimizes $E_{Q(\cdot \mid s)}\left(\left(\tilde{T}(s) - t\right)^2\right)$ for each s. By Theorem 8.1.1 this is minimized by taking $\tilde{T}(s) = E_{Q(\cdot \mid s)}(t)$, the posterior predictive mean of t.

8.4 Decision Theory

Exercises

8.4.1 The model is given by the collection of probability functions $\{\theta^{n\bar{x}}(1-\theta)^{n-n\bar{x}} : \theta \in [0,1]\}$ on the set of all sequences $(x_1, \ldots, x_n)$ of 0's and 1's. The action space is $\mathcal{A} = [0,1]$, the correct action function is $A(\theta) = \theta$, and the loss function is $L(\theta, a) = (\theta - a)^2$.

The risk function for T is given by $R_T(\theta) = E_\theta\left((\theta - \bar{x})^2\right) = \text{Var}_\theta(\bar{x}) = \theta(1-\theta)/n$. This is plotted below.

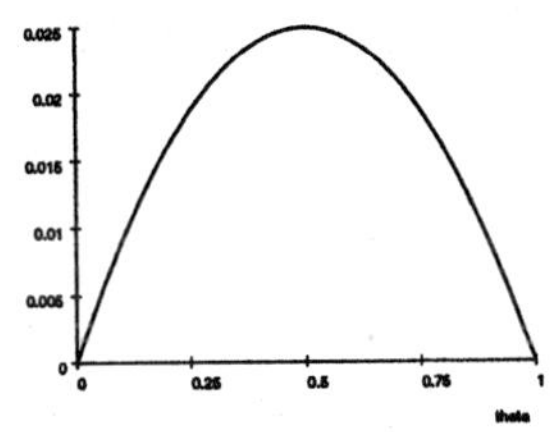

8.4.2 The model is given by the collection of probability functions

$$\left\{\left(\prod_{i=1}^{n} x_i!\right))^{-1}\lambda^{n\bar{x}}e^{-n\lambda} : \lambda \geq 0\right\}$$

on the set of all sequences $(x_1, \ldots, x_n)$ of nonnegative integers. The action space is $\mathcal{A} = [0, \infty)$, the correct action function is $A(\lambda) = \lambda$, and the loss function is $L(\lambda, a) = (\lambda - a)^2$.

The risk function for T is given by $R_T(\lambda) = E_\lambda((\lambda - \bar{x})^2) = \text{Var}_\lambda(\bar{x}) = \lambda/n$. This is plotted below for $n = 25$.

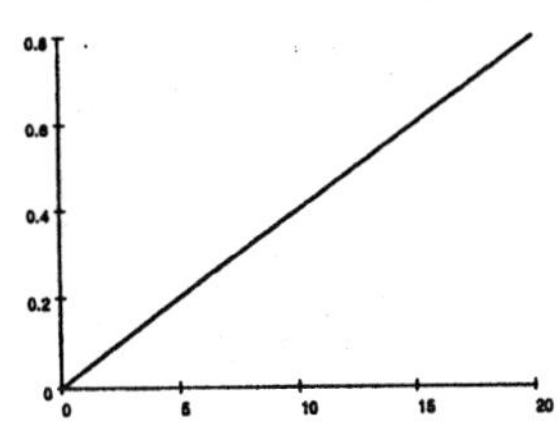

8.4.3 The model is given by the collection of density functions

$$\left\{\frac{1}{\sqrt{2\pi}\sigma_0}\exp\left\{-\frac{1}{2\sigma_0^2}\sum_{i=1}^{n}(x_i-\mu)^2\right\}:\mu\in R^1\right\}$$

on the set of all sequences $(x_1,\ldots,x_n)$ of real numbers. The action space is $\mathcal{A}=R^1$, the correct action function is $A(\mu)=\mu$, and the loss function is $L(\mu,a)=(\mu-a)^2$.

The risk function for T is given by $R_T(\mu)=E_\mu\left((\mu-\bar{x})^2\right)=\mathrm{Var}_\mu(\bar{x})=\frac{\sigma_0^2}{n}$. This is plotted below for $n=25$ and $\sigma_0^2=2$ (note $2/25=0.08$).

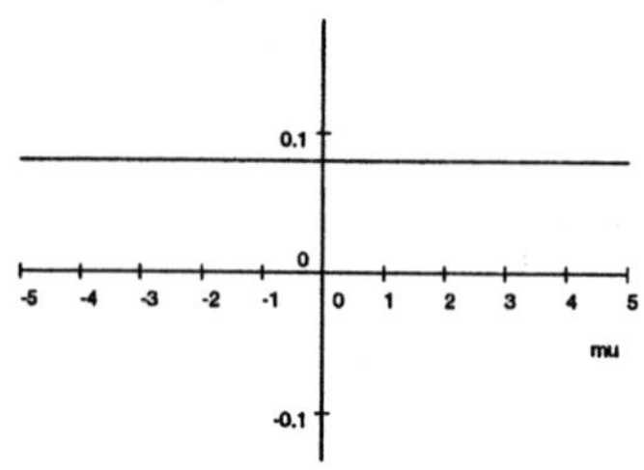

8.4.4 The model is given by the collection of probability functions $\left\{\theta^{n\bar{x}}(1-\theta)^{n-n\bar{x}}:\theta\in[0,1]\right\}$ on the set of all sequences $(x_1,\ldots,x_n)$ of 0's and 1's. The action space is $\mathcal{A}=\{H_0,H_a\}$, where $H_0:\theta=1/2$, the correct action function is

$$A(\theta)=\begin{cases}H_0 & \theta=1/2\\ H_a & \theta\neq 1/2\end{cases}$$

and the loss function is

$$L(\theta,a)=\begin{cases}0 & \theta=1/2, a=H_0 \text{ or } \theta\neq 1/2, a=H_a\\ 1 & \theta=1/2, a=H_a \text{ or } \theta\neq 1/2, a=H_0.\end{cases}$$

The test function φ is given by

$$\varphi(n\bar{x})=\begin{cases}0 & n\bar{x}\notin\{0,1,n-1,n\}\\ 1 & n\bar{x}\in\{0,1,n-1,n\}.\end{cases}$$

The risk function for φ is given by

$$\begin{aligned}R_\varphi(\theta)&=P_\theta(\varphi(n\bar{x})=1)=P_\theta(\{0,1,n-1,n\})\\&=\binom{n}{0}(1-\theta)^n+\binom{n}{1}\theta(1-\theta)^{n-1}+\binom{n}{n-1}\theta^{n-1}(1-\theta)+\binom{n}{n}\theta^n.\end{aligned}$$

A plot of R_φ, when $n=10$ (the power equals 2.1484×10^{-2} at $\theta=1/2$) follows.

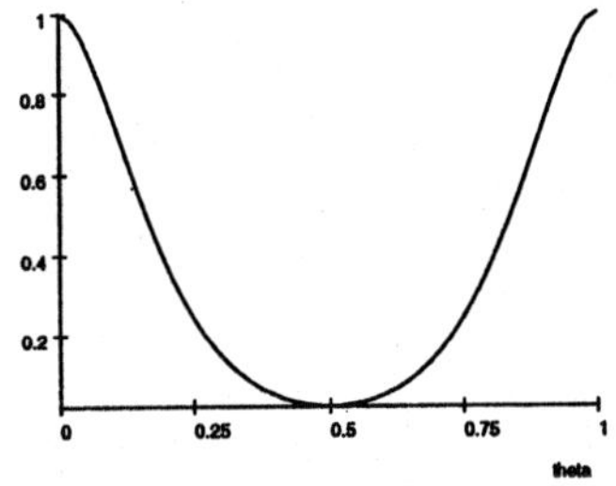

8.4.5
(a) The risk function is given by

$$\begin{aligned} R_d(a) &= E_a(L(a, d(s))) \\ &= \frac{1}{4}L(a, d(1)) + \frac{1}{4}L(a, d(2)) + 0L(a, d(3)) + \frac{1}{2}L(a, d(4)) \\ &= \frac{1}{4}L(a, a) + \frac{1}{4}L(a, a) + \frac{1}{2}L(a, b) = \frac{1}{2}L(a, b) = \frac{1}{2}, \end{aligned}$$

$$\begin{aligned} R_d(b) &= E_b(L(b, d(s))) \\ &= \frac{1}{2}L(b, d(1)) + 0L(b, d(2)) + \frac{1}{4}L(b, d(3)) + \frac{1}{4}L(b, d(4)) \\ &= \frac{1}{2}L(b, a) + \frac{1}{4}L(b, a) + \frac{1}{4}L(b, b) = \frac{1}{2}L(b, a) + \frac{1}{4}L(b, a) = \frac{3}{4}. \end{aligned}$$

(b) Consider the risk function of the decision function d^* given by $d^*(1) = b, d^*(2) = a, d^*(3) = b, d^*(4) = a$. The risk function is given by

$$\begin{aligned} R_{d^*}(a) &= E_a(L(a, d^*(s))) \\ &= \frac{1}{4}L(a, d^*(1)) + \frac{1}{4}L(a, d^*(2)) + 0L(a, d^*(3)) + \frac{1}{2}L(a, d^*(4)) \\ &= \frac{1}{4}L(a, b) + \frac{1}{4}L(a, a) + \frac{1}{2}L(a, a) = \frac{1}{4}L(a, b) = \frac{1}{4}, \end{aligned}$$

$$\begin{aligned} R_{d^*}(b) &= E_b(L(b, d^*(s))) \\ &= \frac{1}{2}L(b, d^*(1)) + 0L(b, d^*(2)) + \frac{1}{4}L(b, d^*(3)) + \frac{1}{4}L(b, d^*(4)) \\ &= \frac{1}{2}L(b, b) + \frac{1}{4}L(b, b) + \frac{1}{4}L(b, a) = \frac{1}{4}L(b, a) = \frac{1}{4}, \end{aligned}$$

so $R_{d^*}(a) < R_d(a), R_{d^*}(b) < R_d(b)$ and d is not optimal.

8.4.6 The model is given by the collection of probability functions

$$\{(\prod_{i=1}^{n} x_i!)^{-1}\lambda^{n\bar{x}}e^{-n\lambda} : \lambda \geq 0\}$$

on the set of all sequences $(x_1, \ldots, x_n)$ of nonnegative integers. The action space is $\mathcal{A} = \{H_0, H_a\}$, where $H_0 : \lambda \leq \lambda_0$. The correct action function is

$$A(\lambda) = \begin{cases} H_0 & \lambda \leq \lambda_0 \\ H_a & \lambda > \lambda_0 \end{cases}$$

and the loss function is

$$L(\lambda, a) = \begin{cases} 0 & \lambda \leq \lambda_0, a = H_0 \text{ or } \lambda > \lambda_0, a = H_a \\ 1 & \lambda \leq \lambda_0, a = H_a \text{ or } \lambda > \lambda_0, a = H_0. \end{cases}$$

The test function φ is given by

$$\varphi(x_1, \ldots, x_n) = \begin{cases} 0 & n\bar{x} < \lfloor n\lambda_0 + 2\sqrt{n\lambda_0} \rfloor \\ 1/2 & n\bar{x} = \lfloor n\lambda_0 + 2\sqrt{n\lambda_0} \rfloor \\ 1 & n\bar{x} > \lfloor n\lambda_0 + 2\sqrt{n\lambda_0} \rfloor. \end{cases}$$

The power function for φ is given by (using $n\bar{x} \sim$ Poisson(λ))

$$\begin{aligned} \beta_\varphi(\lambda) &= P_\lambda(\varphi(x_1, \ldots, x_n) = 1) \\ &= \frac{1}{2} P_\lambda\left(n\bar{x} = \lfloor n\lambda_0 + 2\sqrt{n\lambda_0} \rfloor\right) + P_\lambda\left(n\bar{x} > \lfloor n\lambda_0 + 2\sqrt{n\lambda_0} \rfloor\right) \\ &= \frac{1}{2} \frac{(n\lambda)^{\lfloor n\lambda_0 + 2\sqrt{n\lambda_0} \rfloor}}{\left(\lfloor n\lambda_0 + 2\sqrt{n\lambda_0} \rfloor\right)!} \exp\{-n\lambda\} + \sum_{k=\lfloor n\lambda_0 + 2\sqrt{n\lambda_0} \rfloor + 1}^{\infty} \frac{(n\lambda)^k}{k!} \exp\{-n\lambda\}. \end{aligned}$$

When $\lambda_0 = 1$ and $n = 5$ then $\lfloor n\lambda_0 + 2\sqrt{n\lambda_0} \rfloor = 9$, so the power function is given by

$$\begin{aligned} R_\varphi(\lambda) &= \frac{1}{2} \frac{(10\lambda)^9}{(9)!} \exp\{-5\lambda\} + \sum_{k=10}^{\infty} \frac{(5\lambda)^k}{k!} \exp\{-5\lambda\} \\ &= \frac{1}{2} \frac{(5\lambda)^9}{(9)!} \exp\{-5\lambda\} + 1 - \sum_{k=0}^{9} \frac{(5\lambda)^k}{k!} \exp\{-5\lambda\}. \end{aligned}$$

This is plotted below.

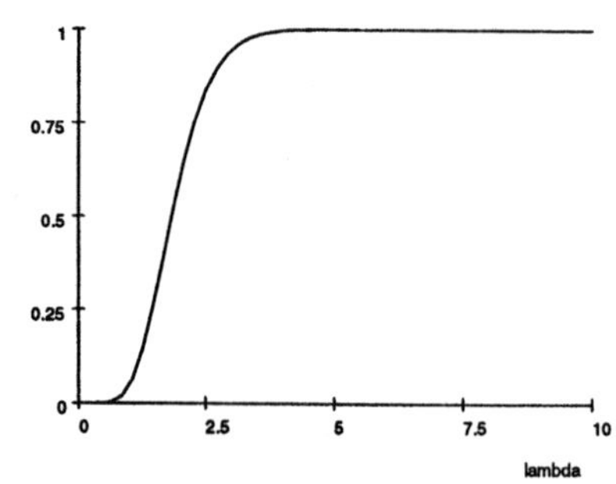

8.4.7 The model is given by the collection of density functions

$$\left\{\frac{1}{\sqrt{2\pi}\sigma_0}\exp\left\{-\frac{1}{2\sigma_0^2}\sum_{i=1}^{n}(x_i-\mu)^2\right\}:\mu\in R^1\right\}$$

on the set of all sequences $(x_1,\ldots,x_n)$ of real numbers. The action space is $\mathcal{A}=\{H_0,H_a\}$, where $H_0:\mu=\mu_0$. The correct action function is

$$A(\lambda)=\begin{cases} H_0 & \lambda\le\lambda_0 \\ H_a & \lambda>\lambda_0 \end{cases}$$

and the loss function is

$$L(\lambda,a)=\begin{cases} 0 & \mu=\mu_0, a=H_0 \text{ or } \mu\ne\mu_0, a=H_a \\ 1 & \mu=\mu_0, a=H_a \text{ or } \mu\ne\mu_0, a=H_0. \end{cases}$$

The test function φ is given by

$$\varphi(x_1,\ldots,x_n)=\begin{cases} 0 & \bar{x}\in\left[\mu_0-2\sigma_0/\sqrt{n},\mu_0+2\sigma_0/\sqrt{n}\right] \\ 1 & \bar{x}\notin\left[\mu_0-2\sigma_0/\sqrt{n},\mu_0+2\sigma_0/\sqrt{n}\right]. \end{cases}$$

The power function for φ is given by (using $\bar{x}\sim N(\mu,\sigma_0^2/n)$)

$$\begin{aligned}&\beta_\varphi(\mu)=P_\mu(\varphi(x_1,\ldots,x_n)=1)=1-P_\mu\left(\bar{x}\in\left[\mu_0-2\sigma_0/\sqrt{n},\mu_0+2\sigma_0/\sqrt{n}\right]\right)\\ &=1-P_\mu\left(\frac{\mu_0-\mu}{\sigma_0/\sqrt{n}}-2<\frac{\bar{x}-\mu}{\sigma_0/\sqrt{n}}<\frac{\mu_0-\mu}{\sigma_0/\sqrt{n}}+2\right)\\ &=1-\left(\Phi\left(\frac{\mu_0-\mu}{\sigma_0/\sqrt{n}}+2\right)-\Phi\left(\frac{\mu_0-\mu}{\sigma_0/\sqrt{n}}-2\right)\right).\end{aligned}$$

When $\mu_0=0,\sigma_0=3,n=10$ the power function is

$$\beta_\varphi(\mu)=1-\left(\Phi\left(-\frac{\mu}{3/\sqrt{10}}+2\right)-\Phi\left(-\frac{\mu}{3/\sqrt{10}}-2\right)\right).$$

This is plotted below (power equals 0.0455 at $\mu=0$).

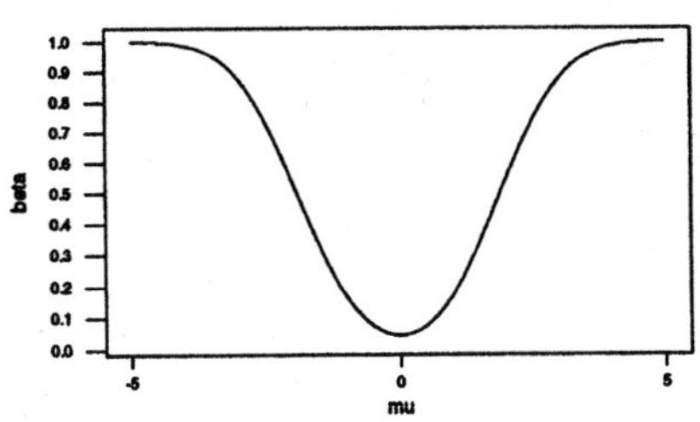

Problems

8.4.8 Suppose we have that $\delta(s,\cdot)$ is degenerate at $d(s)$ for each s. Then clearly $d : S \to \mathcal{A}$.

Now suppose we have $d : S \to \mathcal{A}$ and define

$$\delta(s,B) = \begin{cases} 1 & d(s) \in B \\ 0 & \text{otherwise} \end{cases}$$

for $B \subset \mathcal{A}$. Then $\delta(s,\mathcal{A}) = 1$ and, if $B_1, B_2, \ldots$ are mutually disjoint subsets of $\mathcal{A}$, then $d(s) \in B_i$ for one i (and only one) if and only if $d(s) \in \cup_{j=1}^{\infty} B_j$, so $\delta\left(s, \cup_{j=1}^{\infty} B_j\right) = \sum_{j=1}^{\infty} \delta(s, B_j)$. Therefore, $\delta(s,\cdot)$ is a probability measure for each s and δ is a decision function.

Now, using the fact that $\delta(s,\cdot)$ is a discrete probability measure degenerate at $d(s)$, we have that $R_\delta(\theta) = E_\theta\left(E_{\delta(s,\cdot)}(L(\theta,a))\right) =$
$E_\theta(\delta(s,\{d(s)\})(L(\theta,d(s))) = E_\theta(L(\theta,d(s)))$ since $\delta(s,\{d(s)\}) = 1$.

8.4.9
(a) Consider the decision function $d_{\theta_0}(s) \equiv A(\theta_0)$. Then note that $R_{d_{\theta_0}}(\theta_0) = 0$. Then, if δ is optimal, we must have that $R_\delta(\theta_0) \le R_{d_{\theta_0}}(\theta_0)$ for every θ_0, so $R_\delta(\theta) \equiv 0$. But this implies that $E_{\delta(s,\cdot)}(L(\theta,a)) = 0$ at every s, where $P_\theta(\{s\}) > 0$. Since $L(\theta,a) \ge 0$, then Challenge 3.3.10 implies that $\delta(s,\{L(\theta,a)=0\}) = 1$ and, since $L(\theta,a) = 0$ if and only if $a = A(\theta)$, this implies that $\delta(s,\cdot)$ is degenerate at $A(\theta)$ for each s for which $P_\theta(\{s\}) > 0$.
(b) Part (a) proved that, for an optimal δ, $\delta(s,\cdot)$ is degenerate at $A(\theta)$ for each s for which $P_\theta(\{s\}) > 0$. But if there exists s such that $P_{\theta_1}(\{s\}) > 0$ and $P_{\theta_2}(\{s\}) > 0$ and $A(\theta_1) \ne A(\theta_2)$, then this cannot happen and so no optimal δ can exist.

8.4.10 Suppose δ is not minimax. Then there exists decision function δ^* such that $\sup_\theta R_{\delta^*}(\theta) < \sup_\theta R_\delta(\theta)$. But since $R_\delta(\theta)$ is constant in θ this implies that $R_{\delta^*}(\theta) < R_\delta(\theta)$ for every θ and so δ is not admissible contradicting the hypothesis. Therefore, δ must be minimax.

Challenge

8.4.11 We have that $R_d(\theta_0) = E_{\theta_0}(L(\theta_0,d(s))) = 0$. Now suppose that d is not admissible. Then there exists decision function δ such that $R_\delta(\theta) \le R_d(\theta)$ for every θ and $R_\delta(\theta) < R_d(\theta)$ for some θ. But this implies that $0 = R_\delta(\theta_0) = E_{\theta_0}\left(E_{\delta(s,\cdot)}(L(\theta_0,a))\right) = 0$ and then Challenge 3.3.10 implies that the set $C = \left\{s : E_{\delta(s,\cdot)}(L(\theta_0,a)) > 0\right\}$ satisfies $P_{\theta_0}(C) = 0$. But by hypothesis this implies that $P_\theta(C) = 0$ for every θ. This in turn implies that $R_\delta(\theta) = 0$ for every θ. This says δ is optimal and contradicts the hypothesis that no such decision function exists.

In most practical problems, there does not exist an optimal decision function. So this result says that, in the typical decision problem, constants are admissible, i.e., decision functions that completely ignore the data are admissible. So the property of admissibility for a decision function is not a very strong one.

Chapter 9

Model Checking

9.1 Checking the Sampling Model

Exercises

9.1.1 The observed discrepancy statistic is given by $D(r) = \frac{1}{\sigma_0^2}\sum_{i=1}^{n}(x_i - \bar{x})^2$ $= \frac{19}{4}4.79187 = 22.761$. Now $D(R) \sim \chi^2(19)$ distribution, so the P-value is then given by $P(D(R) > 22.761) = .248$, which doesn't suggest evidence against the model.

9.1.2 **(a)** The plot of the standardized residuals is given below.

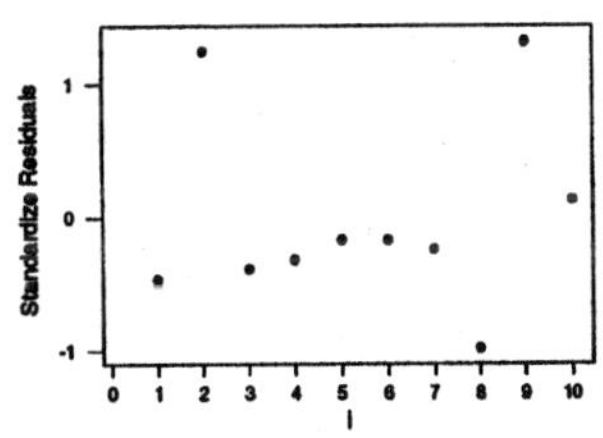

(b) The normal probability plot of the standardized residuals is given below.

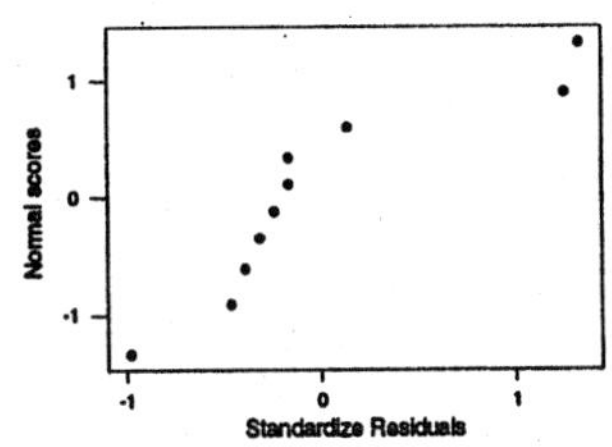

(c) The preceding plots suggest that the sample is probably not from a normal distribution.

9.1.3 (a) The plot of the standardized residuals is given below.

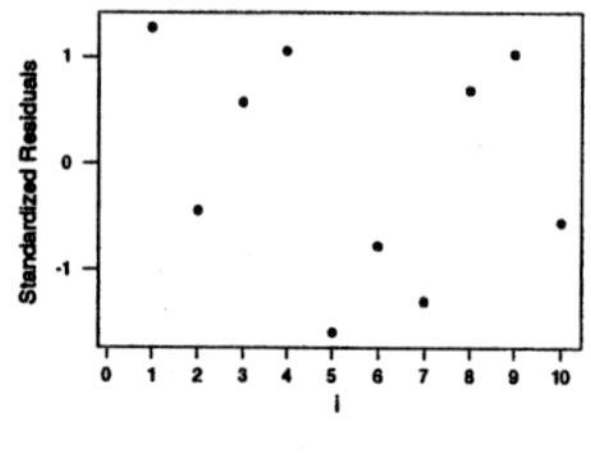

(b) The normal probability plot of the standardized residuals is given below.

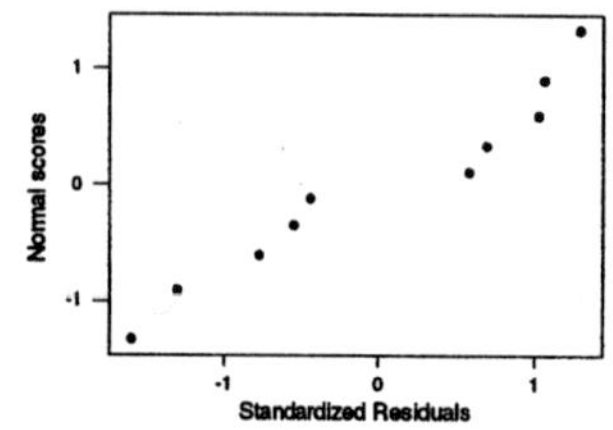

(c) The preceding plots suggest that the normal assumption seems reasonable.

9.1.4 We have $f_{.1} = 5$ conservative, $f_{1.} = 3$ males, and $f_{11} = 2$ conservative males. The Hypergeometric(10, 5, 3) probability function is given by the following table.

i	0	1	2	3
$p(i)$	0.083	0.417	0.417	0.083

The P-value is then equal to 1. Hence, we have no evidence against the model of independence between gender and political orientation.

9.1.5 By grouping the data into five equal intervals each having length 0.2, the expected counts for each interval are $np_i = 4$, and the observed counts are given in the following table.

Interval	Count
$(0.0, 0.2]$	4
$(0.2, 0.4]$	7
$(0.4, 0.6]$	3
$(0.6, 0.8]$	4
$(0.8, 1]$	2

The Chi-squared statistic is equal to 3.50 and the P-value is given by ($X^2 \sim \chi^2(4)$) $P(X^2 \geq 3.5) = 0.4779$ Therefore, we have no evidence against the Uniform model being correct.

9.1.6 First note that if the die is fair, the expected number of counts for each possible outcome is 166.667. The Chi-squared statistic is equal to 9.5720 and the P-value is given by ($X^2 \sim \chi^2(5)$) $P(X^2 \geq 9.5720) = .08831$. Therefore, we have some evidence that the die might not be fair. The standardized residuals are given in the following table.

i	1	2	3	4	5	6
r_i	−0.069541	0.214944	−0.467818	−0.316093	0.309772	0.328737

None of these look unusual.

Computer Exercises

9.1.7 The plot is given below.

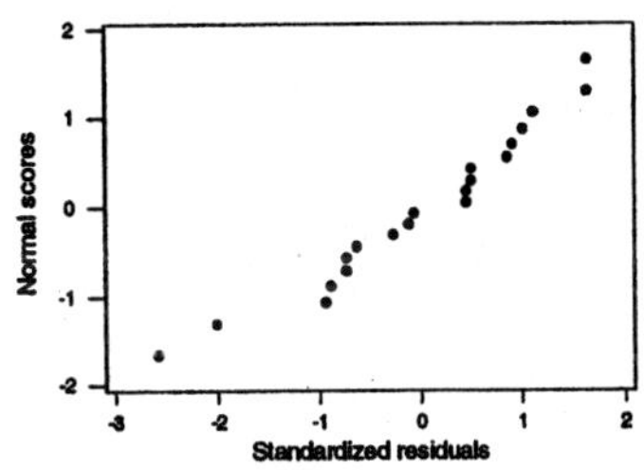

From this we have no evidence against the normality assumption.

9.1.8 Not all the graphs look like straight lines. With a small sample size like $n = 10$, we should expect a fairly wide variety of shapes.

9.1.9 We have $X_{\cdot 1} = 56$ conservative, $X_{1\cdot} = 35$ males and $X_{11} = 20$ conservative males. Using the Hypergeometric(100, 56, 35) probability function to calculate the probability of observing a value with probability less than or equal to $P(X_{11} = 20 \,|\, |X_{1\cdot}, X_{\cdot 1}) = 0.164941$, we obtain that the P-value is 1. Therefore, we have no evidence against the model of independence between gender and political orientation.

9.1.10 The Binomial(10, .2) distribution gives rise to the following cell probabilities and cell expected numbers.

0	0.107374	1.07374
1	0.268435	2.68435
2	0.301990	3.01990
3	0.201327	2.01327
4	0.088080	0.88080

5	0.026424	0.26424
6	0.005505	0.05505
7	0.000786	0.00786
8	0.000074	0.00074
9	0.000004	0.00004
10	0.000000	0.00000

So we grouped the last 7 cells, to ensure $E(X_i) = np_i \geq 1$, to obtain

0	0.107374	1.07374
1	0.268435	2.68435
2	0.301990	3.01990
3	0.201327	2.01327
4	0.120874	1.20874

The Chi-squared statistic obtained for the simulated sample of 1000 was equal to 1.7233 with P-value equal to .94331. Hence, there is some evidence that the sample is not from this distribution as this indicates that the observed value of the Chi-squared statistic is surprisingly small.

Problems

9.1.11 We have $E(a_1Y_1 + \cdots + a_kY_k) = a_1\mu_1 + \cdots + a_k\mu_k$, so $E(Y_i) = \mu_i$ by taking $a_i = 1$ and $a_j = 0$ whenever $j \neq i$. By Theorem 3.3.3 (b) we have $\text{Var}(a_1Y_1 + \cdots + a_kY_k) = a_1^2\,\text{Var}(Y_1) + \cdots + a_k^2\,\text{Var}(Y_k) + 2\sum_{i<j} a_ia_j\,\text{Cov}(Y_i, Y_j) = \sum_{i=1}^k \sum_{j=1}^k a_ia_j\sigma_{ij}$. Putting $a_i = 1$ and $a_j = 0$ whenever $i \neq j$, we obtain $\text{Var}(Y_i) = \sigma_{ii}$ and this implies that $Y_i \sim N(\mu_i, \sigma_{ii})$.

Putting $a_i = a_j = 1$ and $a_l = 0$ whenever $l \notin \{i, j\}$, we obtain $\text{Var}(Y_i + Y_j) = \sigma_{ii} + \sigma_{jj} + 2\sigma_{ij} = \text{Var}(Y_i) + \text{Var}(Y_j) + 2\,\text{Cov}(Y_i, Y_j)$. This immediately implies that $\text{Cov}(Y_i, Y_i) = \sigma_{ij}$.

9.1.12 Using Theorem 4.6.1, we have that

$$\sum_{i=1}^n a_iR_i = \sum_{i=1}^n a_i\left(X_i - \bar{X}\right) = \sum_{i=1}^n a_iX_i - \frac{1}{n}\left(\sum_{i=1}^n a_i\right)\sum_{i=1}^n X_i = \sum_{i=1}^n (a_i - \bar{a})\,X_i$$
$$\sim N\left(\sum_{i=1}^n (a_i - \bar{a})\,\mu, \sigma_0^2 \sum_{i=1}^n (a_i - \bar{a})^2\right) = N\left(0, \sigma_0^2 \sum_{i=1}^n (a_i - \bar{a})^2\right).$$

Therefore, by Problem 9.1.11, R is multivariate normal with mean vector given by $(0, \ldots, 0)$ and variance matrix given by $\Sigma = (\text{Cov}(R_i, R_j))$ and $\sum_{i=1}^k \sum_{j=1}^k a_ia_j\,\text{Cov}(R_i, R_j) = \sigma_0^2 \sum_{i=1}^n (a_i - \bar{a})^2$. Putting $a_i = 1$ and $a_j = 0$ whenever $i \neq j$, we have that $\text{Var}(R_i) = \sigma_0^2\left\{(1 - 1/n)^2 + (n-1)/n^2\right\} = \sigma_0^2(1 - 1/n)$. Putting $a_i = a_j = 1$ and $a_l = 0$ whenever $l \notin \{i, j\}$, we obtain $\text{Var}(R_i) + \text{Var}(R_j) + 2\,\text{Cov}(R_i, R_j) = \sigma_0^2 \sum_{i=1}^n (a_i - \bar{a})^2 = 2\sigma_0^2(1 - 2/n)^2 + \sigma_0^2(n-2)4/n^2 = \sigma_0^2\left(2 - 8/n + 8/n^2 + 4/n - 8/n^2\right) = \sigma_0^2(2 - 4/n)$. Therefore, $\text{Cov}(R_i, R_j) = \sigma_0^2(1 - 2/n - 1 + 1/n) = -\sigma_0^2/n$.

9.1.13 We have that (arguing as in the solution to Problem 9.1.12)

$$\begin{aligned}\operatorname{Cov}\left(\bar{X}, \sum_{i=1}^{n} a_i R_i\right) &= \operatorname{Cov}\left(\sum_{i=1}^{n} \frac{1}{n} X_i, \sum_{i=1}^{n} (a_i - \bar{a}) X_i\right) \\ &= \sum_{i=1}^{n} \sum_{j=1}^{n} \frac{1}{n} (a_j - \bar{a}) \operatorname{Cov}(X_i, X_j) = \frac{1}{n} \sum_{i=1}^{n} (a_i - \bar{a}) \operatorname{Cov}(X_i, X_i) \\ &= \frac{1}{n} \sum_{i=1}^{n} (a_i - \bar{a}) \operatorname{Var}(X_i) = \frac{\sigma_0^2}{n} \sum_{i=1}^{n} (a_i - \bar{a}) = 0.\end{aligned}$$

Theorem 4.6.2 gives the result.

9.1.14 The likelihood function is given by

$$L(\alpha_1, \beta_1) = \alpha_1^{x_{1\cdot}} (1 - \alpha_1)^{n - x_{1\cdot}} \beta_1^{x_{\cdot 1}} (1 - \beta_1)^{n - x_{\cdot 1}}.$$

The log-likelihood function is then $l(\alpha_1, \beta_1) = x_{1\cdot} \ln(\alpha_1) + (n - x_{1\cdot}) \ln(1 - \alpha_1) + x_{\cdot 1} \ln(\beta_1) + (n - x_{\cdot 1}) \ln(1 - \beta_1)$. If we fix β_1, then the partial derivative with respect to α_1 is

$$\frac{x_{1\cdot}}{\alpha_1} - \frac{n - x_{1\cdot}}{1 - \alpha_1}$$

with second derivative

$$-\frac{x_{1\cdot}}{\alpha_1^2} - \frac{n - x_{1\cdot}}{(1 - \alpha_1)^2}.$$

Solving

$$\frac{x_{1\cdot}}{\alpha_1} - \frac{n - x_{1\cdot}}{1 - \alpha_1} = 0$$

leads to $\hat{\alpha}_1 = x_{1\cdot}/n$. That this is a maximum is seen from the second derivative as it is negative at this point. Since it does not involve β_1, this is the MLE of α_1. A similar argument leads to the value $\hat{\beta}_1 = x_{\cdot 1}/n$ as the MLE of β_1.

9.1.15 Consider the set of all sequences of ordered pairs $((a_1, b_1), \ldots, (a_n, b_n))$ where $x_{1\cdot}$ of the a_i equal 1 (with rest equal to 2), $x_{\cdot 1}$ of the b_i equal to 1 (with the rest equal to 2). We are required to count the number of such sequences.

We can select $x_{1\cdot}$ of these pairs to have $a_i = 1$ in $\binom{n}{x_{1\cdot}}$ ways. Let us suppose that we have made these choices.

Now let i denote the number of pairs where $a_i = 1$ and $b_i = 1$. Clearly $\max\{0, x_{1\cdot} + x_{\cdot 1} - n\} \le i \le \min\{x_{1\cdot}, x_{\cdot 1}\}$. We can pick i of the pairs where $a_i = 1$ so that $b_i = 1$ in $\binom{x_{1\cdot}}{i}$ ways and then choose the remaining pairs that will have $a_i = 2$ and $b_i = 1$ in $\binom{n - x_{1\cdot}}{x_{\cdot 1} - i}$ ways. The multiplication principle then implies that there are $\binom{n}{x_{1\cdot}}\binom{x_{1\cdot}}{i}\binom{n - x_{1\cdot}}{x_{\cdot 1} - i}$ such sequences. Therefore, the number of samples satisfying the constrains (9.1.2) is equal to

$$\binom{n}{x_{1\cdot}} \sum_{i=\max\{0, x_{1\cdot} + x_{\cdot 1} - n\}}^{\min\{x_{1\cdot}, x_{\cdot 1}\}} \binom{x_{1\cdot}}{i} \binom{n - x_{1\cdot}}{x_{\cdot 1} - i}.$$

Using the fact the probability function of Hypergeometric$(n, f_{1\cdot}, f_{\cdot 1})$ is given by

$$P(X_{11} = i) = \frac{\binom{x_{1\cdot}}{i}\binom{n-x_{1\cdot}}{x_{\cdot 1}-i}}{\binom{n}{x_{\cdot 1}}},$$

for $\max\{0, x_{1\cdot} + x_{\cdot 1} - n\} \le i \le \min\{x_{1\cdot}, x_{\cdot 1}\}$, we get that the number of such samples is equal to

$$\binom{n}{x_{1\cdot}}\binom{n}{x_{\cdot 1}} \sum_{i=\max\{0, x_{1\cdot}+x_{\cdot 1}-n\}}^{\min\{x_{1\cdot}, x_{\cdot 1}\}} \frac{\binom{x_{1\cdot}}{i}\binom{n-x_{1\cdot}}{x_{\cdot 1}-i}}{\binom{n}{x_{\cdot 1}}} = \binom{n}{x_{1\cdot}}\binom{n}{x_{\cdot 1}}$$

as claimed.

Computer Problems

9.1.16 A density histogram of a sample of 10^4 form the distribution of $D(R) = -\frac{1}{n}\sum_{i=1}^{10} \ln\left(\frac{R_i^2}{n-1}\right)$, when the model is correct, is given below.

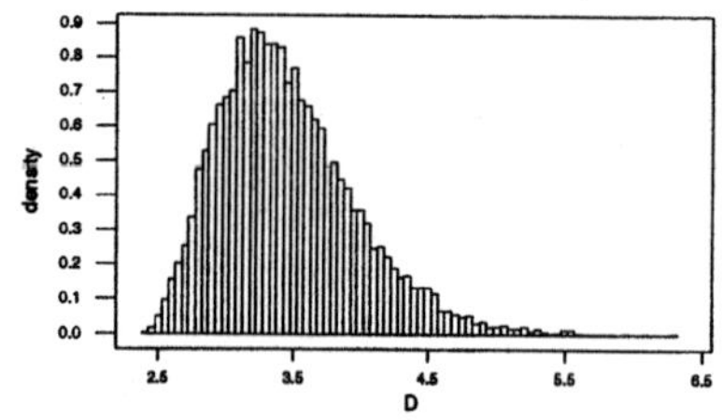

For the data of Exercise 9.1.3 we obtained the value $D(r) = -\frac{1}{n}\sum_{i=1}^{n} \ln\left(\frac{r_i^2}{n-1}\right) = 2.60896$, and the proportion of sample values of D in the simulation that were greater is 0.9864. This can be viewed as evidence that the normal location-scale model is not correct as the observed value of D is surprisingly small.

The following code was used for this simulation.

```
gmacro
goodnessoffit
base 34256734
note - generated sample is stored in c1
note - residuals are placed in c2
note - value of D(r) are placed in c3
note - k1 = size of data set
let k1=10
do k2=1:10000
random k1 c1
let k3=mean(c1)
let k4=sqrt(k1-1)*stdev(c1)
```

```
let c2=((c1-k3)/k4)**2
let c2=loge(c2)
let k5=-sum(c2)/k1
let c3(k2)=k5
enddo
endmacro
```

9.1.17 We get the following histogram (using the code below) when we are sampling from a normal distribution.

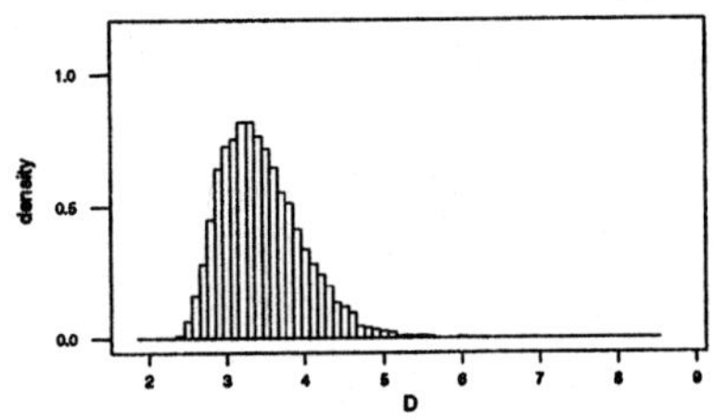

We get the following histogram (using the code below) when we are sampling from a Cauchy distribution.

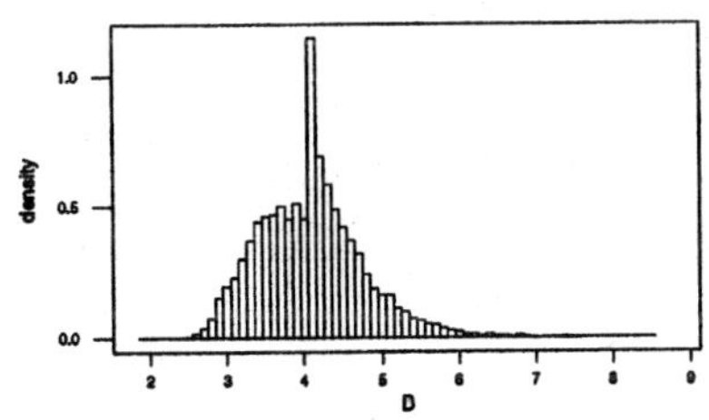

We see that the distribution of D is quite different under a normal model than a Cauchy model. The distribution when sampling under Cauchy sampling has a longer right tail and a sharp peak at its mode. Note that a larger sample size than 10^4 is required to get a smoother histogram.

```
gmacro
goodnessoffit
base 34256734
note - generated sample is stored in c1
note - residuals are placed in c2
note - value of D(r) are placed in c3
note - k1 = size of data set
let k1=10
do k2=1:10000
```

```
random k1 c1
let k3=mean(c1)
let k4=sqrt(k1-1)*stdev(c1)
let c2=((c1-k3)/k4)**2
let c2=loge(c2)
let k5=-sum(c2)/k1
let c3(k2)=k5
random k1 c1;
student 1.
let k3=mean(c1)
let k4=sqrt(k1-1)*stdev(c1)
let c2=((c1-k3)/k4)**2
let c2=loge(c2)
let k5=-sum(c2)/k1
let c4(k2)=k5
enddo
endmacro
```

9.1.18 The interval counts are $10, 3, 1, 2, 1, 3$. The likelihood function is then given by

$$L(\theta \mid f_1, ..., f_6) = \left(1 - e^{-2\theta}\right)^{10} \left(e^{-2\theta} - e^{-4\theta}\right)^3 \left(e^{-4\theta} - e^{-6\theta}\right) \left(e^{-6\theta} - e^{-8\theta}\right)^2 \times \left(e^{-8\theta} - e^{-10\theta}\right) \left(e^{-10\theta}\right)^3,$$

so the log-likelihood is given by

$$10 \ln \left(1 - e^{-2\theta}\right) + 3 \ln \left(e^{-2\theta} - e^{-4\theta}\right) + \ln \left(e^{-4\theta} - e^{-6\theta}\right) + 2 \ln \left(e^{-6\theta} - e^{-8\theta}\right) + \ln \left(e^{-8\theta} - e^{-10\theta}\right) - 30\theta$$

which is plotted below.

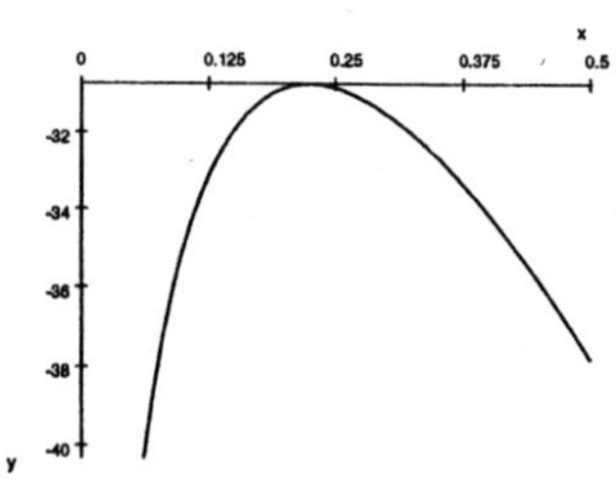

By successively plotting the log-likelihood over smaller and smaller intervals, the MLE was determined to be $\hat{\theta} = .22448$. Accordingly, we get the following expected counts $20\left(1 - e^{-2\hat{\theta}}\right) = 7.2342$, $20\left(e^{-2\hat{\theta}} - e^{-4\hat{\theta}}\right) = 4.6175$, $20\left(e^{-4\hat{\theta}} - e^{-6\hat{\theta}}\right) = 2.9473$, $20\left(e^{-6\hat{\theta}} - e^{-8\hat{\theta}}\right) = 1.8812$, $20\left(e^{-8\hat{\theta}} - e^{-10\hat{\theta}}\right) =$

$1.2008, 20\left(e^{-10\hat{\theta}}\right) = 2.1190$, and the Chi-squared statistic equals

$$\begin{aligned} X_0^2 &= \frac{(7.2342-10)^2}{7.2342} + \frac{(4.6175-3)^2}{4.6175} + \frac{(2.9473-1)^2}{2.9473} + \frac{(1.8812-2)^2}{1.8812} \\ &\quad + \frac{(1.2008-1)^2}{1.2008} + \frac{(2.1190-3)^2}{2.1190} \\ &= 3.3180 \end{aligned}$$

The P-value equals ($X^2 \sim \chi^2(1)$) $P(X^2 \geq 3.3180) = 1 - .4939 = 0.5061$. Hence, we don't have evidence against the model.

9.1.19

(a) We have that

$P((-\infty, 600]) = \Phi\left(\frac{600-\mu}{500}\right)$,
$P((600, 1200]) = \Phi\left(\frac{1200-\mu}{500}\right) - \Phi\left(\frac{600-\mu}{500}\right)$,
$P((1200, 1800]) = \Phi\left(\frac{1800-\mu}{500}\right) - \Phi\left(\frac{1200-\mu}{500}\right)$,
$P((1800, \infty)) = 1 - \Phi\left(\frac{1800-\mu}{500}\right)$,

so the log likelihood is given by

$$\begin{aligned} &9 \ln \Phi\left(\frac{600-\mu}{500}\right) + 20 \ln\left(\Phi\left(\frac{1200-\mu}{500}\right) - \Phi\left(\frac{600-\mu}{500}\right)\right) + \\ &7 \ln\left(\Phi\left(\frac{1800-\mu}{500}\right) - \Phi\left(\frac{1200-\mu}{500}\right)\right) + 2 \ln\left(1 - \Phi\left(\frac{1800-\mu}{500}\right)\right). \end{aligned}$$

This is plotted below.

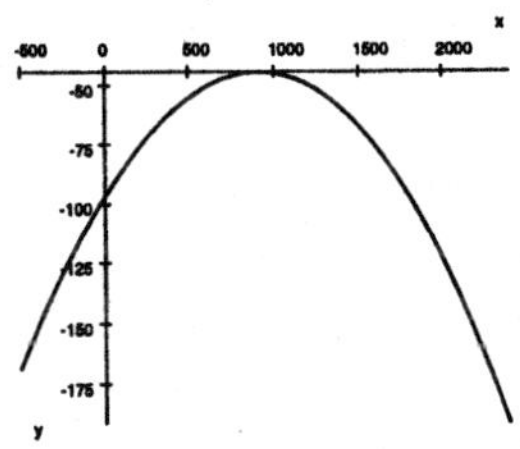

Plotting the log-likelihood over successively smaller intervals, we obtain the MLE as $\hat{\mu} = 914.3$. This leads to the expected counts

$38\,\text{NormalDist}\left(\frac{600-\hat{\mu}}{500}\right) = 10.063$,
$38\left(\text{NormalDist}\left(\frac{1200-\hat{\mu}}{500}\right) - \text{NormalDist}\left(\frac{600-\hat{\mu}}{500}\right)\right) = 17.151$,
$38\left(\text{NormalDist}\left(\frac{1800-\hat{\mu}}{500}\right) - \text{NormalDist}\left(\frac{1200-\hat{\mu}}{500}\right)\right) = 9.333$,
$38\left(1 - \text{NormalDist}\left(\frac{1800-\hat{\mu}}{500}\right)\right) = 1.453$,

and the Chi-squared statistic is given by

$$X_0^2 = \frac{(10.063-9)^2}{10.063} + \frac{(17.151-20)^2}{17.151} + \frac{(9.333-7)^2}{9.333} + \frac{(1.453-2)^2}{1.453} = 1.375.$$

The P-value in this case is given by ($X^2 \sim \chi^2(2)$) $P(X^2 \geq 1.375) = 1 - .4972 = 0.5028$, so we have no evidence against the model.
(b) The overall MLE of μ, namely without grouping, is $\hat{\mu} = \bar{x} = 900$, so there is a difference.
(c) We have that
$P((-\infty, 600]) = \Phi\left(\frac{600-\mu}{\sigma}\right)$,
$P((600, 1200]) = \Phi\left(\frac{1200-\mu}{\sigma}\right) - \Phi\left(\frac{600-\mu}{\sigma}\right)$,
$P((1200, 1800]) = \Phi\left(\frac{1800-\mu}{\sigma}\right) - \Phi\left(\frac{1200-\mu}{\sigma}\right)$,
$P((1800, \infty)) = 1 - \Phi\left(\frac{1800-\mu}{\sigma}\right)$,
so the log likelihood is given by

$$9\ln\Phi\left(\frac{600-\mu}{\sigma}\right) + 20\ln\left(\Phi\left(\frac{1200-\mu}{\sigma}\right) - \Phi\left(\frac{600-\mu}{\sigma}\right)\right)$$
$$+ 7\ln\left(\Phi\left(\frac{1800-\mu}{\sigma}\right) - \Phi\left(\frac{1200-\mu}{\sigma}\right)\right) + 2\ln\left(1 - \Phi\left(\frac{1800-\mu}{\sigma}\right)\right).$$

Challenge

9.1.20 We have that $X_i = \mu + \sigma Z_i$ where $Z_1, \ldots, Z_n$ is a sample from f. Now observe that $\bar{x} = \mu + \sigma\bar{z}$ and $\sum_{i=1}^n (x_i - \bar{x})^2 = \sum_{i=1}^n (\mu + \sigma z_i - \mu + \sigma\bar{z})^2 = \sigma^2 \sum_{i=1}^n (z_i - \bar{z})^2$. Therefore,

$$r(x_1, \ldots, x_n) = \left(\frac{x_1 - \bar{x}}{s}, \ldots, \frac{x_n - \bar{x}}{s}\right)$$
$$= \left(\frac{z_1 - \bar{z}}{\sqrt{\sum_{i=1}^n (z_i - \bar{z})^2}}, \ldots, \frac{z_n - \bar{z}}{\sqrt{\sum_{i=1}^n (z_i - \bar{z})^2}}\right),$$

so is a function of the z_i. This implies that the distribution of R is independent of (μ, σ) and so is ancillary.

9.2 Checking the Bayesian Model

Exercises

9.2.1
(a) The probability of obtaining $s = 2$ from f_1 is $1/3$, which is a reasonable value, so we have no evidence against the model $\{f_1, f_2\}$.
(b) The prior predictive M distribution is given by

$$m(1) = \frac{3}{10}\frac{1}{3} + \frac{7}{10}\frac{1}{3} = \frac{1}{3},$$
$$m(2) = \frac{3}{10}\frac{1}{3} + \frac{7}{10}0 = \frac{1}{10},$$
$$m(3) = \frac{3}{10}\frac{1}{3} + \frac{7}{10}\frac{2}{3} = \frac{17}{30}.$$

So the probability of a data set occurring with probability as small or smaller than $m(2)$ is $1/10$, so the observation 2 is not very surprising. Accordingly, there is no evidence of a prior-data conflict.
(c) The prior predictive M now is given by

$$m(1) = \frac{1}{100}\frac{1}{3} + \frac{99}{100}\frac{1}{3} = \frac{1}{3},$$
$$m(2) = \frac{1}{100}\frac{1}{3} + \frac{99}{100}0 = \frac{1}{300},$$
$$m(3) = \frac{1}{100}\frac{1}{3} + \frac{99}{100}\frac{2}{3} = \frac{199}{300}.$$

So the probability of a data set occurring with probability as small or smaller than $m(2)$ is $1/300$, and the observation 2 is surprising. Accordingly, there is some evidence of a prior-data conflict.

9.2.2 The prior predictive probability function for the minimal sufficient statistic $y = n\bar{x} = \sum_{i=1}^{n} x_i$ is given by

$$m(y) = \frac{\Gamma(7)}{\Gamma(12)} \frac{\Gamma(6)}{(\Gamma(3))^2} \frac{\Gamma(y+3)\Gamma(9-y)}{\Gamma(y+1)\Gamma(7-y)}.$$

A tabulation and a plot of this is given below.

y	m(y)
0	0.060606
1	0.136364
2	0.194805
3	0.216450
4	0.194805
5	0.136364
6	0.060606

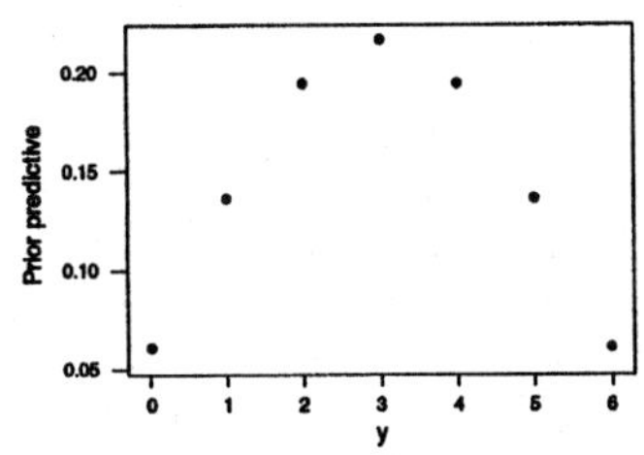

Using the symmetry of the prior predictive, the probability of obtaining a value with probability of occurrence no greater than $y = n\bar{x} = 2$ is equal to $m(0) + m(1) + m(2) + m(4) + m(5) + m(6) = 2(0.060606) + 2(0.194805) + 2(0.136364) = 0.78355$. Therefore, the observation $y = n\bar{x} = 2$ is not surprising and we conclude that there isn't any prior-data conflict.

Computer Exercises

9.2.3 The prior predictive probability function for the minimal sufficient statistics $y = n\bar{x} = \sum_{i=1}^{n} x_i$ is given by

$$m(y) = \frac{\Gamma(21)}{\Gamma(26)} \frac{\Gamma(6)}{(\Gamma(3))^2} \frac{\Gamma(y+3)\Gamma(23-y)}{\Gamma(y+1)\Gamma(21-y)}$$

A tabulation and plot of this is given below.

y	m(y)
0	0.0043478
1	0.0118577
2	0.0214568
3	0.0321852
4	0.0431959
5	0.0537549
6	0.0632411
7	0.0711462
8	0.0770751
9	0.0807453
10	0.0819876
11	0.0807453
12	0.0770751
13	0.0711462
14	0.0632411
15	0.0537549
16	0.0431959
17	0.0321852
18	0.0214568
19	0.0118577
20	0.0043478

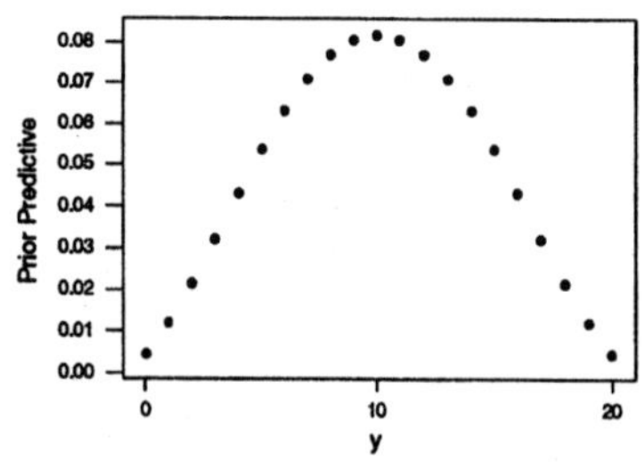

Using the symmetry of the prior predictive, the probability of obtaining a value probability of occurrence no greater than $y = n\bar{x} = 6$ equals $2m(0) + 2m(1) + 2m(2) + 2m(3) + 2m(4) + 2m(5) + 2m(6) = 0.460079$. Therefore, the observation

$y = n\bar{x} = 6$ is not surprising and we conclude that there isn't any prior-data conflict.

Problems

9.2.4 First, by Corollary 4.6.1 we have $\bar{X} \sim N(\mu, \sigma_0^2/n)$. Then we can write $\bar{X}$ as $\bar{X} = \mu + Z/\sqrt{n}$, where $Z \sim N(0, \sigma_0^2)$ is independent of $\mu \sim N(\mu_0, \tau_0^2)$. Hence, by Theorem 4.6.1 we have, that the prior predictive distribution of $\bar{X}$ is the $N(\mu_0, \tau_0^2 + \sigma_0^2/n)$ distribution.

9.2.5 We have that $Y = n\bar{X} \sim \text{Gamma}(n, \theta)$ so the prior predictive distribution of $Y$is given by

$$\begin{aligned} m(y) &= \int_0^\infty \frac{\theta^n}{\Gamma(n)} y^{n-1} \exp(-y\theta) \frac{\beta_0^{\alpha_0} \theta^{\alpha_0 - 1} e^{-\beta_0 \theta}}{\Gamma(\alpha_0)} \, d\theta \\ &= \frac{\beta_0^{\alpha_0}}{\Gamma(\alpha_0)\Gamma(n)} y^{n-1} \int_0^\infty \theta^{n+\alpha_0-1} \exp(-(\beta_0 + y)\theta) \, d\theta \\ &= \frac{\beta_0^{\alpha_0}}{\Gamma(\alpha_0)\Gamma(n)} \frac{\Gamma(\alpha_0 + n)}{(\beta_0 + y)^{n+\alpha_0}} y^{n-1} = \frac{\Gamma(\alpha_0 + n)}{\Gamma(\alpha_0)\Gamma(n)} \left(\frac{y}{\beta_0}\right)^{n-1} \left(1 + \frac{y}{\beta_0}\right)^{-(\alpha_0+n)} \frac{1}{\beta_0}. \end{aligned}$$

Making the transformation $\bar{x} = y/n$, we see that the prior predictive density of $\bar{X}$ is given by

$$m(\bar{x}) = \frac{\Gamma(\alpha_0 + n)}{\Gamma(\alpha_0)\Gamma(n)} \left(\frac{n\bar{x}}{\beta_0}\right)^{n-1} \left(1 + \frac{n\bar{x}}{\beta_0}\right)^{-(\alpha_0+n)} \frac{n}{\beta_0}$$

and from this we deduce that the prior predictive of $\alpha_0\bar{X}/\beta_0$ is $F(n, \alpha_0)$.

9.2.6 We know that $(x_1, ..., x_k) \sim \text{Multinomial}(n, \theta_1, \ldots, \theta_k)$. Therefore, the prior predictive distribution of $(x_1, ..., x_k)$ is given by

$$\begin{aligned} m(x_1, ..., x_k) &= \binom{n}{x_1 \ldots x_k} \frac{\Gamma(\alpha_1 + \cdots + \alpha_k)}{\Gamma(\alpha_1) \cdots \Gamma(\alpha_k)} \int_0^1 \cdots \int_0^{1-\theta_2-\cdots-\theta_{k-1}} \theta_1^{\alpha_1 + x_1 - 1} \cdots \\ &\quad \times (1 - \theta_1 - \cdots - \theta_{k-1})^{\alpha_k + x_k - 1} \, d\theta_1 \cdots d\theta_{k-1} \\ &= \binom{n}{x_1 \ldots x_k} \frac{\Gamma(\alpha_1 + \cdots + \alpha_k)}{\Gamma(\alpha_1) \cdots \Gamma(\alpha_k)} \frac{\Gamma(\alpha_1 + x_1) \cdots \Gamma(\alpha_k + x_k)}{\Gamma(\alpha_1 + \cdots + \alpha_k + n)}. \end{aligned}$$

9.2.7 When $X_1, \ldots, X_n$ is a sample from the Uniform$[0, \theta]$ distribution then $X_{(n)}$ has density given by $n\left(x_{(n)}\right)^{n-1}/\theta^n$ for $0 < x_{(n)} < \theta$. Therefore, the prior predictive density of $X_{(n)}$ is given by

$$\begin{aligned} m\left(x_{(n)}\right) &= \int_0^\infty \frac{n}{\theta^n}\left(x_{(n)}\right)^{n-1} I_{[x_{(n)},\infty)}(\theta)\,\frac{\theta^{-\alpha} I_{[\beta,\infty)}(\theta)}{(\alpha-1)\,\beta^{\alpha-1}}\,d\theta \\ &= \frac{n\left(x_{(n)}\right)^{n-1}}{(\alpha-1)\,\beta^{\alpha-1}}\int_0^\infty \theta^{-\alpha-n} I_{[\max\{x_{(n)},\beta\},\infty)}(\theta)\,d\theta \\ &= \frac{n\left(x_{(n)}\right)^{n-1}}{(\alpha-1)\,\beta^{\alpha-1}}\int_{\max\{x_{(n)},\beta\}}^\infty \theta^{-\alpha-n}\,d\theta \\ &= \frac{n\left(x_{(n)}\right)^{n-1}}{(\alpha+n-1)(\alpha-1)\,\beta^{\alpha-1}}\left(\max\{x_{(n)},\beta\}\right)^{-\alpha-n+1}. \end{aligned}$$

Challenge

9.2.8 The prior predictive distribution is given by the joint density of $(\bar{X}, S^2, \mu, 1/\sigma^2)$ divided by the posterior density of $(\mu, 1/\sigma^2)$. The joint density of $(\bar{X}, S^2, \mu, 1/\sigma^2)$, using the fact $\bar{X} \sim N(\mu, \sigma^2/n)$ independent of $(n-1)\,S^2/\sigma^2 \sim \chi^2(n-1)$, is given by

$$\left\{\frac{n^{1/2}}{\sqrt{2\pi}\sigma}\exp\left(-\frac{n}{2\sigma^2}(\bar{x}-\mu)^2\right)\right\}\left\{\begin{array}{c}\frac{1}{\Gamma((n-1)/2)}\left(\frac{n-1}{2\sigma^2}\right)^{\frac{n-1}{2}}\left(s^2\right)^{\frac{n-1}{2}-1} \\ \times\exp\left(-\frac{n-1}{2\sigma^2}s^2\right)\end{array}\right\}\times$$
$$\left\{\frac{1}{\sqrt{2\pi}\tau_0\sigma}\exp\left(-\frac{1}{2\tau_0^2\sigma^2}(\mu-\mu_0)^2\right)\right\}\left\{\frac{\beta_0^{\alpha_0}}{\Gamma(\alpha_0)}\left(\frac{1}{\sigma^2}\right)^{\alpha_0-1}\exp\left(-\frac{\beta_0}{\sigma^2}\right)\right\}.$$

Then using the same algebraic manipulations as carried out in Section 7.5, we have that this joint density equals

$$\left\{\frac{n^{1/2}}{\sqrt{2\pi}}\frac{(n-1)^{\frac{n-1}{2}}\left(s^2\right)^{\frac{n-1}{2}-1}}{\Gamma((n-1)/2)}\right\}\left\{\frac{1}{\tau_0}\frac{\beta_0^{\alpha_0}}{\Gamma(\alpha_0)}\right\}\left\{\left(n+\frac{1}{\tau_0^2}\right)^{-1/2}\frac{\Gamma(\alpha_0+n/2)}{\beta_x^{\alpha_0+n/2}}\right\}$$
$$\times\frac{1}{\sqrt{2\pi}}\left(n+\frac{1}{\tau_0^2}\right)^{1/2}\left(\frac{1}{\sigma^2}\right)^{1/2}\exp\left(-\frac{1}{2\sigma^2}\left(n+\frac{1}{\tau_0^2}\right)(\mu-\mu_x)^2\right)$$
$$\times\frac{\beta_x^{\alpha_0+n/2}}{\Gamma(\alpha_0+n/2)}\left(\frac{1}{\sigma^2}\right)^{\alpha_0+n/2-1}\exp\left(-\beta_x\frac{1}{\sigma^2}\right).$$

where $\mu_x = (n+1/\tau_0^2)^{-1}(\mu_0/\mu_0 + n\bar{x})$ and

$$\beta_x = \beta_0 + \frac{n}{2}\bar{x}^2 + \frac{\mu_0^2}{2\tau_0^2} + \frac{n-1}{2}s^2 - \frac{1}{2}\left(n+\frac{1}{\tau_0^2}\right)^{-1}\left(\frac{\mu_0}{\tau_0^2}+n\bar{x}\right)^2.$$

Since we know that that the posterior distribution of (μ, σ^2) is given by $\mu \mid \sigma^2, x \sim N(\mu_x, (n+1/\tau_0^2)^{-1}\sigma^2)$ and $1/\sigma^2 \mid x \sim \text{Gamma}(\alpha_0+n/2, \beta_x)$, this implies that the prior predictive density of $(\bar{X}, S^2)$ is given by

$$\left\{\frac{n^{1/2}}{\sqrt{2\pi}}\frac{(n-1)^{\frac{n-1}{2}}\left(s^2\right)^{\frac{n-1}{2}-1}}{\Gamma((n-1)/2)}\right\}\left\{\frac{1}{\tau_0}\frac{\beta_0^{\alpha_0}}{\Gamma(\alpha_0)}\right\}\left\{\left(n+\frac{1}{\tau_0^2}\right)^{-1/2}\frac{\Gamma(\alpha_0+n/2)}{\beta_x^{\alpha_0+n/2}}\right\}.$$

Chapter 10

Relationships among Variables

10.1 Related Variables

Exercises

10.1.1 From the definitions we know that if the conditional distribution of Y given X does not change as we change X, then X and Y are unrelated and then for any x_1, x_2, (that occur with positive probability) and y we have $P(Y = y \mid X = x_1) = P(Y = y \mid X = x_2)$. Hence,

$$\frac{P(X = x_1, Y = y)}{P(X = x_1)} = \frac{P(X = x_2, Y = y)}{P(X = x_2)}$$

so $P(X = x_1, Y = y) = P(X = x_2, Y = y)\, P(X = x_1) / P(X = x_2)$. Summing this over x_1 leads to $P(X = x_2, Y = y) = P(X = x_2)\, P(Y = y)$, and this implies that X and Y are statistically independent. Conversely, if X and Y are statistically independent, then for all x and y we have

$$P(Y = y \mid X = x) = \frac{P(X = x, Y = y)}{P(X = x)} = \frac{P(X = x)\, P(Y = y)}{P(X = x)} = P(Y = y),$$

so the conditional distribution of Y given X does not change as we change X, and therefore X and Y are unrelated.

10.1.2 Suppose there exists $x_1 \neq x_2$ such that $g(x_1) \neq g(x_2)$ and $f_X(x_1) \neq 0, f_X(x_2) \neq 0$, where f_X is the relative frequency function for X. Then we must have

$$f_Y(y \mid X = x_i) = \frac{f_{X,Y}(x_i, y)}{f_X(x_i)} = \begin{cases} 0 & y \neq g(x_i) \\ 1 & y = g(x_i) \end{cases}$$

for $i = 1, 2$. Since $g(x_1) \neq g(x_2)$, this implies that the conditional distribution of Y changes as we change X, and therefore X and Y are related.

If $g(x) = c$ for all x, we have $g(x_1) = g(x_2) = c$, so $f_Y(y \mid X = x1) = f_Y(y \mid X = x_2)$ for all y, i.e., the conditional distribution of Y given X does not changes as we change x and so they are not related.

10.1.3 To check whether or not a relationship exists between Y and X we calculate the conditional distributions of Y given X. These are given in the following table.

	$Y=1$	$Y=2$	$Y=3$
$X=1$	$0.15/0.73 = .20548$	$0.18/0.73 = .24658$	$0.40/0.73 = .54795$
$X=2$	$0.12/0.27 = .44444$	$0.09/0.27 = .33333$	$0.06/0.27 = .22222$

The conditional distribution of Y given $X = x$ does change as we change x, so we conclude that X and Y are related.

10.1.4 To check whether or not a relationship exists between Y and X we calculate the conditional distribution of Y given X. These are given in the following table.

	$Y=1$	$Y=2$	$Y=3$
$X=1$	$\frac{1/6}{2/3}=\frac{1}{4}$	$\frac{1/6}{2/3}=\frac{1}{4}$	$\frac{1/3}{2/3}=\frac{1}{2}$
$X=2$	$\frac{1/12}{1/3}=\frac{1}{4}$	$\frac{1/12}{1/3}=\frac{1}{4}$	$\frac{1/6}{1/3}=\frac{1}{2}$

As we can see, the conditional distribution of Y given $X = x$ does not change at all as we change x, so we conclude that X and Y are unrelated.

10.1.5 Suppose that $P(X = x) > 0$. We have that

$$P(Y = y \mid X = x) = \frac{P(X = x, X^2 = y)}{P(X = x)} = \begin{cases} 0 & y \neq x^2 \\ 1 & y = x^2 \end{cases}$$

and so the conditional distributions will change with x whenever X is not degenerate.

10.1.6 This cannot be claimed to be a cause-effect relationship because we can't assign the value birth-weight at birth.

10.1.7 If the conditional distribution of life-length given various smoking habits changes, then we can conclude that these two variables are related. However, we can't assign the value of smoking habit (perhaps different amount of smoking), and there might be many other confounding variables that should be take into account, e.g., exercise habits, eating habits, sleeping habits, etc. So we can't conclude that this relationship is a cause-effect relationship.

10.1.8 The teacher should conduct an experiment in which a random sample is drawn from the population of students. Then half of this sample should be randomly selected to write the exam with open book, while the other half writes it with closed book. Then a comparison should be made of the conditional distributions of the response variable Y (the grade obtained) given the predictor X (closed or open book) using the samples to make inference about these distributions.

10.1.9 The researcher should draw a random sample from the population of voters and ask them to measure their attitude towards a particular political party on a scale from favorably disposed to unfavorably disposed. Then the researcher should randomly select half of this sample to be exposed to a negative ad (an ad that points out various negative attributes about the opponents), while the other half is exposed to a positive ad (one that points out various positive attributes of the party). They should all then be asked to measure their attitude towards the particular political party on the same scale. Then compare the conditional distribution of the response variable Y (the change in attitude from before seeing the ad to after) given the predictor X (type of ad exposed to) using the samples to make inference about these distributions.

10.1.10 Recall that the correlation of any two random variables is non-zero only if the covariance of them is non-zero. This immediately implies that the two variables are not independent, else $\mathrm{Cov}(X, Y) = 0$. Therefore, the two variables are related.

10.1.11
(a) First, let $x_1 = 0$ denote usual diet and $x_2 = 1$ denote new diet. The experimental design is given by $\{(0, 100), (1, 100)\}$.
(b) There are several concerns about the conduct of this study. First, we have not taken a sample from the population of interest. The individuals involved in the study have volunteered and, as a group, they might be very different than the full population, e.g., in their ability to stick to the diet. Second, the sample size might be too small relative to the population size, so inference may be inconclusive.
(c) We should group the individuals according to their initial weight W into homogenous groups (blocks) and then randomly apply the treatments to the individuals in each block and compare the conditional distribution of the response given the two predictors, type of diet and initial weight. This will make the comparisons more accurate by reducing variability.

10.1.12
(a) There are 10 conditional distributions since the factor W has two levels and the factor X has 5 levels and so there $5(2) = 10$ combinations.
(b) The predictor variable W (gender) is a categorical variable, while both the response variable Y and the predictor variable X (age in years) are quantitative variables.
(c) To have a balanced design we should allocate 200 individuals to each combination of the factors.
(d) A relationship between the response and the predictors can't be claimed to be a cause-effect relationship since we can't assign the values of the predictor variables.

(e) We should use family income as a blocking variable having, say, two levels, namely low and high. Then look at the conditional distributions of the response given the blocking variable and the two predictors.

10.1.13
(a) The response variable could be whether or not they have watched the particular program. Predictor variables might be the number of members in each household, whether or not they received the brochure, number of televisions in the house, number of children, etc.
(b) We cannot claim for a cause-effect relationship since for many of the predictor variables it would be impossible to assign the different values it can take.

10.1.14 Given a fixed value of X, the conditional distribution of Y given W and X does not change as W changes and is given by the $N(3,5)$ distribution for $X = 0$ and the $N(4,5)$ distribution for $X = 1$. Therefore, we can conclude that W does not have a relationship with Y. However, for a fixed value of W, the conditional distribution of Y given W and X changes as X changes from the $N(3,5)$ distribution for $X = 0$ to the $N(4,5)$ distribution for $X = 1$. Therefore, we can conclude that X does have a relationship with Y.

10.1.15 Given the value $X = 1$, the conditional distribution of Y given W and X does not change as W changes and is given by the $N(4,5)$ distribution. While given the value $X = 0$, the conditional distribution of Y given W and X changes as W changes from the $N(2,5)$ distribution for $W = 0$ to the $N(3,5)$ distribution for $W = 1$. Therefore, we conclude that W does have a relationship with Y. Now, the conditional distribution of Y given W and X changes as X changes, for a fixed value of W, and we can conclude that X does have a relationship with Y.

10.1.16 In Exercise 10.1.14 the predictors do not interact since the changes in the conditional distribution of Y given W and X, as we change X, does not depend on the value of W. While in Exercise 10.1.15 the changes in the conditional distribution of Y given W and X, as we change W, depend on the value of X, so the predictors interact.

Problems

10.1.17 The situation is somewhat simpler when the predictors do not interact because we can ignore the other predictor when studying the effects of changing just one predictor as the change is the same no matter what value the other predictor takes. Typically, the experimenter cannot control whether or not the predictors interact.

10.1.18 If X and Y are related, then there exist x_1, x_2, y such that $f_{Y|X}(y\,|\,x_1) \neq f_{Y|X}(y\,|\,x_2)$. Now suppose that $f_{X|Y}(x\,|\,y) = f_{X|Y}(x\,|\,y')$ for all x, y, y', i.e., that the conditional distribution of X given Y does not change as we change Y. Then $f_{X|Y}(x\,|\,y) = f_{X,Y}(x,y)/f_Y(y) = f_{X,Y}(x,y')/f_Y(y') = f_{X|Y}(x\,|\,y')$ which implies $f_{X,Y}(x,y) = (f_{X,Y}(x,y')/f_Y(y'))f_Y(y)$ for all x, y, y', which in turn implies

$$f_{X,Y}(x,y) = \sum_{y'} f_{X,Y}(x,y)\, f_Y\left(y'\right) = \sum_{y'} \frac{f_{X,Y}\left(x,y'\right)}{f_Y(y')} f_Y(y)\, f_Y\left(y'\right)$$
$$= \left(\sum_{y'} f_{X,Y}\left(x,y'\right)\right) f_Y(y) = f_X(x)\, f_Y(y)$$

for every x, y. But this implies $f_{Y|X}(y\,|\,x_1) = f_X(x_1)\, f_Y(y)\,/ f_X(x_1) = f_Y(y) = f_{Y|X}(y\,|\,x_2)$, which is a contradiction. Therefore we must have that $f_{X|Y}(x\,|\,y) \neq f_{X|Y}(x\,|\,y')$ for all x, y, y', i.e., that the conditional distribution of X given Y changes as we change Y, which implies that Y and X are related variables (by Definition 10.1).

10.1.19 We have that

$$\begin{aligned}
\operatorname{Cov}(U,V) &= E(UV) - E(U)E(V) \\
&= E((X+Z)(Y+Z)) - E(X+Z)E(Y+Z) \\
&= E(XY + XZ + YZ + Z^2) - (E(X) + E(Z))(E(Y) + E(Z)) \\
&= E(XY) + E(XZ) + E(YZ) + E\left(Z^2\right) - 0 \\
&= E(X)E(Y) + E(X)E(Z) + E(Y)E(Z) + E\left(Z^2\right) = 1,
\end{aligned}$$

so U and V are not independent and so must be related.

10.1.20 First note that the joint probability distribution function of X and Y is given by $P(X = x, Y = y) = \binom{n}{x}\binom{n-x}{y}\left(\frac{1}{3}\right)^x\left(\frac{1}{3}\right)^y\left(\frac{1}{3}\right)^{n-x-y}$. Since $X \sim$ Binomial$(n, 1/3)$, the marginal probability function of X is given by $P(X = x) = \binom{n}{x}\left(\frac{1}{3}\right)^x\left(\frac{2}{3}\right)^{n-x}$. Therefore, the conditional distribution of Y given $X = x$ has probability function

$$\frac{\binom{n}{x}\binom{n-x}{y}\left(\frac{1}{3}\right)^x\left(\frac{1}{3}\right)^y\left(\frac{1}{3}\right)^{n-x-y}}{\binom{n}{x}\left(\frac{1}{3}\right)^x\left(\frac{2}{3}\right)^{n-x}} = \binom{n-x}{y}\left(\frac{1}{3}\Big/\frac{2}{3}\right)^y\left(\frac{1}{3}\Big/\frac{2}{3}\right)^{n-x-y}$$
$$= \binom{n-x}{y}\left(\frac{1}{2}\right)^y\left(\frac{1}{2}\right)^{n-x-y}$$

and this is the Binomial$(n - x, 1/2)$ distribution and this changes with x. Therefore, X and Y are related.

10.1.21 By Problem 2.8.17 X and Y are independent if and only if $\rho = 0$ and $\operatorname{Corr}(X,Y) = \rho$.

10.1.22 If the conditional distribution of Y given $X = x$ and $Z = z$ changes as we change x for some value z then X and Y are related. If the conditional distribution of Y given $X = x$ and $Z = z$ never changes as we change z for each fixed value of x, then Z and Y are not related.

Now $p_{X,Y,Z}(x,y,z) = p_{Y|X,Z}(y\,|\,x,z)\,p_{X,Z}(x,z) =$ $p_{Y|X,Z}(y\,|\,x,z)\,p_{X|Z}(x\,|\,z)p_Z(z)$ and, because $p_{Y|X,Z}(y\,|\,x,z)$ is constant in z, we must have that

$$\begin{aligned} p_{Y|X}(y\,|\,x) &= \frac{p_{X,Y}(x,y)}{p_X(x)} = \frac{\sum_z p_{Y|X,Z}(y\,|\,x,z)\,p_{X|Z}(x\,|\,z)p_Z(z)}{p_X(x)} \\ &= \frac{\sum_z p_{X|Z}(x\,|\,z)p_Z(z)}{p_X(x)} = p_{Y|X,Z}(y\,|\,x,z). \end{aligned}$$

Therefore, $p_{X,Y,Z}(x,y,z) = p_{Y|X}(y\,|\,x)\,p_{X|Z}(x\,|\,z)p_Z(z)$.

10.2 Categorical Response and Predictors

Exercises

10.2.1 First, note that the predictor variable, X-year, is not random. The estimated conditional distribution of Y given X are recorded in the following table.

	June	July	August
Year 1	$60/240 = .25$	$100/240 = .41667$	$80/240 = .33333$
Year 2	$80/240 = .33333$	$100/240 = .41667$	$60/240 = .25$

Under the null hypothesis of no difference in the distributions of thunderstorms between the two years, the MLEs are given by

$$\hat\theta_1 = \frac{140}{480} = .29167,\ \hat\theta_2 = \frac{200}{480} = .41667,\ \hat\theta_3 = \frac{140}{480} = .29167.$$

Then the estimates of the expected counts $n_i\theta_j$ are given in the following table.

	June	July	August
Year 1	70	100	70
Year 2	70	100	70

The Chi-squared statistic is then equal to $X_0^2 = 5.7143$ and, with $X^2 \sim \chi^2(2)$, the P-value equals $P\left(X^2 > 5.7143\right) = .05743$. Therefore, we don't have evidence against the null hypothesis of no difference in the distributions of thunderstorms between the two years, at least at the .05 level.

10.2.2 First note that the predictor variable, X (received vitamin C or not), is deterministic. The estimated conditional distributions of Y given X are recorded in the following table.

	No cold	Cold
Placebo	.22143	.77857
Vitamin C	.12230	.87770

Under the null hypothesis of no relationship between taking vitamin C and the incidence of the common cold, the MLEs are given by

$$\hat{\theta}_1 = \frac{48}{279} = .17204, \ \hat{\theta}_2 = \frac{231}{279} = .82796.$$

Then the estimates of the expected counts $n_i\theta_j$ are given in the following table.

	No cold	Cold
Placebo	24.086	115.91
Vitamin C	23.914	115.09

The Chi-squared statistic is equal to $X_0^2 = 4.8105$ and, with $X^2 \sim \chi^2(1)$, the P-value equals $P(X^2 > 4.8105) = .02829$. Therefore, we have evidence against the null hypothesis of no relationship between taking vitamin C and the incidence of the common cold.

10.2.3 The estimated conditional distributions of Y (second digit) given X (first digit) are recorded in the following table.

	Second digit 0	Second digit 1
First digit 0	0.489796	0.510204
First digit 1	0.500000	0.500000

Under the null hypothesis of no relationship between the digits, the MLEs are given by

$$\hat{\theta}_{.1} = \frac{495}{1000} = .495, \ \hat{\theta}_{.2} = \frac{505}{1000} = .505$$

for the Y probabilities and

$$\hat{\theta}_{1.} = \frac{490}{1000} = .49, \ \hat{\theta}_{.2} = \frac{510}{1000} = .51$$

for the X probabilities. Then the estimates of the expected counts $n_i\theta_{i.}\theta_{.j}$ are given in the following table.

	Second digit 0	Second digit 1
First digit 0	242.55	247.45
First digit 1	252.45	257.55

The Chi-squared statistic is then equal to $X_0^2 = .10409$ and, with $X^2 \sim \chi^2(1)$, the P-value equals $P(X^2 > 4.8105) = .74698$. Therefore, we have no evidence against the null hypothesis of no relationship between the two digits.

10.2.4 First, note that the predictor variable, X (university), is not random. The estimated conditional distributions of Y given X are recorded in the following table.

	Fail	Pass
University 1	0.187500	0.812500
University 2	0.077193	0.922807

Under the null hypothesis of no relationship between calculus grades and university, the MLE's are given by

$$\hat{\theta}_1 = \frac{55}{461} = .11931,\ \hat{\theta}_2 = \frac{406}{461} = .88069.$$

Then the estimates of the expected counts $n_i\theta_j$ are given in the following table.

	Fail	Pass
University 1	20.999	155.0
University 2	34.003	251.0

The Chi-squared statistic is then equal to $X_0^2 = 12.598$ and, with $X^2 \sim \chi^2(1)$, the P-value equals $P(X^2 > 12.598) = .00039$. Therefore, we have strong evidence against the null hypothesis of no relationship between the calculus grades and university.

10.2.5

(a) First, note that the predictor variable, X (gender), is not random. The estimated conditional distributions of Y given X are given in the following table.

	$Y =$ fair	$Y =$ red	$Y =$ medium	$Y =$ dark	$Y =$ jet black
$X =$ m	0.281905	0.0566667	0.404286	0.240000	0.0171429
$X =$ f	0.305104	0.0544027	0.379697	0.252944	0.0078519

Under the null hypothesis of no relationship between hair color and gender, the MLEs are given by

$$\hat{\theta}_1 = \frac{1136}{3883} = .292557,\ \hat{\theta}_2 = \frac{216}{3883} = .055627,\ \hat{\theta}_3 = \frac{1526}{3883} = .0.392995,$$
$$\hat{\theta}_4 = \frac{955}{3883} = .245944,\ \hat{\theta}_5 = \frac{50}{3883} = 0.012877.$$

Then the estimates of the expected counts $n_i\theta_j$ are given in the following table.

	$Y =$ fair	$Y =$ red	$Y =$ medium	$Y =$ dark	$Y =$ jet black
$X =$ m	614.370	116.817	825.290	516.482	27.041
$X =$ f	521.630	99.183	700.710	438.518	22.959

The Chi-squared statistic is then equal to $X_0^2 = 10.4674$ and, with $X^2 \sim \chi^2(4)$, the P-value equals $P(X^2 > 10.4674) = .03325$. Therefore, we have some evidence against the null hypothesis of no relationship between hair color and gender.

(b) The appropriate bar plots are the two conditional distributions and these are plotted as follows for males and then females.

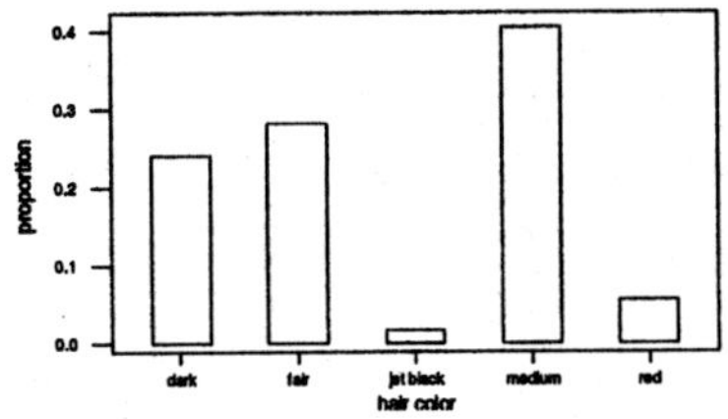

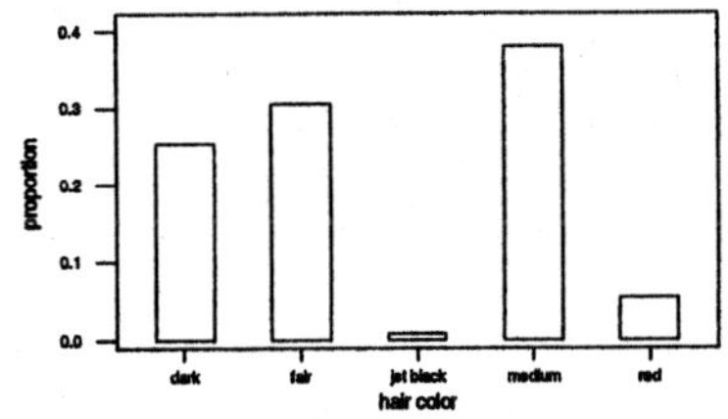

(b) The standardized residuals are given in the following table. They all look reasonable, so nothing stands out as an explanation of why the model of independence doesn't fit. Overall, it looks like a large sample size has detected a small difference.

	$Y = \text{fair}$	$Y = \text{red}$	$Y = \text{medium}$	$Y = \text{dark}$	$Y = \text{jet black}$
$X = \text{m}$	-1.07303	0.20785	1.05934	-0.63250	1.73407
$X = \text{f}$	1.16452	-0.22557	-1.14966	0.68642	-1.88191

10.2.6

(a) First, note that the predictor variable X, is not random. The estimated conditional distributions of Y given X are given in the following table.

	$X = 1$	$X = 2$	$X = 3$	$X = 4$
$Y = 0$	0.48	0.40	0.64	0.56
$Y = 1$	0.52	0.60	0.36	0.44

Under the null hypothesis of no relationship between X and Y, the MLEs are given by

$$\hat{\theta}_1 = \frac{52}{100} = .52,\ \hat{\theta}_2 = \frac{48}{100} = .48.$$

Then the estimates of the expected counts $n_i\theta_j$ are given in the following table.

	$X = 1$	$X = 2$	$X = 3$	$X = 4$
$Y = 0$	13	13	13	13
$Y = 1$	12	12	12	12

The Chi-squared statistic is then equal to $X^2 = 3.20513$ and, with $X^2 \sim \chi^2(3)$, the P-value equals $P(X^2 > 3.20513) = .36107$. Therefore, we don't have any evidence against the null hypothesis of no cause-effect relationship between X and Y.

(b) If a relationship had been detected, this would be evidence of a cause-effect relationship because we have assigned the value of X to each sample element.

10.2.7 We should first generate a value for $X_1 \sim$ Dirichlet$(1,3)$. Then generate U_2 from the Beta$(1,2)$ distribution and set $X_2 = (1 - X_1)\, U_2$. Then generate U_3 from the Beta$(1,1)$ distribution and set $X_3 = (1 - X_1 - X_2)\, U_3$. Then set $X_4 = 1 - X_1 - X_2 - X_3$.

Problems

10.2.8 We place a Dirichlet$(1,1,1,1,1,1,1,1,1,1,1,1)$ prior distribution on $(\theta_{11}, \theta_{21}, \theta_{31}, \theta_{12}, \theta_{22}, \theta_{32}, \theta_{13}, \theta_{23}, \theta_{33}, \theta_{14}, \theta_{24})$, so the posterior is proportional to (using $\theta_{34} = 1-$ the other parameters) $\theta_{11}^{17}\theta_{21}^{17}\theta_{31}^{12}\theta_{12}^{11}\theta_{22}^{9}\theta_{32}^{13}\theta_{13}^{11}\theta_{23}^{8}\theta_{33}^{19}\theta_{14}^{14}$ $\times\theta_{24}^{7}\theta_{34}^{28}$. Therefore, the posterior distribution is Dirichlet$(18,18,13,12,10,14,$ $12,9,20,15,8,29)$.

10.2.9 We place a Dirichlet$(1,1,1)$ prior on $\left(\theta_{O|X=j}, \theta_{A|X=j}\right)$ for $j = P, G, C$, and we assume that these three distributions are independent. Therefore, the posterior is proportional to

$$\left(\theta_{O|X=P}\right)^{983}\left(\theta_{A|X=P}\right)^{679}\left(\theta_{B|X=P}\right)^{134} \times \left(\theta_{O|X=G}\right)^{383}\left(\theta_{A|X=G}\right)^{416}\left(\theta_{B|X=G}\right)^{84}$$
$$\times \left(\theta_{O|X=C}\right)^{2892}\left(\theta_{A|X=C}\right)^{2625}\left(\theta_{B|X=C}\right)^{570},$$

so $\left(\theta_{O|X=P}, \theta_{A|X=P}\right) \mid$ data $\sim$ Dirichlet$(984, 680, 135)$, $\left(\theta_{O|X=G}, \theta_{A|X=G}\right) \mid$ data $\sim$ Dirichlet$(384, 417, 85)$, $\left(\theta_{O|X=C}, \theta_{A|X=C}\right) \mid$ data $\sim$ Dirichlet$(2893, 2626,$ $571)$ and they are independent.

10.2.10 Consider the following 2×2 table.

	$Y=1$	$Y=2$	$P(X=x)$
$X=1$	θ_{11}	θ_{12}	$\theta_{11}+\theta_{12}$
$X=2$	θ_{21}	θ_{22}	$\theta_{21}+\theta_{22}$
$P(Y=y)$	$\theta_{11}+\theta_{21}$	$\theta_{12}+\theta_{22}$	1

Now$X$ and Y independent implies that

$$\theta_{11} = (\theta_{11}+\theta_{12})(\theta_{11}+\theta_{21}), \quad \theta_{12} = (\theta_{11}+\theta_{12})(\theta_{12}+\theta_{22})$$
$$\theta_{21} = (\theta_{21}+\theta_{22})(\theta_{11}+\theta_{21}), \quad \theta_{22} = (\theta_{21}+\theta_{22})(\theta_{12}+\theta_{22}).$$

and this implies that

$$\frac{\theta_{11}\theta_{22}}{\theta_{12}\theta_{21}} = \frac{(\theta_{11}+\theta_{12})(\theta_{11}+\theta_{21})(\theta_{21}+\theta_{22})(\theta_{12}+\theta_{22})}{(\theta_{11}+\theta_{12})(\theta_{12}+\theta_{22})(\theta_{21}+\theta_{22})(\theta_{11}+\theta_{21})} = 1.$$

Now $\theta_{11}\theta_{22}/\theta_{12}\theta_{21} = 1$ implies that $\theta_{11}\theta_{22} = \theta_{12}\theta_{21}$ and so $(\theta_{11}+\theta_{12})\,\theta_{22} = \theta_{12}(\theta_{21}+\theta_{22})$ and $(\theta_{11}+\theta_{12})(\theta_{12}+\theta_{22}) = \theta_{12}(\theta_{21}+\theta_{22}=\theta_{11}+\theta_{12}) = \theta_{12}$. Also, $\theta_{11}\theta_{22} = \theta_{12}\theta_{21}$ implies $(\theta_{11}+\theta_{21})\,\theta_{22} = (\theta_{12}+\theta_{22})\,\theta_{21}$ and so $(\theta_{11}+\theta_{21})$ $\times(\theta_{21}+\theta_{22}) = (\theta_{12}+\theta_{22}+\theta_{11}+\theta_{21})\,\theta_{21} = \theta_{21}$. Similarly, $\theta_{22} = (\theta_{21}+\theta_{22})$ $\times(\theta_{12}+\theta_{22})$ and $\theta_{11} = (\theta_{11}+\theta_{12})(\theta_{11}+\theta_{21})$, so X and Y are independent.

10.2.11 When sampling with replacement from the population, we can think of the sample as an i.i.d. sample from this population, so each observation has probability θ_{ij} of falling in the (i, j) category, namely θ_{ij}. Then when f_{ij} sample elements fall in this cell the likelihood takes the form $\prod_{i=1}^{a}\prod_{j=1}^{b}\theta_{ij}^{f_{ij}}$ as claimed.

10.2.12 First, note that there are only $ab-1$ free parameters, so we place $\theta_{ab} = 1 - \sum_{(i,j)\neq(a,b)}\theta_{ij}$. The likelihood function is given by $L(\theta_{11}, ..., \theta_{ab} \mid (x_1, y_1), ..., (x_n, y_n)) = \prod_{i=1}^{a}\prod_{j=1}^{b}\theta_{ij}^{f_{ij}}$. The log-likelihood function is given by $l(\theta_{11}, ..., \theta_{a(b-1)} \mid (x_1, y_1), ..., (x_n, y_n)) = \sum_{i=1}^{a}\sum_{j=1}^{b} f_{ij}\ln\theta_{ij}$.

The score function is then given by

$$S(\theta_{11}, ..., \theta_{a(b-1)} \mid (x_1, y_1), ..., (x_n, y_n)) = \begin{pmatrix} \frac{f_{11}}{\theta_{11}} - \frac{f_{ab}}{\theta_{ab}} \\ \frac{f_{12}}{\theta_{12}} - \frac{f_{ab}}{\theta_{ab}} \\ \vdots \end{pmatrix}.$$

Setting this equal to 0 and solving leads to $\theta_{ij} = (f_{ij}/f_{ab})\,\theta_{ab}$. Then summing both sides over all $(i, j) \neq (a, b)$ leads to $1 - \theta_{ab} = (n - f_{ab})\,\theta_{ab}/f_{ab}$ or $\theta_{ab} = f_{ab}/n$, and this implies that $\theta_{ij} = f_{ij}/n$ gives a unique solution to the score equations.

Now the log-likelihood takes the value $-\infty$ whenever any $\theta_{ij} = 0$, so the log-likelihood does not attain its maximum at such a point. Therefore, the log-likelihood is maximized at some point for which all $\theta_{ij} \neq 0$, and the log-likelihood is continuously differentiable at such a point. Since the unique solution to the score equations is such a point, it must be the MLE.

10.2.13 We let $\theta_{1\cdot}, \ldots, \theta_{(a-1)\cdot}, \theta_{\cdot 1}, \ldots, \theta_{\cdot(b-1)}$ be the free parameters since $\theta_{a\cdot} = 1 - \sum_{i=1}^{a-1}\theta_{i\cdot}$ and $\theta_{\cdot b} = 1 - \sum_{j=1}^{b-1}\theta_{\cdot j}$. The likelihood function is then given by $L(\theta_{1\cdot}, \ldots, \theta_{(a-1)\cdot}, \theta_{\cdot 1}, \ldots, \theta_{\cdot(b-1)} \mid (x_1, y_1), ..., (x_n, y_n)) = \prod_{i=1}^{a}\prod_{j=1}^{b}(\theta_{i\cdot}\theta_{\cdot j})^{f_{ij}}$ $= \prod_{i=1}^{a}(\theta_{i\cdot})^{f_{i\cdot}}\prod_{j=1}^{b}(\theta_{\cdot j})^{f_{\cdot j}}$. The log-likelihood function is given by $l(\theta_{1\cdot}, \ldots, \theta_{(a-1)\cdot}, \theta_{\cdot 1}, \ldots, \theta_{\cdot(b-1)} \mid (x_1, y_1), ..., (x_n, y_n)) =$ $\sum_{i=1}^{a} f_{i\cdot}\ln\theta_{i\cdot} + \sum_{j=1}^{b} f_{\cdot j}\ln\theta_{\cdot j}$. The score function is then given by

$$S(\theta_{1\cdot}, \ldots, \theta_{(a-1)\cdot}, \theta_{\cdot 1}, \ldots, \theta_{\cdot(b-1)} \mid (x_1, y_1), ..., (x_n, y_n)) = \begin{pmatrix} \frac{f_{1\cdot}}{\theta_{1\cdot}} - \frac{f_{a\cdot}}{\theta_{a\cdot}} \\ \vdots \\ \frac{f_{\cdot 1}}{\theta_{\cdot 1}} - \frac{f_{\cdot b}}{\theta_{\cdot b}} \\ \vdots \end{pmatrix}.$$

Setting this equal to 0 and solving leads to

$$\theta_{i\cdot} = \frac{f_{i\cdot}}{f_{a\cdot}}\theta_{a\cdot},\ \theta_{\cdot j} = \frac{f_{\cdot j}}{f_{\cdot b}}\theta_{\cdot b}.$$

Summing these over $i = 1, \ldots, a-1$ and $j = 1, \ldots, b-1$ leads to the equations

$$1 - \theta_{a\cdot} = \frac{n - f_{a\cdot}}{f_{a\cdot}}\theta_{a\cdot} \text{ and } 1 - \theta_{\cdot b} = \frac{n - f_{\cdot b}}{f_{\cdot b}}\theta_{\cdot b}.$$

Therefore, $\theta_{a\cdot} = f_{a\cdot}/n, \theta_{\cdot b} = f_{\cdot b}/n$, and this implies that $\theta_{i\cdot} = f_{i\cdot}/n, \theta_{\cdot j} = f_{\cdot j}/n$ gives a unique solution to the score equations.

Now the log-likelihood takes the value $-\infty$ whenever any $\theta_{i\cdot} = 0$ or $\theta_{\cdot j} = 0$, so the log-likelihood does not attain its maximum at such a point. Therefore, the log-likelihood is maximized at some point for which all $\theta_{i\cdot} \neq 0$ or $\theta_{\cdot j} \neq 0$, and the log-likelihood is continuously differentiable at such a point. Since the unique solution to the score equations is such a point, it must be the MLE.

10.2.14 There are $a\,(b-1)$ free parameters because $\theta_{b|X=i} = 1 - \sum_{j=1}^{b-1} \theta_{j|X=i}$ for $i = 1, \ldots, a$. The likelihood function is given by

$$L\left(\theta_{1|X=1}, \ldots, \theta_{b-1|X=1}, \ldots, \theta_{b-1|X=a} \,|\, (x_1, y_1), \ldots, (x_n, y_n)\right) = \prod_{i=1}^{a}\prod_{j=1}^{b} \left(\theta_{j|X=i}\right)^{f_{ij}}.$$

The log-likelihood function is given by

$$l\left(\theta_{1|X=1}, \ldots, \theta_{b-1|X=1}, \ldots, \theta_{b-1|X=a} \,|\, (x_1, y_1), \ldots, (x_n, y_n)\right) = \sum_{i=1}^{a}\sum_{j=1}^{b} f_{ij} \ln \theta_{j|X=i}.$$

The score function is then given by

$$S\left(\theta_{1|x=1}, \theta_{1|x=2} \,|\, (x_1, y_1), \ldots, (x_n, y_n)\right) = \begin{pmatrix} \frac{f_{11}}{\theta_{1|X=1}} - \frac{f_{1b}}{\theta_{b|X=1}} \\ \vdots \end{pmatrix}.$$

Setting this equal to 0 and solving leads to $\theta_{j|X=i} = (f_{ij}/f_{ib})\,\theta_{b|X=i}$. Summing both sides over $j = 1, \ldots, b-1$ leads to

$$1 - \theta_{b|X=i} = \frac{n_{i\cdot} - f_{ib}}{f_{ib}} \theta_{b|X=i}$$

and this implies that $\theta_{b|X=i} = f_{ib}/n_i$ further implying that $\theta_{j|X=i} = f_{ij}/n_i$ gives a unique solution to the score equations.

Now the log-likelihood takes the value $-\infty$ whenever any $\theta_{j|X=i} = 0$, so the log-likelihood does not attain its maximum at such a point. Therefore, the log-likelihood is maximized at some point for which all $\theta_{j|X=i} \neq 0$, and the log-likelihood is continuously differentiable at such a point. Since the unique solution to the score equations is such a point, it must be the MLE.

10.2.15 There are $b-1$ free parameters because $\theta_b = 1 - \sum_{j=1}^{b-1} \theta_j$. The likelihood function is given by $L\left(\theta_1, \ldots, \theta_{b-1} \,|\, (x_1, y_1), \ldots, (x_n, y_n)\right) = \prod_{i=1}^{a}\prod_{j=1}^{b} \theta_i^{f_{ij}} = \prod_{j=1}^{b} \theta_j^{f_{\cdot j}}$. The log-likelihood function is given by $l\left(\theta_1, \ldots, \theta_{b-1} \,|\, (x_1, y_1), \ldots, (x_n, y_n)\right) = \sum_{j=1}^{b} f_{\cdot j} \ln \theta_j$. The score function is then given by

$$S(\theta_1, \ldots, \theta_{b-1} \mid (x_1, y_1), ..., (x_n, y_n)) = \begin{pmatrix} \frac{f_{\cdot 1}}{\theta_1} - \frac{f_{\cdot b}}{\theta_b} \\ \vdots \\ \frac{f_{\cdot (b-1)}}{\theta_{b-1}} - \frac{f_{\cdot b}}{\theta_b} \end{pmatrix}.$$

Setting this equal to 0 gives $\theta_j = (f_{\cdot j}/f_{\cdot b})\,\theta_b$ and summing this over $j = 1, \ldots, b-1$ gives $1 - \theta_b = (n - f_{\cdot b})\,\theta_b/f_{\cdot b}$. This implies that $\theta_b = f_{\cdot b}/n$, further implying that $\theta_j = f_{\cdot j}/n$ gives a unique solution to the score equations.

Now the log-likelihood takes the value $-\infty$ whenever any $\theta_j = 0$, so the log-likelihood does not attain its maximum at such a point. Therefore the log-likelihood is maximized at some point for which all $\theta_j \neq 0$, and the log-likelihood is continuously differentiable at such a point. Since the unique solution to the score equations is such a point, it must be the MLE.

10.2.16 First, note that the density of Dirichlet$(\alpha_1, ..., \alpha_k)$ density is given by $\frac{\Gamma(\alpha_1+\cdots+\alpha_k)}{\Gamma(\alpha_1)\cdots\Gamma(\alpha_k)} x_1^{\alpha_1-1} x_2^{\alpha_2-1} \cdots x_k^{\alpha_k-1}$. Therefore,

$$E\left(X_1^{l_1} \cdots X_k^{l_k}\right) = \int_0^1 \cdots \int_0^{1-x_2-\cdots-x_{k-1}} x_1^{l_1} \cdots (1 - x_1 - \cdots - x_{k-1})^{l_k}$$

$$\times \frac{\Gamma(\alpha_1 + \cdots + \alpha_k)}{\Gamma(\alpha_1) \cdots \Gamma(\alpha_k)} x_1^{\alpha_1 - 1} x_2^{\alpha_2 - 1} \cdots (1 - x_1 - \cdots - x_{k-1})^{\alpha_k - 1} dx_1 \cdots dx_{k-1}$$

$$= \frac{\Gamma(\alpha_1 + \cdots + \alpha_k)}{\Gamma(\alpha_1) \cdots \Gamma(\alpha_k)} \frac{\Gamma(\alpha_1 + l_1) \cdots \Gamma(\alpha_k + l_k)}{\Gamma(\alpha_1 + \cdots + \alpha_k + l_1 + \cdots + l_k)}.$$

Computer Problems

10.2.17 The following code generates the sample in c2, c3, c4, c5.

```
gmacro
dirichlet
note - the base command sets the seed for the random number
       generator (so you can repeat a simulation).
base 34256734
note - here we provide the algorithm for generating from a
       Dirichlet(k1,k2,k3,k4) distribution.
note - assign the values of the parameters.
let k1=1
let k2=1
let k3=1
let k4=1
let k5=K2+k3+k4
let k6=k3+k4
note - generate the sample with i-th sample in i-th row of
       c2, c3, c4, c5, ....
do k10=1:10000
```

```
random 1 c1;
beta k1 k5.
let c2(k10)=c1(1)
random 1 c1;
beta k2 k6.
let c3(k10)=(1-c2(k10))*c1(1)
random 1 c1;
beta k3 k4.
let c4(k10)=(1-c2(k10)-c3(k10))*c1(1)
let c5(k10)= 1-c2(k10)-c3(k10)-c4(k10)
enddo
endmacro
```

Based on the output, the following commands calculate the estimates of the expectations.

```
MTB > let k1=mean(c2)
MTB > let k2=mean(c3)
MTB > let k3=mean(c4)
MTB > let k4=mean(c5)
MTB > print k1-k4
Data Display
K1 0.247073
K2 0.251701
K3 0.251028
K4 0.250198
```

From Appendix C the exact values of each of these expectations is given by $1/(1+1+1+1) = .25$.

10.2.18 From Problem 10.2.8 we need to generate from a Dirichlet$(18, 18, 13, 12, 10, 14, 12, 9, 20, 15, 8, 29)$ distribution. The code below generates the sample.

```
gmacro
dirichlet
note - the base command sets the seed for the random number
       generator (so you can repeat a simulation).
base 34256734
note - here we provide the algorithm for generating from a
       Dirichlet(k1,k2,k3,k4) distribution.
note - assign the values of the parameters.
let k1=18
let k2=18
let k3=13
let k4=12
let k5=10
let k6=14
let k7=12
let k8=9
let k9=20
```

```
let k10=15
let k11=8
let k12=29
let k20=K2+k3+k4+k5+k6+k7+k8+k9+k10+k11+k12
let k21=k3+k4+k5+k6+k7+k8+k9+k10+k11+k12
let k22=k4+k5+k6+k7+k8+k9+k10+k11+k12
let k23=k5+k6+k7+k8+k9+k10+k11+k12
let k24=k6+k7+k8+k9+k10+k11+k12
let k25=k7+k8+k9+k10+k11+k12
let k26=k8+k9+k10+k11+k12
let k27=k9+k10+k11+k12
let k28=k10+k11+k12
let k29=k11+k12
let k30=k12
note - generate the sample with i-th sample in i-th row of
       c2, c3, c4, c5, ....
do k100=1:10000
random 1 c1;
beta k1 k20.
let c2(k100)=c1(1)
random 1 c1;
beta k2 k21.
let c3(k100)=(1-c2(k100))*c1(1)
random 1 c1;
beta k3 k22.
let c4(k100)=(1-c2(k100)-c3(k100))*c1(1)
random 1 c1;
beta k4 k23.
let c5(k100)=(1-c2(k100)-c3(k100)-c4(k100))*c1(1)
random 1 c1;
beta k5 k24.
let c6(k100)=(1-c2(k100)-c3(k100)-c4(k100)-c5(k100))*c1(1)
random 1 c1;
beta k6 k25.
let c7(k100)=(1-c2(k100)-c3(k100)-c4(k100)-c5(k100)
   -c6(k100))*c1(1)
random 1 c1;
beta k7 k26.
let c8(k100)=(1-c2(k100)-c3(k100)-c4(k100)-c5(k100)
   -c6(k100)-c7(k100))*c1(1)
random 1 c1;
beta k8 k27.
let c9(k100)=(1-c2(k100)-c3(k100)-c4(k100)-c5(k100)
   -c6(k100)-c7(k100)-c8(k100))*c1(1)
random 1 c1;
beta k9 k28.
```

```
let c10(k100)=(1-c2(k100)-c3(k100)-c4(k100)-c5(k100)-c6(k100)
   -c7(k100)-c8(k100)-c9(k100))*c1(1)
random 1 c1;
beta k10 k29.
let c11(k100)=(1-c2(k100)-c3(k100)-c4(k100)-c5(k100)-c6(k100)
   -c7(k100)-c8(k100)-c9(k100)-c10(k100))*c1(1)
random 1 c1;
beta k11 k30.
let c12(k100)=(1-c2(k100)-c3(k100)-c4(k100)-c5(k100)-c6(k100)
   -c7(k100)-c8(k100)-c9(k100)-c10(k100)-c11(k100))*c1(1)
let c13(k100)=(1-c2(k100)-c3(k100)-c4(k100)-c5(k100)-c6(k100)
   -c7(k100)-c8(k100)-c9(k100)-c10(k100)-c11(k100)-c12(k100))
enddo
endmacro
```

Once the sample is generated, the following code generates the estimates.

```
MTB > let k1=mean(c2)
MTB > let k2=mean(c3)
MTB > let k3=mean(c4)
MTB > let k4=mean(c5)
MTB > let k5=mean(c6)
MTB > let k6=mean(c7)
MTB > let k7=mean(c8)
MTB > let k8=mean(c9)
MTB > let k9=mean(c10)
MTB > let k10=mean(c11)
MTB > let k11=mean(c12)
MTB > let k12=mean(c13)
MTB > print k1-k12
Data Display
K1 0.101230
K2 0.101019
K3 0.0728538
K4 0.0675903
K5 0.0562978
K6 0.0785378
K7 0.0675277
K8 0.0507912
K9 0.112003
K10 0.0844297
K11 0.0449191
K12 0.162800
```

From Appendix C the exact posterior expected values are given by (where $s = 18 + 18 + 13 + 12 + 10 + 14 + 12 + 9 + 20 + 15 + 8 + 29 = 178$) and

$(18/s, 18/s, 13/s, 12/s, 10/s, 14/s, 12/s, 9/s, 20/s, 15/s, 8/s, 29/s)$. So the estimates are as recorded in the following table.

i	Estimate of posterior mean of α_i
1	0.10112
2	0.10112
3	7.3034×10^{-2}
4	6.7416×10^{-2}
5	5.6180×10^{-2}
6	7.8652×10^{-2}
7	6.7416×10^{-2}
8	5.0562×10^{-2}
9	0.112 36
10	8.4270×10^{-2}
11	4.4944×10^{-2}
12	0.16292

Challenge

10.2.17 We have that $U_1, U_2, \ldots, U_{k-1}$ are independent, with $U_i \sim \text{Beta}(\alpha_i, \alpha_{i+1} + \cdots + \alpha_k)$ and

$$X_1 = U_1, X_2 = (1 - X_1)\, U_2, \ldots, X_{k-1} = (1 - X_1 - \cdots - X_{k-1})\, U_{k-1},$$

so

$$U_1 = X_1, U_2 = X_2/\,(1 - X_1)\,, \ldots, U_{k-1} = X_{k-1}/\,(1 - X_1 - \cdots - X_{k-1})\,.$$

From this we deduce that the matrix of partial derivatives of this transformation is lower triangular and the ith element along the diagonal is $\partial U_i/\partial X_i = 1/\,(1 - X_1 - \cdots - X_{i-1})$. Therefore, the Jacobian derivative is given by $\prod_{i=2}^{k-1} (1 - X_1 - \cdots - X_{i-1})^{-1}$. Now the joint density of $(U_1, U_2, \ldots, U_{k-1})$ proportional to

$$u_1^{\alpha_1 - 1} (1 - u_1)^{\alpha_2 + \cdots + \alpha_k - 1}\, u_2^{\alpha_2 - 1} (1 - u_2)^{\alpha_3 + \cdots + \alpha_k - 1} \cdots u_{k-1}^{\alpha_{k-1} - 1} (1 - u_{k-1})^{\alpha_k - 1}.$$

Therefore, the joint density of $(X_1, X_2, \ldots, X_{k-1})$ is proportional to

$$\left\{ \begin{array}{l} x_1^{\alpha_1 - 1} (1 - x_1)^{\alpha_2 + \cdots + \alpha_k - 1} \left(\frac{x_2}{1 - x_1}\right)^{\alpha_2 - 1} \left(1 - \frac{x_2}{1 - x_1}\right)^{\alpha_3 + \cdots + \alpha_k - 1} \\ \quad \cdots \left(\frac{x_{k-1}}{1 - x_1 - \cdots - x_{k-2}}\right)_{k-1}^{\alpha_{k-1} - 1} \left(1 - \frac{x_{k-1}}{1 - x_1 - \cdots - x_{k-2}}\right)^{\alpha_k - 1} \end{array} \right\}$$

$$\times \prod_{i=2}^{k-1} (1 - x_1 - \cdots - x_{i-1})^{-1}$$

$$= x_1^{\alpha_1 - 1} x_2^{\alpha_2 - 1} \cdots x_{k-1}^{\alpha_{k-1} - 1} (1 - x_1 - \cdots - x_{k-1})^{\alpha_k - 1}$$

$$\times (1 - x_1)^{\alpha_2 + \cdots + \alpha_k - 1} (1 - x_1)^{1 - \alpha_2 - (\alpha_3 + \cdots + \alpha_k - 1)} (1 - x_1)^{-1}$$

$$\times (1 - x_1 - x_2)^{\alpha_3 + \cdots + \alpha_k - 1} (1 - x_1 - x_2)^{1 - \alpha_3 - (\alpha_4 + \cdots + \alpha_k - 1)} (1 - x_1 - x_2)^{-1} \times \cdots$$

$$= x_1^{\alpha_1 - 1} x_2^{\alpha_2 - 1} \cdots x_{k-1}^{\alpha_{k-1} - 1} (1 - x_1 - \cdots - x_{k-1})^{\alpha_k - 1},$$

so $(X_1, X_2, \ldots, X_{k-1}) \sim \text{Dirichlet}(\alpha_1, \alpha_2, \ldots, \alpha_k)$.

10.3 Quantitative Response and Predictors

Exercises

10.3.1 Since $\bar{x} \in [0, 1]$ with probability 1, we have that $\bar{x}$ is the least-squares estimate of the mean θ.

10.3.2 Since $\bar{x} \in [0, \theta] \subset [0, \infty)$ with probability 1, we have that $\bar{x}$ is the least-squares estimate of the mean $\theta/2 \in [0, \infty)$.

10.3.3 Since $\bar{x} \in (0, \infty)$ with probability 1, we have that $\bar{x}$ is the least-squares estimate of the mean $1/\theta \in (0, \infty)$.

10.3.4
(a) A scatter plot is given below.

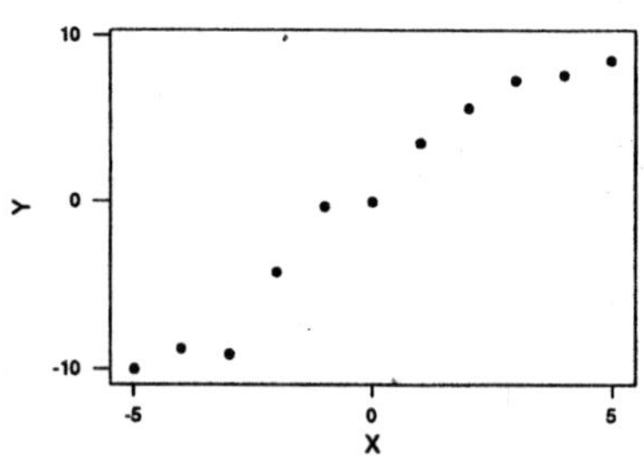

(b) The least-squares estimates of β_1 and β_2 are given by $b_2 = 2.1024$ and $b_1 = \bar{y} = -0.00091$, so the least-squares line is given by $y = -0.00091 + 2.1024x$. A scatter plot of the data together with a plot of the least-squares line follows.

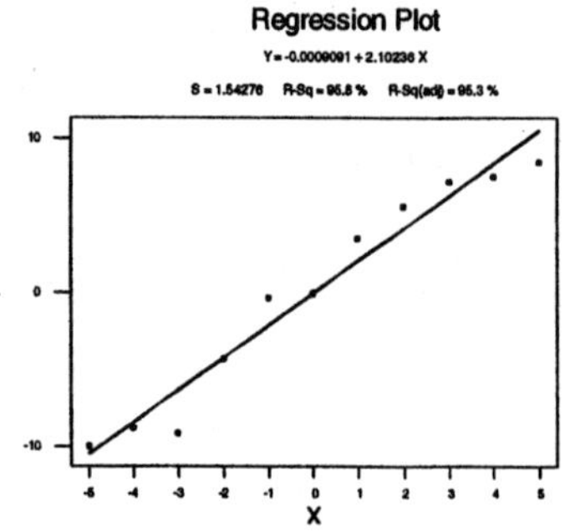

(c) The plot of the standardized residuals against X follows.

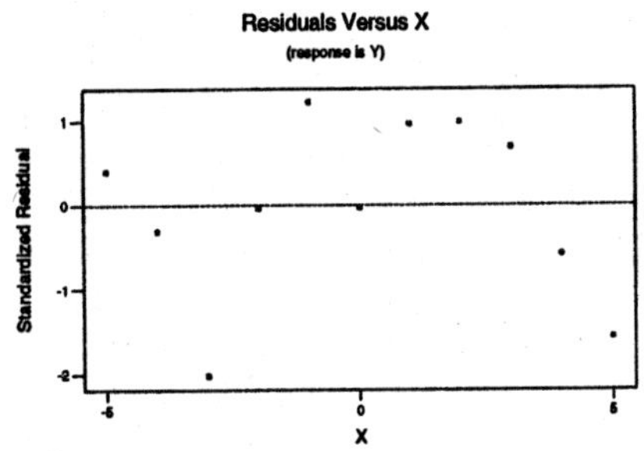

(d) A normal probability plot of the standardized residuals is given below.

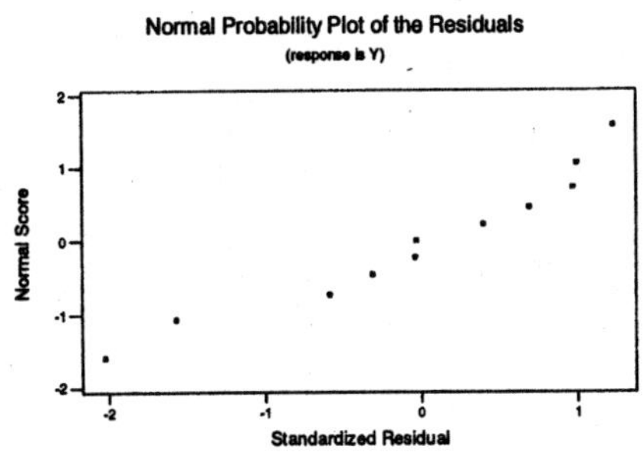

(e) Both graphs indicate that the normal simple linear regression model is reasonable.
(f) A .95-confidence interval for the intercept is given by

$$-0.00091 \pm 0.4652\,(2.2622) = (-1.0533, 1.0515)$$

and a .95-confidence interval for the slope is given by $2.1024 \pm 0.1471 \cdot 2.2622 = (1.7696, 2.4352)$.
(g) The ANOVA table is follows.

Source	Df	SS	MS
X	1	486.19	486.19
Error	9	21.42	2.38
Total	10	507.61	

The F statistic for testing $H_0 : \beta_2 = 0$ is given by $F = 486.19/2.38 = 204.28$ and, since $F \sim F(1,9)$ under H_0, the P-value is given by $P(F > 204.28) = .000$, so we reject the null hypothesis of no effect between X and Y.
(h) The proportion of the observed variation in the response that is being explained by changes in the predictor is given by the coefficient of determination $R^2 = 486.19/507.61 = .9578$.
(i) The prediction is given by $y = -0.00091 + 2.1024\,(0) = -0.00091$. This is an interpolation because 0.0 is in the range of observed X values. The standard

error of this prediction is, since $\bar{x} = 0$ (using Corollary 10.3.1), $(2.38/11)^{1/2} = 0.46515$.

(j) The prediction is given by $y = -0.00091 + 2.1024\,(6) = 12.613$. This is an extrapolation because 6 is not in the range of observed X values. The standard error of this prediction is, since $\bar{x} = 0$ (using Corollary 10.3.1),

$$(2.38)^{1/2}\left(\frac{1}{11} + \frac{(6-0)^2}{110}\right)^{1/2} = 0.99763.$$

(k) The prediction is given by $y = -0.00091 + 2.1024\,(20) = 42.047$. This is an extrapolation because 12 is not in the range of observed X values. The standard error of this prediction is, since $\bar{x} = 0$ (using Corollary 10.3.1),

$$(2.38)^{1/2}\left(\frac{1}{11} + \frac{(20-0)^2}{110}\right)^{1/2} = 2.978\,4.$$

The standard errors get larger as we move away from the observed X values.

10.3.5

(a) A scatter plot of the data follows.

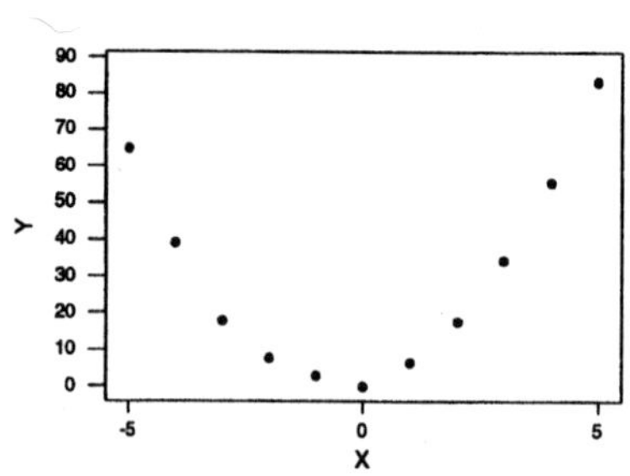

(b) The least-squares estimates of β_1 and β_2 are given by $b_2 = 2.10236$ and $b_1 = 29.9991$. The least-squares line is then given by $y = 29.9991 + 2.10236x$. A scatter plot of the data together with a plot of the least-squares line follows.

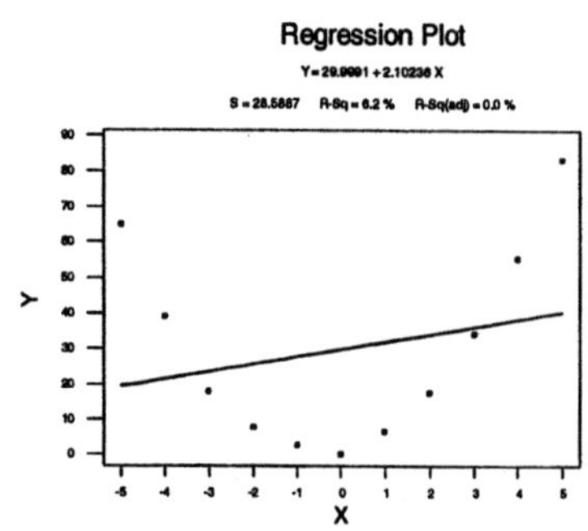

(c) The plot of the standardized residuals against X follows.

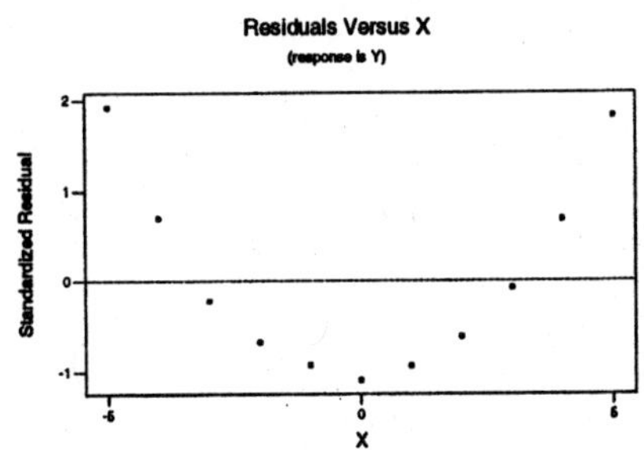

(d) A normal probability plot of the standardized residuals follows.

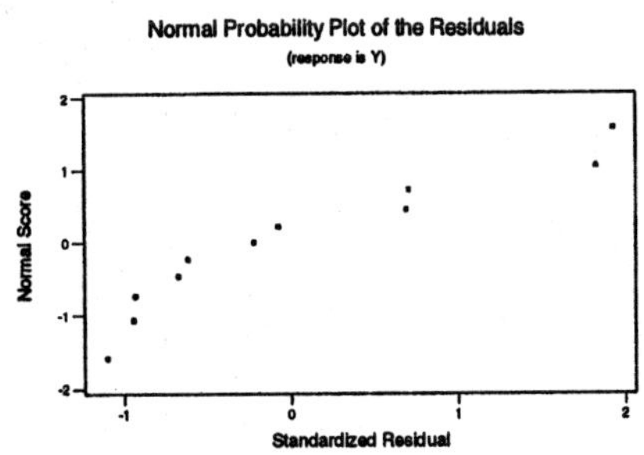

(e) The plot of the standardized residuals against X indicates very clearly that there is a problem with this model.
(f) Based on (e), it is not appropriate to calculate confidence intervals for the intercept and slope.
(g) Nothing can be concluded about the relationship between Y and X based on this model as we have determined that it is inappropriate.
(h) The proportion of the observed variation in the response that is being explained by changes in the predictor is given by the coefficient of determination $R^2 = 486.193/7842.01 = 0.062$, which is very low.

10.3.6
(a) A scatter plot of the data is given below.

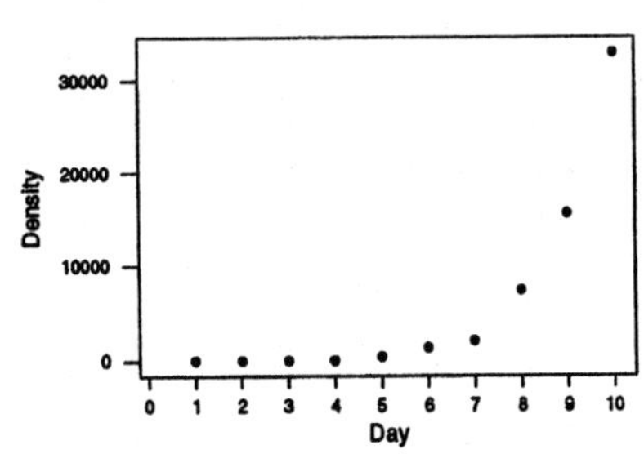

(b) The least-squares estimates of β_1 and β_2 are given by $b_2 = 2732.67$ and $b_1 = -9033.28$, respectively. The least-squares line is then given by $y = -9033.28 + 2732.67x$. A scatter plot of the data together with a plot of the least-squares line follows.

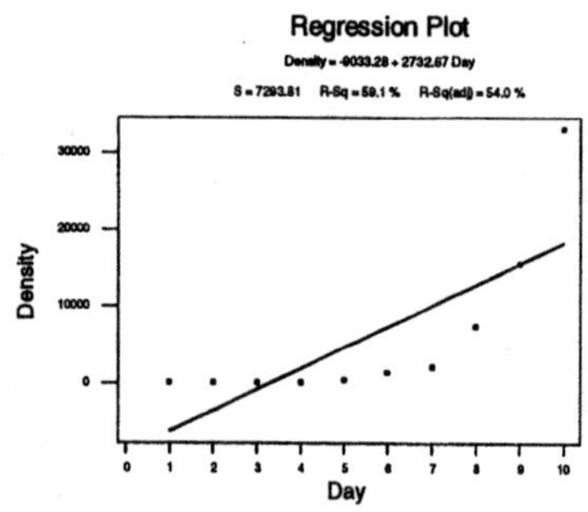

(c) A plot of the standardized residuals against X follows.

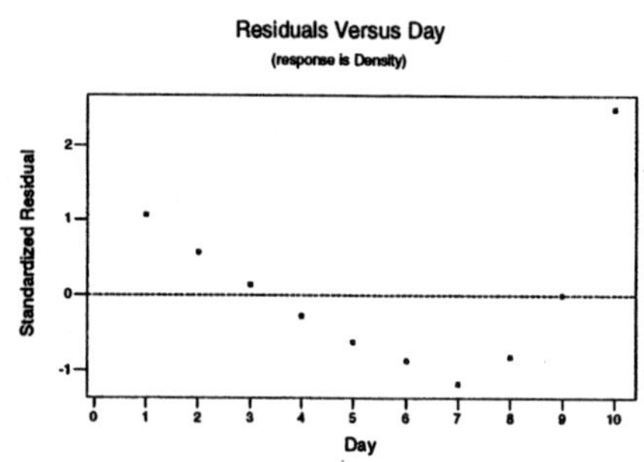

(d) A normal probability plot of the standardized residuals follows.

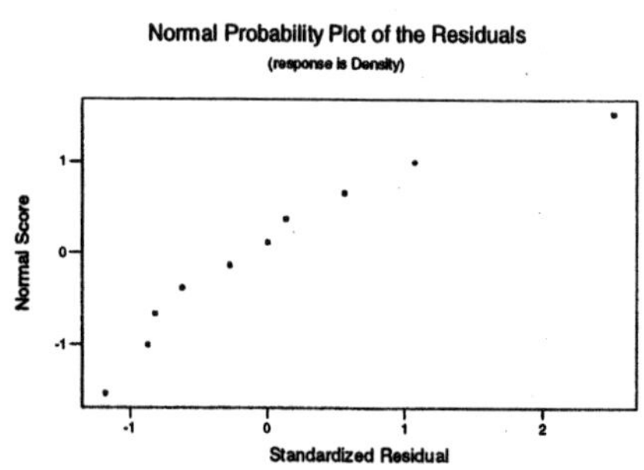

(e) The plot of the standardized residuals against X indicates very clearly that there is a problem with this model.

(f) Taking the logarithm of the response, we obtain the least-squares line given by $\ln(y) = 0.169155 + 1.06500x$. A scatter plot of the data together with the least-squares line follows

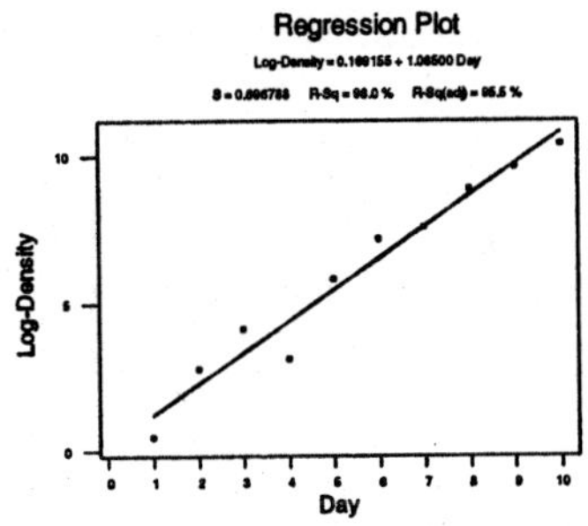

A plot of the standardized residuals against X follows.

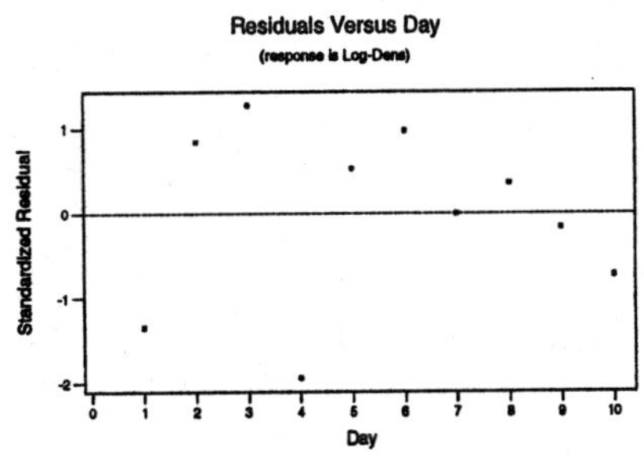

A normal probability plot of the standardized residuals follows.

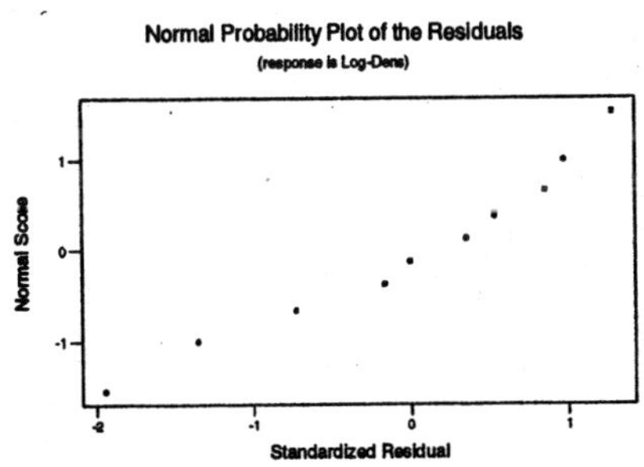

Both graphs above look reasonable and therefore indicate no evidence against the normal linear model for the transformed response.

(g) As we can see from the scatter plot in part (a), the relationship between X and Y is definitely non-linear, and therefore it is not appropriate to calculate confidence intervals for the intercept and slope. However, after transforming the response, the relationship looks quite linear, so for this model a .95-confidence intervals for the intercept and the slope are given by $0.169155 \pm 0.4760\,(2.306) = (-.9285, 1.2668)$ and $1.065 \pm 0.07671\,(2.306) = (.88811, 1.2419)$, respectively.

(h) The ANOVA table based on the transformed data (in part f) is given below.

Source	Df	SS	MS
X	1	93.573	93.573
Error	8	3.884	0.486
Total	9	97.458	

The F statistic for testing $H_0 : \beta_2 = 0$ for this model is then given by $F = 93.573/3.884 = 24.092$ and, since $F \sim F(1,8)$ under H_0, the P-value is $P(F > 24.092) = 0.000$. Therefore, we have strong evidence against the null hypothesis of no relationship between $\ln Y$ and X.

(i) Yes, we can conclude that there is a relationship. We can then express the relationship between X and Y as $E(\ln Y \mid X = x) = 0.169155 + 1.06500x$.

(j) The proportion of the observed variation in the response that is being explained by changes in the predictor is given by the coefficient of determination $R^2 = 616068769/1.042E{+}09 = 59.1$ for the first model, which is quite low, and $R^2 = 93.573/97.458 = .96014$ for the second model (as in part f),which is quite high.

(k) The prediction of $\ln Y$ at $X = 12$ is given by $0.169155 + 1.06500(12) = 12.949$. The prediction of Y is then given by $\exp(12.949) = 4.2042 \times 10^5$. This is an extrapolation as 12 lies outside the range of observed X values.

Computer Exercises

10.3.7

(a) A scatter plot of the data is given below.

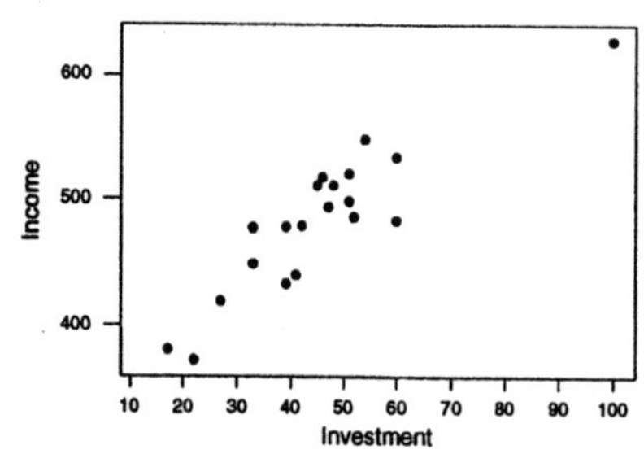

(b) The least-squares estimates of β_1 and β_2 are given by $b_2 = 3.04845$ and $b_1 = 344.703$. The least-squares line is then given by $y = 344.703 + 3.04845x$. A scatter plot of the data together with a plot of the least-squares line follows.

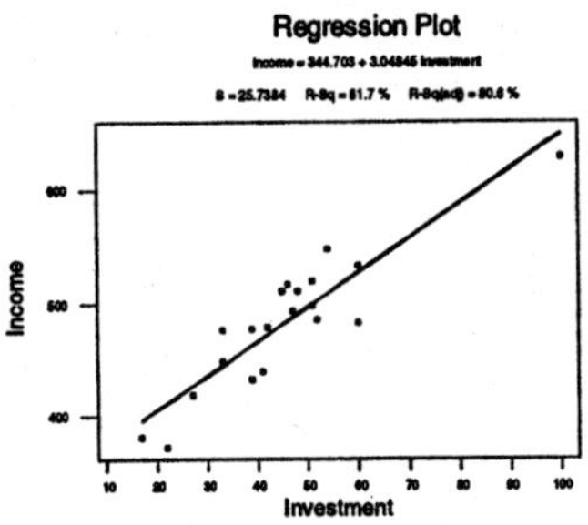

(c) The plot of the standardized residuals against X follows.

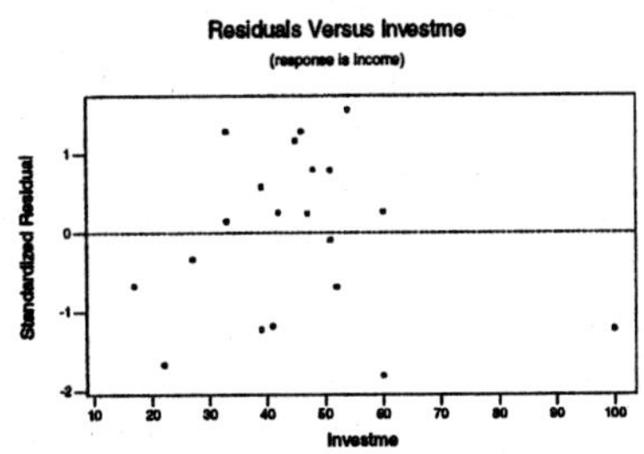

(d) A normal probability plot of the standardized residuals follows.

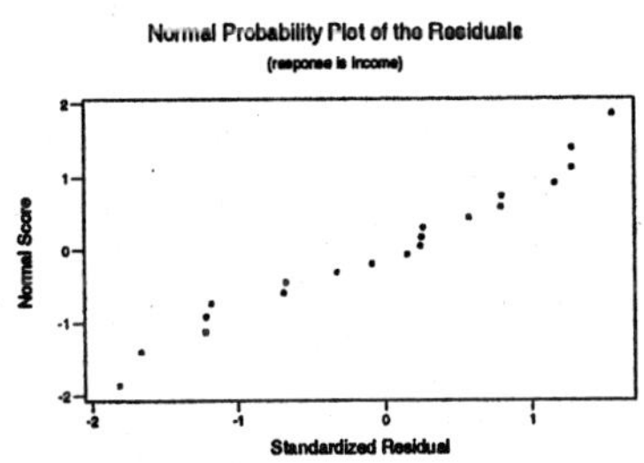

(e) Both plots above indicate that the model assumptions are reasonable.
(f) A .95-confidence interval for the intercept is given by $344.703 \pm 16.48\,(2.1009) = (310.08, 379.33)$and a .95-confidence interval for the slope is given by $3.04845 \pm 0.3406\,(2.1009) = (2.3329, 3.764)$.
(g) The F statistics for testing $H_0 : \beta_2 = 0$ is given by $F = 53069/662 = 80.165$ and, since $F \sim F\,(1, 18)$ under H_0, the P-value is $P(F > 80.165) = 0.000$, indicating strong evidence against the null hypothesis of no linear relationship. Since we have accepted the model as appropriate, this leads us to conclude that a relationship between Y and X exists.
(h) The proportion of the observed variation in the response that is being explained by changes in the predictor is given by the coefficient of determination $R^2 = 53069/64993 = .81653$, which is reasonably high.

10.3.8

(a) A scatter plot of the data follows.

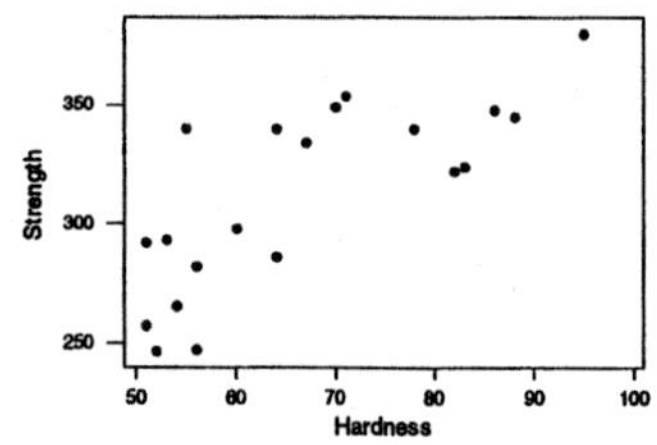

(b) The least-squares estimates of β_1 and β_2 are given by $b_2 = 2.14440$ and $b_1 = 168.854$ respectively. The least-squares line is then given by $y = 168.854 + 2.14440x$. A scatter plot of the data together with a plot of the least-squares line follows.

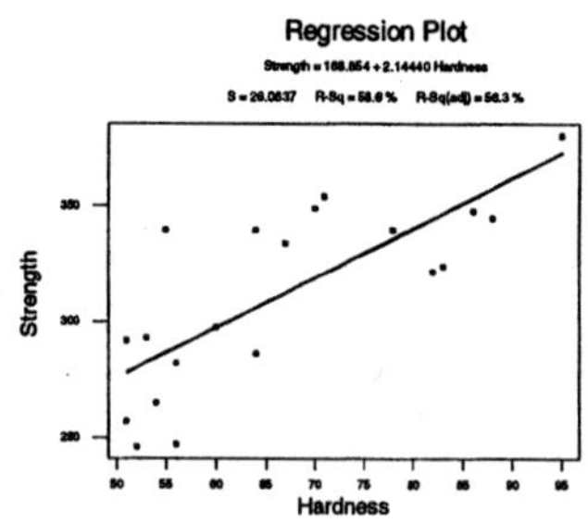

(c) A plot of the standardized residuals against X follows.

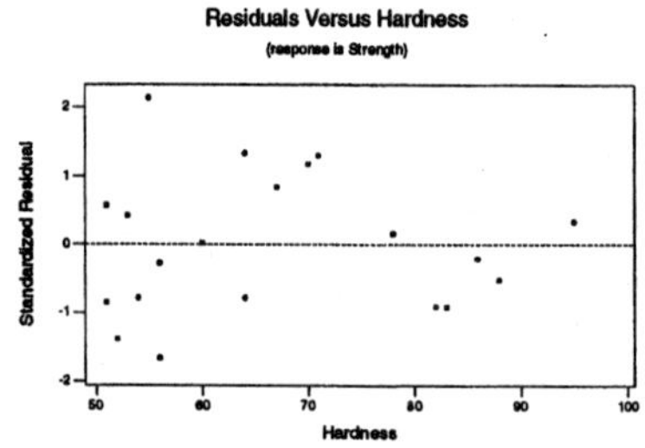

(d) A normal probability plot of the standardized residuals follows.

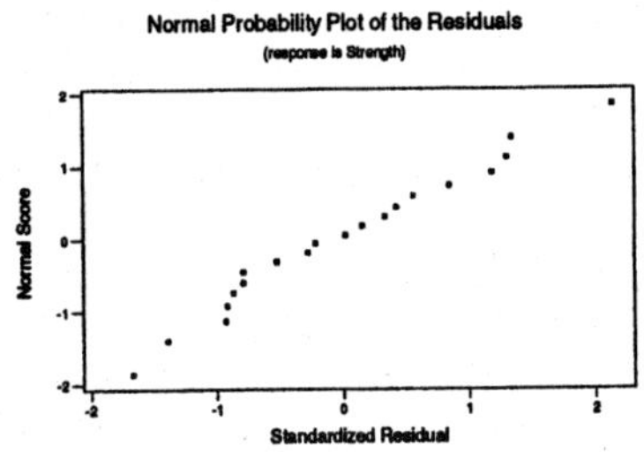

(e) Both plots look reasonable. The first plot might reveal some trend indicating a possible violation of the assumption of equal variances.
(f) A .95-confidence interval for the intercept and the slope are given by $168.854 \pm 28.98 \cdot (2.1009) = (107.97, 229.74)$ and $2.1444 \pm 0.4250\,(2.1009) = (1.2515, 3.0373)$, respectively.
(g) The F statistic for testing $H_0 : \beta_2 = 0$ is given by $F = 17323/680 = 25.475$ and, since $F \sim F\,(1, 18)$ under H_0, the P-value equals $P(F > 25.475) = 0.000$, indicating strong evidence against the null hypothesis of no linear relationship. We conclude that there is a linear relationship between X and Y.
(h) The proportion of the observed variation in the response that is being explained by changes in the predictor is given by the coefficient of determination $R^2 = 17323/29570 = .58583$.

10.3.9
(a) A scatter plot of the response Y against the predictor W (speed) follows.

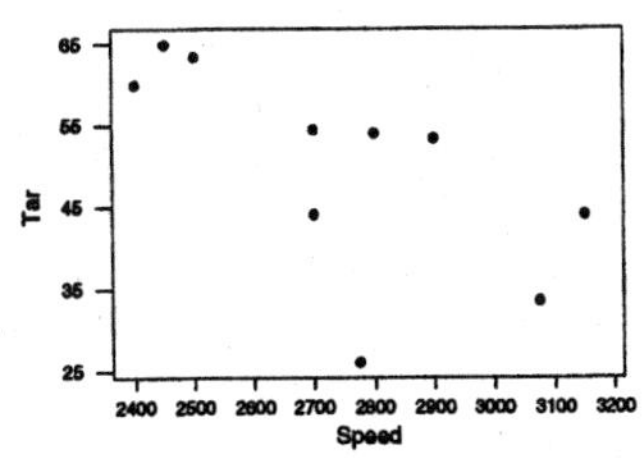

The scatter plot of the response Y against the predictor X (temperature) follows.

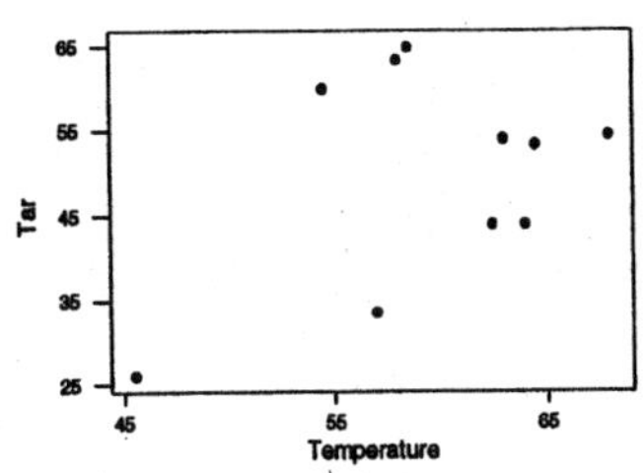

(b) The least-squares estimates of β_1, β_2, and β_3 are given by $b_1 = 87.8$, $b_2 = -0.0406$ and $b_3 = 1.23$. The least-squares equation is then given by $Y = 87.8 - 0.0406W + 1.23X$.
(c) a plot of the standardized residuals against W follows.

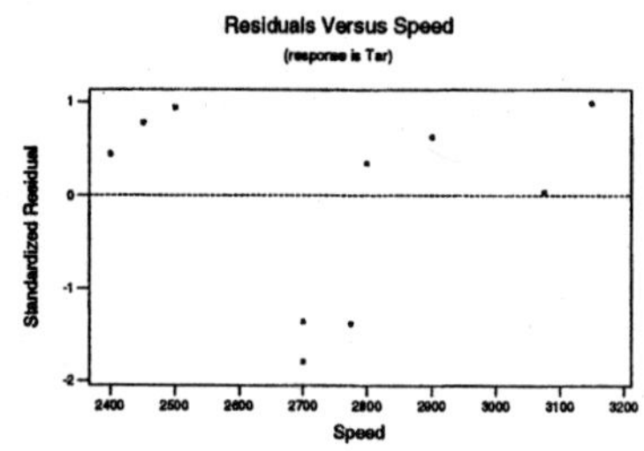

The plot of the standardized residuals against X follows.

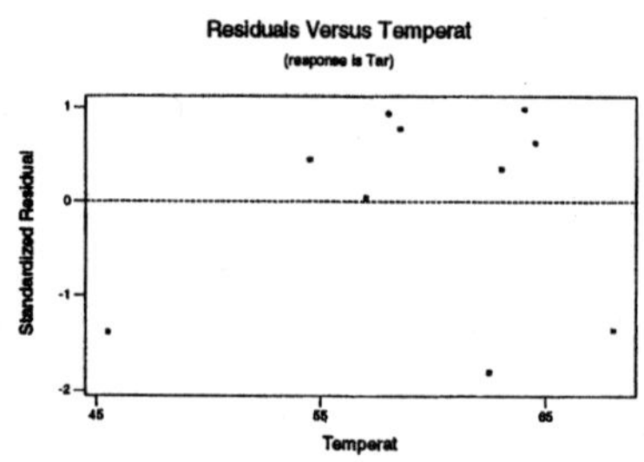

(d) A normal probability plot of the standardized residuals follows.

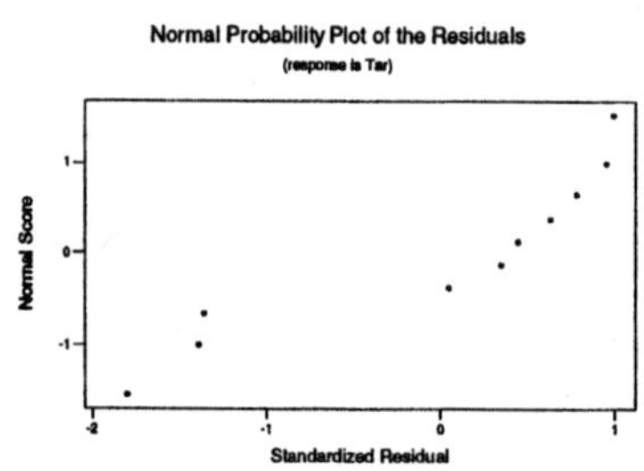

(e) The normal probability plot seems to indicate that the normality assumption is suspect. The other residual plots look reasonable.
(f) The .95-confidence intervals for the regression coefficients are given by
$87.8 \pm 28.98 \cdot (2.3646) = (19.274, 156.33)$ for β_1,
$-0.0406 \pm 0.009142\,(2.3646) = (-0.062217, -0.018983)$for β_2, and
$1.23 \pm 0.3595\,(2.3646) = (.37993, 2.0801)$ for β_3.

(g) The ANOVA table is given below.

Source	Df	SS	MS
W, X	2	1159.29	579.65
Error	7	315.58	45.08
Total	9	1474.87	

The F statistic for testing $H_0 : \beta_2 = \beta_3 = 0$ is given by $F = 579.65/45.08 = 12.858$, and since $F \sim F(2,7)$ under H_0, the P-value equals $P(F > 12.858) = 0.0045$. This provides strong evidence against the null hypothesis of no relationship between the response and the predictors.
(h) The proportion of the observed variation in the response that is being explained by changes in the predictor is given by the coefficient of determination $R^2 = 1159.29/1474.87 = .78603$.
(i) The ANOVA table for testing the null hypothesis $H_0 : \beta_2 = 0$, given that X is in the model, follows.

Source	Df	SS	MS
X	1	271.06	271.06
$W \mid X$	1	888.24	888.24
Error	7	315.58	45.08
Total	9	1474.87	

The F statistic is then $F = 888.24/45.08 = 19.704$, and since $F \sim F(1,7)$ under H_0, the P-value equals $P(F > 19.704) = .00301$, so we have some evidence against the null hypothesis. We conclude that W (speed) has an effect on the response Y (tar), given that X is in the model.
(j) The estimate of the mean of Y when $W = 2750$ and $X = 50.0$ is given by $Y = 87.8 - 0.0406\,(2750) + 1.23\,(50.0) = 37.65$. This is an extrapolation because 50.0 is not in the range of observed X values.

10.3.10
(a) A scatter plot of the response Y against the predictor X follows.

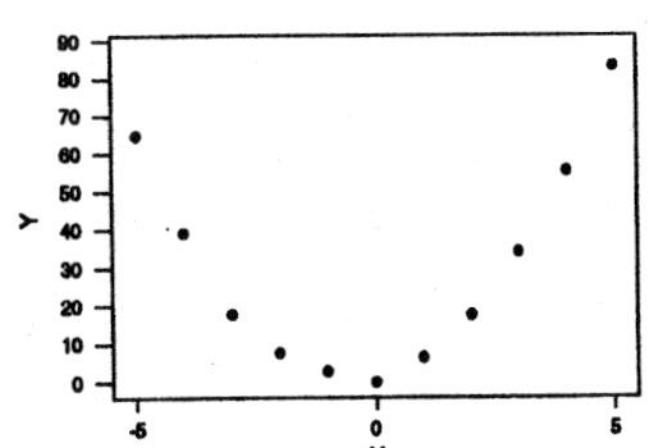

(b) The least-squares estimates of β_1, β_2 and β_3 are given by $b_1 = 0.752$ and $b_2 = 2.10$ and $b_3 = 2.92$. The least-squares line is then given by $Y = 0.752 + 2.10X + 2.92X^2$.
(c) A plot of the standardized residuals against X follows.

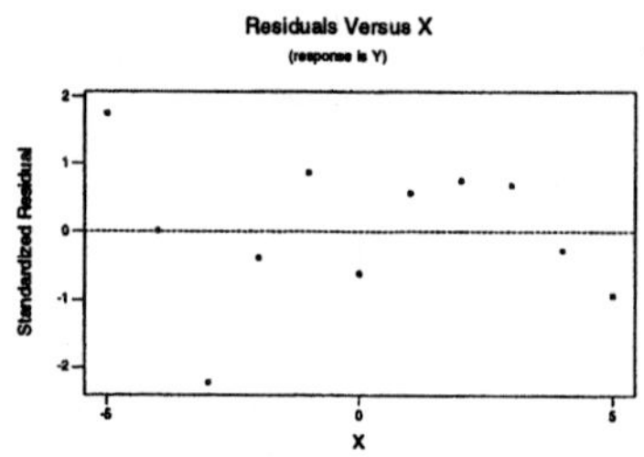

(d) A normal probability plot of the standardized residuals follows.

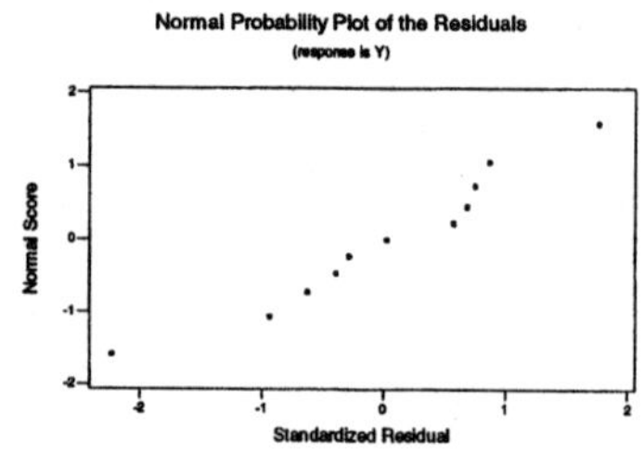

(e) All plots above, for the most part, look reasonable, so the model assumptions seem reasonable.
(f) A .95-confidence intervals for the regression coefficients are given by
$0.752 \pm 0.6553\,(2.306) = (-.75912, 2.2631)$ for β_1,
$2.10 \pm 0.1372\,(2.306) = (1.7836, 2.4164)$ for β_2, and
$2.92 \pm 0.04911\,(2.306) = (2.8068, 3.0332)$ for β_3.
(g) The ANOVA table is given below.

Source	Df	SS	MS
X, X^2	2	7825.5	3912.7
Error	8	16.6	2.1
Total	10	7842.0	

The F statistic for testing $H_0 : \beta_2 = \beta_3 = 0$ is given by $F = 3912.7/2.1 = 1890.54$, and since $F \sim F(2,8)$ under H_0, the P-value equals $P(F > 1890.54) = 0.000$, indicating strong evidence against the null hypothesis of no relationship between the response and the predictors.
(h) The proportion of the observed variation in the response that is being explained by changes in the predictor is given by the coefficient of determination $R^2 = 7825.5/7842.0 = .9979$, which is very high.
(i) The ANOVA table for testing the null hypothesis $H_0 : \beta_3 = 0$ given that X is in the model follows.

Source	Df	SS	MS
X	1	486.2	486.2
$X^2 \mid X$	1	7339.3	7339.3
Error	8	16.6	2.1
Total	10	7842.0	

The F statistic is given by $F = 7339.3/2.1 = 3494.9$, and since $F \sim F(1, 8)$ under H_0, the P-value equals $P(F > 3494.9) = 0.000$, so we have strong evidence against the null hypothesis. We conclude that X^2 has an effect on the response, given that X is in the model.
(j) We predict Y at $X = 6$ by $29.9991 + 2.10236\,(6) = 42.613$ using the simple linear model and by $0.752 + 2.10\,(6) + 2.92\,(6^2) = 118.47$ using the linear model containing the linear and quadratic terms. So there is a substantial difference in these predictions.

Problems

10.3.11 First, note that the mean of this distribution is given by $(1/2)^2 + (1/2)\left((\theta - 2)/2\right) = (\theta - 1)/4$ and that this value is in the interval $(7/4, \infty)$. Therefore, the least-squares estimate is given by $\bar{x}$ whenever $\bar{x} \in (7/4, \infty)$ and is equal to $7/4$ whenever $\bar{x} \le 7/4$.

10.3.12 Since $\sum_{i=1}^{n} (x_i - \bar{x})^2 = 0$, we must have $(x_i - \bar{x})^2 = 0$, so $x_i = \bar{x}$ for every i and all the x_i are equal to the same value, say x. Then we need to estimate the conditional mean of Y at $X = x$ based on a sample $(y_1, \ldots, y_n)$ from this distribution. The model says that this conditional mean is of the form $E(Y \,|\, X = x) = \beta_1 + \beta_2 x$, where $\beta_1, \beta_2 \in R^1$. Therefore, $E(Y \,|\, X = x)$ can be any value in R^1, and the least-squares estimate is given by the sample average $\bar{y}$.

10.3.13 For convenience we write $\text{Cov}(A, B \,|\, X_1 = x_1, \ldots, X_n = x_n) = \text{Cov}(A, B)$. By Theorem 3.3.2 (linearity of covariance) we have

$$\begin{aligned}\text{Cov}(Y_i - B_1 - B_2 x_i, Y_j - B_1 - B_2 x_j) &= \text{Cov}(Y_i, Y_j) - \text{Cov}(Y_i, B_1 + B_2 x_j) \\ &\quad - \text{Cov}(Y_j, B_1 + B_2 x_i) + \text{Cov}(B_1 + B_2 x_i, B_1 + B_2 x_j).\end{aligned}$$

Now $\text{Cov}(Y_i, Y_j) = \sigma^2 \delta_{ij}$, where $\delta_{ij} = 1$ when $i = j$ and is 0 otherwise. Also,

$$\begin{aligned}\text{Cov}(Y_i, B_2) &= \text{Cov}\left(Y_i, \frac{\sum_{j=1}^{n} (Y_j - \bar{Y})(x_j - \bar{x})}{\sum_{j=1}^{n} (x_j - \bar{x})^2}\right) \\ &= \frac{1}{\sum_{i=1}^{n} (x_i - \bar{x})^2} \text{Cov}\left(Y_i, \sum_{j=1}^{n} x_j Y_j - \bar{x} \sum_{j=1}^{n} Y_j - \bar{Y} \sum_{j=1}^{n} x_j + n\bar{x}\bar{Y}\right) \\ &= \frac{1}{\sum_{i=1}^{n} (x_i - \bar{x})^2} \text{Cov}\left(Y_i, \sum_{j=1}^{n} x_j Y_j - \bar{x} n \bar{Y} - n\bar{x}\bar{Y} \sum_{j=1}^{n} x_j + n\bar{x}\bar{Y}\right) \\ &= \frac{1}{\sum_{i=1}^{n} (x_i - \bar{x})^2} \text{Cov}\left(Y_i, \sum_{j=1}^{n} x_j Y_j - \bar{x} n \bar{Y}\right) \\ &= \frac{1}{\sum_{i=1}^{n} (x_i - \bar{x})^2} \left(x_i \sigma^2 - n\bar{x}\, \text{Cov}(Y_i, \bar{Y})\right) \\ &= \frac{1}{\sum_{i=1}^{n} (x_i - \bar{x})^2} \left(x_i \sigma^2 - \bar{x}\sigma^2\right) = \frac{\sigma^2 (x_i - \bar{x})}{\sum_{i=1}^{n} (x_i - \bar{x})^2}.\end{aligned}$$

and $\text{Cov}(Y_i, b_1) = \text{Cov}(Y_i, \bar{Y} - B_2\bar{x}) = \text{Cov}(Y_i, \bar{Y}) - \bar{x}\,\text{Cov}(Y_i, B_2) = \sigma^2/n - \sigma^2 (x_i - \bar{x})\,\bar{x}/\sum_{i=1}^n (x_i - \bar{x})^2$. Therefore,

$$\text{Cov}\left(Y_i, B_1 + B_2 x_j\right) = \frac{\sigma^2}{n} - \frac{\sigma^2 (x_i - \bar{x})\,\bar{x}}{\sum_{i=1}^n (x_i - \bar{x})^2} + \frac{\sigma^2 (x_i - \bar{x})\,x_j}{\sum_{i=1}^n (x_i - \bar{x})^2}$$

$$= \sigma^2 \left(\frac{1}{n} + \frac{(x_i - \bar{x})(x_j - \bar{x})}{\sum_{i=1}^n (x_i - \bar{x})^2} \right) = \text{Cov}\left(Y_j, B_1 + B_2 x_i\right).$$

Also, using Theorem 10.3.3 we have that

$$\begin{aligned}
&\text{Cov}\left(B_1 + B_2 x_i, B_1 + B_2 x_j\right) \\
&= \text{Var}(B_1) + x_i x_j \,\text{Var}(B_2) + (x_i + x_j)\,\text{Cov}(B_1, B_2) \\
&= \sigma^2 \left(\frac{1}{n} + \frac{\bar{x}^2}{\sum_{i=1}^n (x_i - \bar{x})^2} + \frac{x_i x_j}{\sum_{i=1}^n (x_i - \bar{x})^2} - \frac{(x_i + x_j)\,\bar{x}}{\sum_{i=1}^n (x_i - \bar{x})^2} \right) \\
&= \sigma^2 \left(\frac{1}{n} + \frac{(x_i - \bar{x})(x_j - \bar{x})}{\sum_{i=1}^n (x_i - \bar{x})^2} \right).
\end{aligned}$$

All together this implies that

$$\begin{aligned}
&\text{Cov}\left(Y_i - B_1 - B_2 x_i, Y_j - B_1 - B_2 x_j\right) \\
&= \sigma^2 \delta_{ij} - 2\sigma^2 \left(\frac{1}{n} + \frac{(x_i - \bar{x})(x_j - \bar{x})}{\sum_{i=1}^n (x_i - \bar{x})^2} \right) + \sigma^2 \left(\frac{1}{n} + \frac{(x_i - \bar{x})(x_j - \bar{x})}{\sum_{i=1}^n (x_i - \bar{x})^2} \right) \\
&= \sigma^2 \delta_{ij} - \sigma^2 \left(\frac{1}{n} + \frac{(x_i - \bar{x})(x_j - \bar{x})}{\sum_{i=1}^n (x_i - \bar{x})^2} \right).
\end{aligned}$$

10.3.14 We have that

$$\begin{aligned}
&Y_i - (B_1 + B_2 x_i) = Y_i - \left(\bar{Y} - B_2\bar{x} - B_2 x_i\right) = Y_i - \bar{Y} - B_2 (x_i - \bar{x}) \\
&= Y_i - \bar{Y} - (x_i - \bar{x}) \frac{\sum_{j=1}^n (x_j - \bar{x})\left(Y_j - \bar{Y}\right)}{\sum_{j=1}^n (x_j - \bar{x})^2}
\end{aligned}$$

and we note that this is a linear combination of the independent normals $Y_1, \ldots, Y_n$. Therefore, by Theorem 4.6.1 we have that $Y_i - (B_1 + B_2 x_i)$, given $X_1 = x_1, \ldots, X_n = x_n$, is normally distributed with mean

$$\begin{aligned}
&E\left(Y_i - (B_1 + B_2 x_i) \,|\, X_1 = x_1, \ldots, X_n = x_n\right) \\
&= E\left(Y_i \,|\, X_1 = x_1, \ldots, X_n = x_n\right) - E\left(B_1 \,|\, X_1 = x_1, \ldots, X_n = x_n\right) \\
&\quad - E\left(B_2 \,|\, X_1 = x_1, \ldots, X_n = x_n\right) x_i \\
&= \beta_1 + \beta_2 x_i - \beta_1 - \beta_2 x_i = 0,
\end{aligned}$$

and with variance (using Problem 10.3.13 with $i = j$)

$$\begin{aligned}
&\text{Var}\left(Y_i - (B_1 + B_2 x_i) \,|\, X_1 = x_1, \ldots, X_n = x_n\right) \\
&= \sigma^2 \left(1 - \frac{1}{n} - \frac{(x_i - \bar{x})^2}{\sum_{i=1}^n (x_i - \bar{x})^2} \right).
\end{aligned}$$

Therefore,

$$\frac{Y_i - (B_1 + B_2 x_i)}{\sigma\left(1 - \frac{1}{n} - \frac{(x_i-\bar{x})^2}{\sum_{i=1}^n (x_i-\bar{x})^2}\right)^{1/2}} \sim N(0,1)$$

as claimed.

10.3.15 We have that

$$\begin{aligned} Y - (B_1 + B_2 x) &= Y - \left(\bar{Y} - B_2\bar{x} - B_2 x\right) = Y - \bar{Y} - B_2\left(x - \bar{x}\right) \\ &= Y - \bar{Y} - (x_i - \bar{x})\frac{\sum_{j=1}^n (x_j - \bar{x})\left(Y_j - \bar{Y}\right)}{\sum_{j=1}^n (x_j - \bar{x})^2} \end{aligned}$$

and we note that this is a linear combination of the independent normals $Y, Y_1, \ldots, Y_n$. Therefore, by Theorem 4.6.1 we have that $Y - (B_1 + B_2 x)$, given $X = x, X_1 = x_1, \ldots, X_n = x_n$, is normally distributed with mean

$$\begin{aligned} &E\left(Y - (B_1 + B_2 x) \,|\, X = x, X_1 = x_1, \ldots, X_n = x_n\right) \\ &= E\left(Y \,|\, X = x, X_1 = x_1, \ldots, X_n = x_n\right) \\ &\quad - E\left(B_1 \,|\, X = x, X_1 = x_1, \ldots, X_n = x_n\right) \\ &\quad - E\left(B_2 \,|\, X = x, X_1 = x_1, \ldots, X_n = x_n\right) x \\ &= E\left(Y \,|\, X = x\right) - E\left(B_1 \,|\, X_1 = x_1, \ldots, X_n = x_n\right) \\ &\quad - E\left(B_2 \,|\, X_1 = x_1, \ldots, X_n = x_n\right) x \\ &= \beta_1 + \beta_2 x - \beta_1 - \beta_2 x = 0, \end{aligned}$$

and with variance (using Corollary 10.3.1)

$$\begin{aligned} &\text{Var}\left(Y - (B_1 + B_2 x) \,|\, X = x, X_1 = x_1, \ldots, X_n = x_n\right) \\ &= \text{Var}\left(Y \,|\, X = x\right) + \text{Var}\left(B_1 + B_2 x \,|\, X_1 = x_1, \ldots, X_n = x_n\right) \\ &= \sigma^2\left(1 + \frac{1}{n} + \frac{(x - \bar{x})^2}{\sum_{i=1}^n (x_i - \bar{x})^2}\right). \end{aligned}$$

Also, $Y - (B_1 + B_2 x)$ is independent of $(n-2)S^2/\sigma^2 \sim \chi^2(n-2)$, so by Definition 4.6.2

$$\begin{aligned} T &= \frac{Y - (B_1 + B_2 x)}{\sigma\left(1 + \frac{1}{n} + \frac{(x_i-\bar{x})^2}{\sum_{i=1}^n (x_i-\bar{x})^2}\right)^{1/2}} \Big/ \sqrt{\frac{(n-2)S^2}{(n-2)\sigma^2}} \\ &= \frac{Y - (B_1 + B_2 x)}{S\left(1 + \frac{1}{n} + \frac{(x_i-\bar{x})^2}{\sum_{i=1}^n (x_i-\bar{x})^2}\right)^{1/2}} \sim t\,(n-2). \end{aligned}$$

Therefore,

$$\gamma = P\left(-t_{\frac{1+\gamma}{2}}\,(n-2) < T < t_{\frac{1+\gamma}{2}}\,(n-2) \,\Big|\, X = x, X_1 = x_1, \ldots, X_n = x_n\right),$$

so the probability that

$$Y \in \left[B_1 + B_2 x \pm S \left(1 - \frac{1}{n} - \frac{(x_i - \bar{x})^2}{\sum_{i=1}^n (x_i - \bar{x})^2} \right)^{1/2} t_{\frac{1+\gamma}{2}} (n-2) \right]$$

is equal to γ.

10.3.16

(a) Putting $b = \sum_{i=1}^n x_i y_i / \sum_{i=1}^n x_i^2$, we have that

$$\begin{aligned} &\sum_{i=1}^n (y_i - \beta x_i)^2 = \sum_{i=1}^n (y_i - b x_i + b x_i - \beta x_i)^2 \\ &= \sum_{i=1}^n (y_i - b x_i)^2 + 2(b - \beta) \sum_{i=1}^n (y_i - b x_i) x_i + (b - \beta)^2 \sum_{i=1}^n x_i^2 \\ &= \sum_{i=1}^n (y_i - b x_i)^2 + (b - \beta)^2 \sum_{i=1}^n x_i^2 \end{aligned}$$

since $\sum_{i=1}^n (y_i - b x_i) x_i = \sum_{i=1}^n x_i y_i - b \sum_{i=1}^n x_i^2 = 0$, and this is clearly minimized, as a function of β, by b.

(b) We have that

$$\begin{aligned} &E(B \mid X_1 = x_1, \ldots, X_n = x_n) = \frac{\sum_{i=1}^n x_i E(Y_i \mid X_1 = x_1, \ldots, X_n = x_n)}{\sum_{i=1}^n x_i^2} \\ &= \frac{\sum_{i=1}^n x_i (\beta x_i)}{\sum_{i=1}^n x_i^2} = \beta \frac{\sum_{i=1}^n x_i^2}{\sum_{i=1}^n x_i^2} = \beta \end{aligned}$$

and

$$\begin{aligned} \mathrm{Var}(B \mid X_1 = x_1, \ldots, X_n = x_n) &= \frac{\sum_{i=1}^n x_i^2 \,\mathrm{Var}(Y_i \mid X_1 = x_1, \ldots, X_n = x_n)}{\left(\sum_{i=1}^n x_i^2\right)^2} \\ &= \frac{\sigma^2 \sum_{i=1}^n x_i^2}{\left(\sum_{i=1}^n x_i^2\right)^2} = \frac{\sigma^2}{\sum_{i=1}^n x_i^2}. \end{aligned}$$

(c) We have that

$$\begin{aligned} &E(S^2 \mid X_1 = x_1, \ldots, X_n = x_n) \\ &= \frac{1}{n-1} \sum_{i=1}^n E\left((Y_i - B x_i)^2 \mid X_1 = x_1, \ldots, X_n = x_n \right) \end{aligned}$$

and

$$
\begin{aligned}
&E\left((Y_i - Bx_i)^2 \,|\, X_1 = x_1, \ldots, X_n = x_n\right) \\
&= E\left((Y_i - \beta x_i + \beta x_i - Bx_i)^2 \,|\, X_1 = x_1, \ldots, X_n = x_n\right) \\
&= \operatorname{Var}(Y_i \,|\, X_1 = x_1, \ldots, X_n = x_n) + x_i^2 \operatorname{Var}(B \,|\, X_1 = x_1, \ldots, X_n = x_n) \\
&\quad - 2x_i \operatorname{Cov}(Y_i, B \,|\, X_1 = x_1, \ldots, X_n = x_n) \\
&= \sigma^2 + \frac{\sigma^2 x_i^2}{\sum_{i=1}^n x_i^2} - \frac{2x_i^2}{\sum_{i=1}^n x_i^2} \operatorname{Var}(Y_i \,|\, X_1 = x_1, \ldots, X_n = x_n) \\
&= \sigma^2 - \frac{\sigma^2 x_i^2}{\sum_{i=1}^n x_i^2} = \sigma^2 \left(1 - \frac{x_i^2}{\sum_{i=1}^n x_i^2}\right).
\end{aligned}
$$

Combining these, we obtain $E\left(S^2 \,|\, X_1 = x_1, \ldots, X_n = x_n\right) = \sigma^2$.

(d) We have that

$$
\begin{aligned}
&\sum_{i=1}^n y_i^2 = \sum_{i=1}^n (y_i - bx_i + bx_i)^2 = \sum_{i=1}^n (y_i - bx_i)^2 + 2b \sum_{i=1}^n (y_i - bx_i)\, x_i + b^2 \sum_{i=1}^n x_i^2 \\
&= \sum_{i=1}^n (y_i - bx_i)^2 + b^2 \sum_{i=1}^n x_i^2.
\end{aligned}
$$

Here we have that $\sum_{i=1}^n (y_i - bx_i)^2$ is the error sum of squares and $b^2 \sum_{i=1}^n x_i^2$ is the regression sum of squares. The coefficient of determination is then given by $R^2 = b^2 \sum_{i=1}^n x_i^2 / \sum_{i=1}^n y_i^2$ and this is the proportion of the total variation observed in Y (as measured by $\sum_{i=1}^n y_i^2$) due to changes in X.

(e) Since B is a linear combination of independent normal variables we have that B is normally distributed with mean given by (part (b)) β and variance (part (b)) given by $\sigma^2 / \sum_{i=1}^n x_i^2$.

(f) We have that $(B - \beta) / \sigma \left(\sum_{i=1}^n x_i^2\right)^{-1/2} \sim N(0,1)$ independent of $(n-1)S^2/\sigma^2 \sim \chi^2(n-1)$, so $(B - \beta) / S \left(\sum_{i=1}^n x_i^2\right)^{-1/2} \sim t(n-1)$. Now there is no relationship between X and Y if and only if $\beta = 0$, so we test $H_0 : \beta = 0$ by computing the P-value $P(|T| > |b/s \left(\sum_{i=1}^n x_i^2\right)^{-1/2}|)$, where $T \sim t(n-1)$.

(g) We have that $y_i = bx_i + (y_i - bx_i)$ and when the model is correct $y_i - bx_i$ is a value from a distribution with mean 0 and variance (see part (b)) $\sigma^2 \left(1 - x_i^2 / \sum_{i=1}^n x_i^2\right)$. Therefore, the ith standardized residual is given by $(y_i - bx_i) / s \left(1 - x_i^2 / \sum_{i=1}^n x_i^2\right)^{1/2}$. We can plot these in residual plots and normal probability plots to see if they look like samples from the $N(0,1)$ distribution.

10.3.17 First, we should express the β's in terms of the α's as follows $\beta_2 = \alpha_2$ and $\beta_1 = \alpha_1 - \alpha_2 \bar{x}$. Substituting those into the sum of squares, and noting that

$\sum_{i=1}^{n}(x_i - \bar{x}) = \sum(y_i - \bar{y}) = 0$, we get

$$\begin{aligned}
&\sum_{i=1}^{n}(y_i - \beta_1 - \beta_2 x_i)^2 \\
&= \sum_{i=1}^{n}(y_i - \alpha_1 - \alpha_2(x_i - \bar{x}))^2 \\
&= \sum_{i=1}^{n}(y_i - \bar{y} + \bar{y} - \alpha_1 - \alpha_2(x_i - \bar{x}))^2 \\
&= \sum_{i=1}^{n}(y_i - \bar{y})^2 + 2(\bar{y} - \alpha_1)\sum_{i=1}^{n}(y_i - \bar{y}) - 2\alpha_2\sum_{i=1}^{n}(y_i - \bar{y})(x_i - \bar{x}) \\
&\quad + \sum_{i=1}^{n}(\bar{y} - \alpha_1 - \alpha_2(x_i - \bar{x}))^2 \\
&= \sum_{i=1}^{n}(y_i - \bar{y})^2 - 2\alpha_2\sum_{i=1}^{n}(y_i - \bar{y})(x_i - \bar{x}) + n(\bar{y} - \alpha_1)^2 \\
&\quad - 2\alpha_2(\bar{y} - \alpha_1)\sum_{i=1}^{n}(x_i - \bar{x}) + \alpha_2^2\sum_{i=1}^{n}(x_i - \bar{x})^2 \\
&= \sum_{i=1}^{n}(y_i - \bar{y})^2 - 2\alpha_2\sum_{i=1}^{n}(y_i - \bar{y})(x_i - \bar{x}) + n(\bar{y} - \alpha_1)^2 + \alpha_2^2\sum_{i=1}^{n}(x_i - \bar{x})^2
\end{aligned}$$

as claimed.

Clearly, this is minimized for α_1, independently of α_2, by selecting $\alpha_1 = \bar{y}$. Then we must minimize

$$\begin{aligned}
&-2\alpha_2\sum_{i=1}^{n}(y_i - \bar{y})(x_i - \bar{x}) + \alpha_2^2\sum_{i=1}^{n}(x_i - \bar{x})^2 \\
&= \sum_{i=1}^{n}(x_i - \bar{x})^2\left(\alpha_2 - \frac{\sum_{i=1}^{n}(y_i - \bar{y})(x_i - \bar{x})}{\sum_{i=1}^{n}(x_i - \bar{x})^2}\right)^2 - \frac{\left(\sum_{i=1}^{n}(y_i - \bar{y})(x_i - \bar{x})\right)^2}{\sum_{i=1}^{n}(x_i - \bar{x})^2}
\end{aligned}$$

for α_2. Clearly, this is minimized by taking $\alpha_2 = \sum_{i=1}^{n}(y_i - \bar{y})(x_i - \bar{x}) / \sum_{i=1}^{n}(x_i - \bar{x})^2$ as claimed.

10.3.18 The likelihood function is given by

$$\left(2\pi\sigma^2\right)^{-n/2}\exp\left(-\frac{c_y^2 - c_x^2 a^2}{2\sigma^2}\right)\exp\left(-\frac{n}{2\sigma^2}(\alpha_1 - \bar{y})^2\right)\exp\left(-\frac{c_x^2}{2\sigma^2}(\alpha_2 - a)^2\right)$$

The posterior distribution of α_1, given σ^2, is then proportional to

$$\begin{aligned}
&\exp\left(-\frac{n}{2\sigma^2}(\alpha_1-\bar{y})^2-\frac{1}{2\tau_1^2\sigma^2}(\alpha_1-\mu_1)^2\right)\\
&\propto \exp\left(-\frac{1}{2\sigma^2}\left\{\left(n+\frac{1}{\tau_1^2}\right)\alpha_1^2-2\left(n\bar{y}+\frac{\mu_1}{\tau_1^2}\right)\alpha_1\right\}\right)\\
&\propto \exp\left(-\frac{1}{2\sigma^2}\left(n+\frac{1}{\tau_1^2}\right)\left(\alpha_1-\left(n+\frac{1}{\tau_1^2}\right)^{-1}\left(n\bar{y}+\frac{\mu_1}{\tau_1^2}\right)\right)^2\right)
\end{aligned}$$

and we recognize this as being proportional to the density of a $N((n+1/\tau_1^2)^{-1}(n\bar{y}+\mu_1/\tau_1^2), (n+1/\tau_1^2)^{-1}\sigma^2)$ distribution.

Also, the posterior distribution of α_2, given σ^2, is then proportional to

$$\begin{aligned}
&\exp\left(-\frac{c_x^2}{2\sigma^2}(\alpha_2-a)^2-\frac{1}{2\tau_2^2\sigma^2}(\alpha_2-\mu_2)^2\right)\\
&\propto \exp\left(-\frac{1}{2\sigma^2}\left\{\left(c_x^2+\frac{1}{\tau_2^2}\right)\alpha_2^2-\left(c_x^2a+\frac{\mu_2}{\tau_2^2}\right)\alpha_2\right\}\right)\\
&\propto \exp\left(-\frac{1}{2\sigma^2}\left\{\left(c_x^2+\frac{1}{\tau_2^2}\right)\left(\alpha_2-\left(c_x^2+\frac{1}{\tau_2^2}\right)^{-1}\left(c_x^2a+\frac{\mu_2}{\tau_2^2}\right)\right)^2\right\}\right)
\end{aligned}$$

and we recognize this as being proportional to the density of a $N((c_x^2+1/\tau_2^2)^{-1}(c_x^2a+\mu_2/\tau_2^2), (c_x^2+1/\tau_2^2)^{-1}\sigma^2)$ distribution.

Finally, the posterior density of $1/\sigma^2$ is proportional to $(1/\sigma^2)^{\frac{n}{2}+\kappa-1}\exp(-v_{xy}/\sigma^2)$, where

$$v_{xy}=\frac{1}{2}\left\{\begin{array}{c}(c_y^2-a^2c_x^2)+\left[n\bar{y}^2+\frac{\mu_1^2}{\tau_1^2}-\left(n+\frac{1}{\tau_1^2}\right)^{-1}\left(n\bar{y}+\frac{\mu_1}{\tau_1^2}\right)^2\right]\\ +\left[a^2c_x^2+\frac{\mu_2^2}{\tau_2^2}-\left(c_x^2+\frac{1}{\tau_2^2}\right)^{-1}\left(c_x^2a+\frac{\mu_2}{\tau_2^2}\right)^2\right]\end{array}\right\}+v$$

and we recognize this as being proportional to the density of a Gamma$(\kappa+n/2, v_{xy})$ distribution. Therefore, we established that the posterior distributions above are from the same family of distribution as the prior and therefore this prior is conjugate.

10.3.19 When $\tau_1\to\infty, \tau_2\to\infty$ and $\nu\to 0$ and the posterior converges to

$$\alpha_1\,|\,\alpha_2,\sigma^2\sim N(\bar{y},\sigma^2/n),\ \alpha_2\,|\,\sigma^2\sim N(a,\frac{\sigma^2}{c_x^2}),\ 1/\sigma^2\sim \text{Gamma}\left(\kappa+\frac{n}{2},\nu_{xy}\right)$$

where $\nu_{xy}=\left\{c_y^2-a^2c_x^2\right\}/2$. Then the marginal posterior density of α_1 is proportional to

$$\begin{aligned}
&\int_0^\infty\left(\frac{1}{\sigma^2}\right)^{\frac{1}{2}}\exp\left\{-\frac{n}{2\sigma^2}(\alpha_1-\bar{y})^2\right\}\left(\frac{1}{\sigma^2}\right)^{\kappa+\frac{n}{2}-1}\exp\left\{-\frac{\nu_{xy}}{\sigma^2}\right\}d\left(\frac{1}{\sigma^2}\right)\\
&=\int_0^\infty\left(\frac{1}{\sigma^2}\right)^{\kappa+\frac{n}{2}-\frac{1}{2}}\exp\left\{-\left(\nu_{xy}+\frac{n}{2}(\alpha_1-\bar{y})^2\right)\frac{1}{\sigma^2}\right\}d\left(\frac{1}{\sigma^2}\right)
\end{aligned}$$

Making the change of variable $1/\sigma^2 \to w$, where $w = \left(\nu_{xy} + \frac{n}{2}(\alpha_1 - \bar{y})^2\right)/\sigma^2$, in the above integral, shows that the marginal posterior density of α_1 is proportional to $\left(1 + \frac{n}{2\nu_{xy}}(\alpha_1 - \bar{y})^2\right)^{-\frac{2\kappa+n+1}{2}}$. This establishes ((Problem 4.6.14)) that the posterior distribution of α_1 is given by $\sqrt{2\kappa + n}\frac{\alpha_1 - \bar{y}}{\sqrt{2\nu_{xy}/n}} \sim t(2\kappa + n)$.

Challenge

10.3.20 Let μ be the mean and σ^2 be the variance of the distribution of X. By the SLLN we have that $\bar{X} \overset{a.s.}{\to} \mu$ so, of necessity, $\bar{X}^2 \overset{a.s.}{\to} \mu^2$. Further, $\frac{1}{n}\sum_{i=1}^n (X_i - \bar{X})^2 = \frac{1}{n}\sum_{i=1}^n X_i^2 - \bar{X}^2 \overset{a.s.}{\to} \sigma^2 + \mu^2 - \mu^2 = \sigma^2$ since (again by the SSLN) $n^{-1}\sum_{i=1}^n X_i^2 \overset{a.s.}{\to} E(X^2) = \sigma^2 + \mu^2$. Also, for any random variable Y, we have that $Y/\sqrt{n} \overset{a.s.}{\to} 0$. Therefore,

$$\frac{X_i - \bar{X}}{\sqrt{\sum_{i=1}^n (X_i - \bar{X})^2}} = \frac{(X_i - \bar{X})/\sqrt{n}}{\sqrt{\frac{1}{n}\sum_{i=1}^n (X_i - \bar{X})^2}} \overset{a.s.}{\to} \frac{0}{\sigma} = 0.$$

10.4 Quantitative Response and Categorical Predictors

Exercises

10.4.1
(a) A side-by-side boxplot of the data follows.

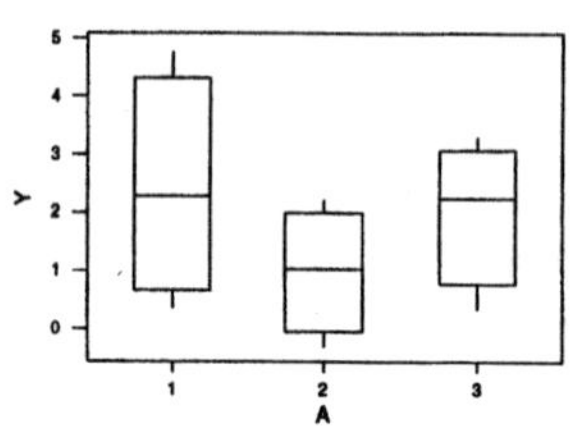

(b) A plot of the standardized residuals against A follows.

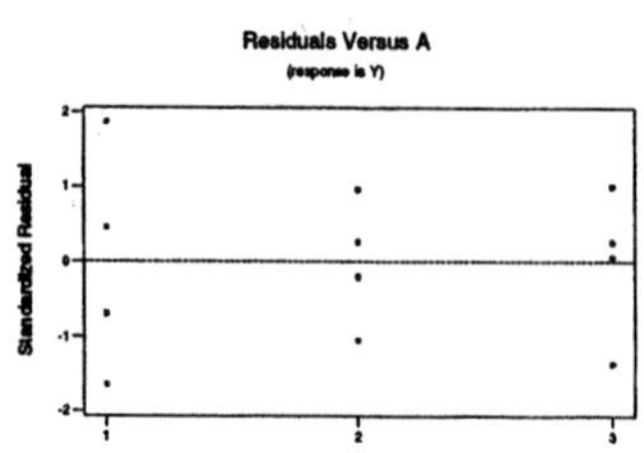

A normal probability plot of the standardized residuals follows.

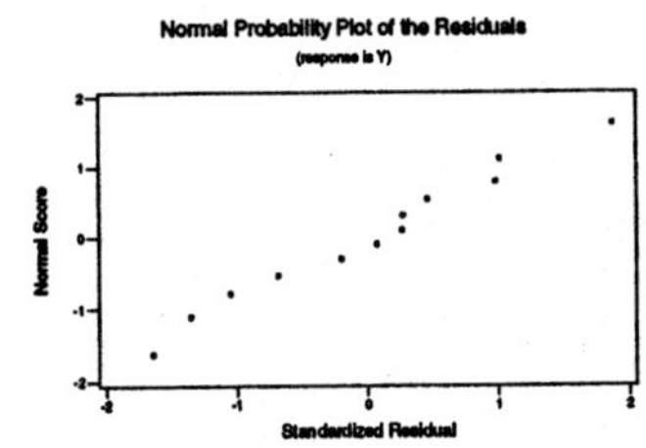

Both plots look reasonable, indicating no serious concerns about the correctness of the model assumptions.
(c) The ANOVA table for testing $H_0 : \beta_1 = \beta_2 = \beta_3$ is given below.

Source	Df	SS	MS
A	2	4.37	2.18
Error	9	18.85	2.09
Total	11	23.22	

The F statistic for testing H_0 is given by $F = 2.18/2.09 = 1.0431$, and since $F \sim F(2,9)$ under H_0, we have P-value $P(F > 1.0431) = .39135$. Therefore, we don't have evidence against the null hypothesis of no difference among the conditional means of Y given X.
(d) Since we didn't find any relationship between Y and X, there is no need to calculate these confidence intervals.

10.4.2
(a) A side-by-side boxplot of the data follows.

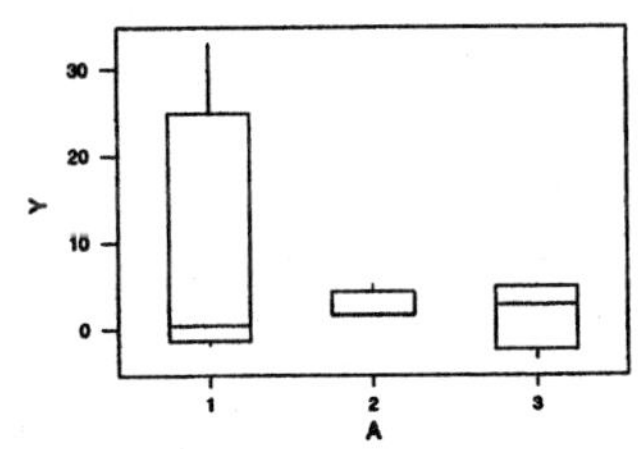

(b) A plot of the standardized residuals against A follows.

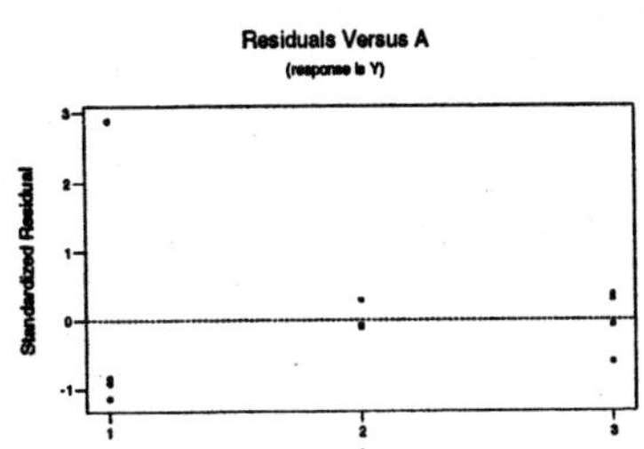

A normal probability plot of the standardized residuals is given below.

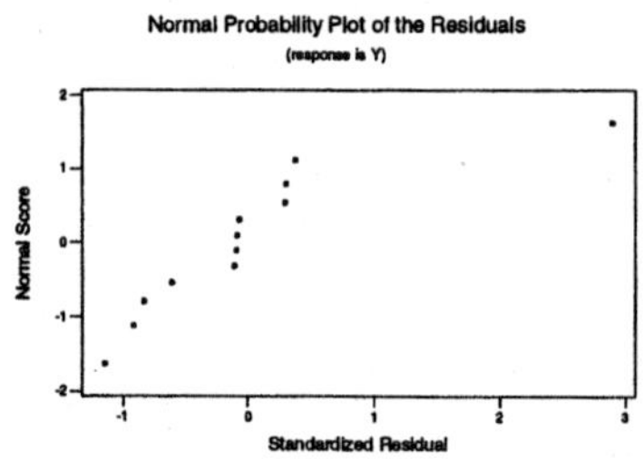

Both plots indicate a problem with the model assumptions.
(c) A possible way to "fix" this problem is to remove the extreme observation in the first category, namely 33.07. After removing this value, we get the following plot of the standardized residuals against A and normal probability plot of the standardized residuals. These look much more reasonable.

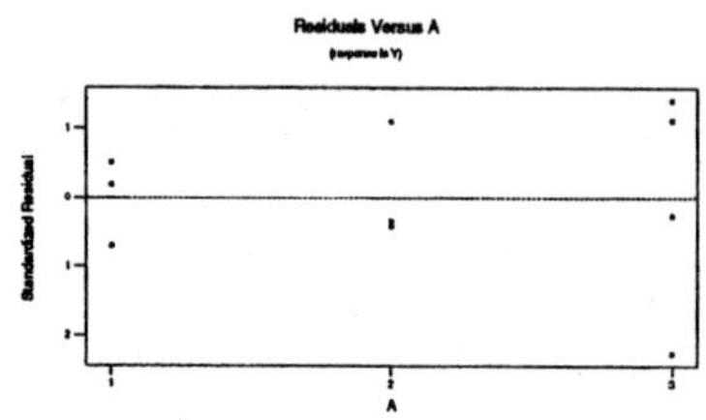

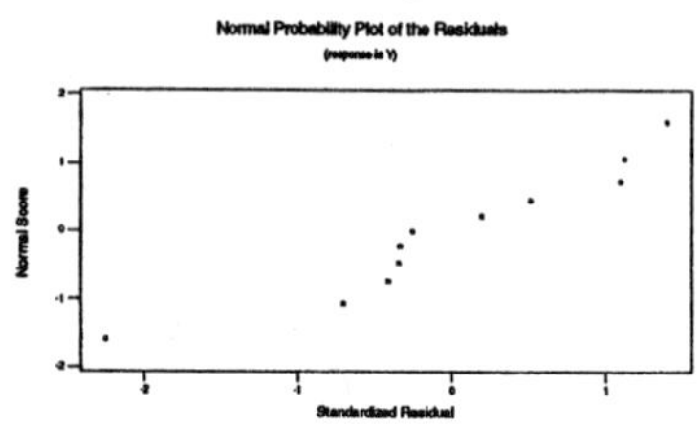

(d) The ANOVA table for testing $H_0 : \beta_1 = \beta_2 = \beta_3$, after removing the outlier, is given below.

Source	Df	SS	MS
A	2	14.840	7.42
Error	8	58.904	7.363
Total	11	73.744	

The F statistics for testing H_0 is given by $F = 7.41/7.363 = 1.01$, and since $F \sim F(2, 8)$ under H_0, the P-value equals $P(F > 1.01) = .407$. Therefore, we have no evidence against the null hypothesis of no difference among the conditional means of Y given X.

(e) There is no need to compute these confidence intervals as we found no evidence of a relationship between the response and the predictor.

10.4.3

(a) A side-by-side boxplot of the data follows.

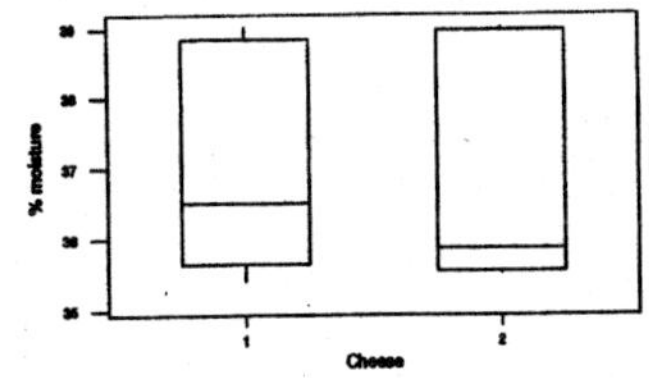

(b) A plot of the standardized residuals against cheese follows.

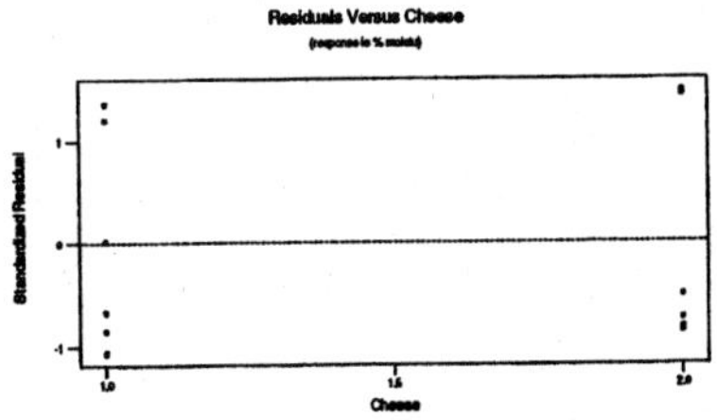

A normal probability plot of the standardized residuals follows.

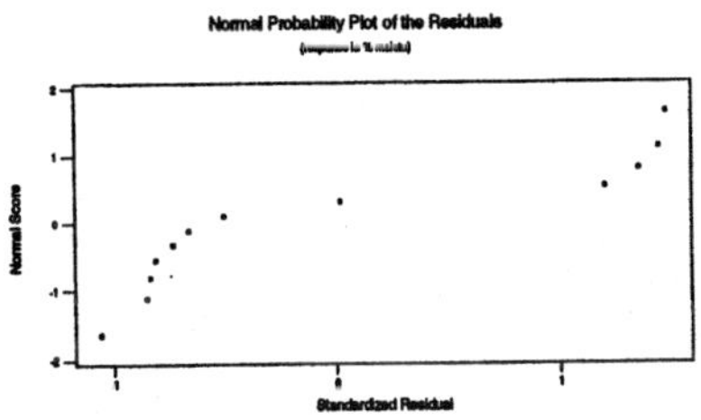

Both plots indicate a possible problem with the model assumptions.

(c) The ANOVA table for testing $H_0 : \beta_1 = \beta_2$ is given below.

Source	Df	SS	MS
Cheese	1	0.114	0.114
Error	10	26.865	2.686
Total	11	26.979	

The F statistic for testing H_0 is given by $F = .114/2.686 = .04$ and, since $F \sim F(1, 10)$ under H_0, the P-value equals $P(F > .04) = .841$. Therefore, we don't have any evidence against the null hypothesis of no difference among the conditional means of Y given Cheese.

10.4.4

(a) A side-by-side boxplot of the data follows.

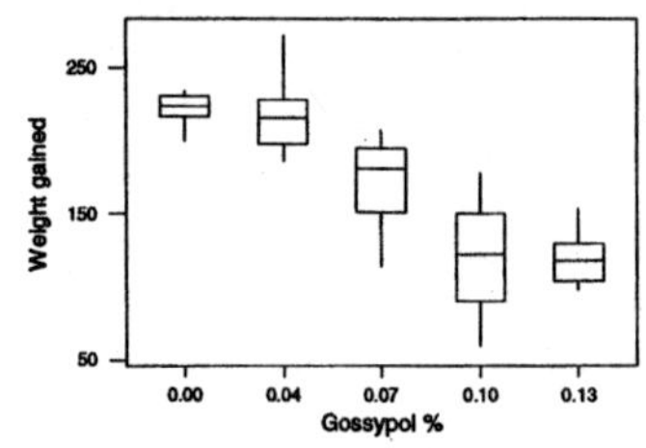

Some of the boxplots don't look very symmetrical, which should be the case for normal samples. So these graphs are some evidence that the normality assumption may not be appropriate.
(b) A normal probability plot of the standardized residuals follows.

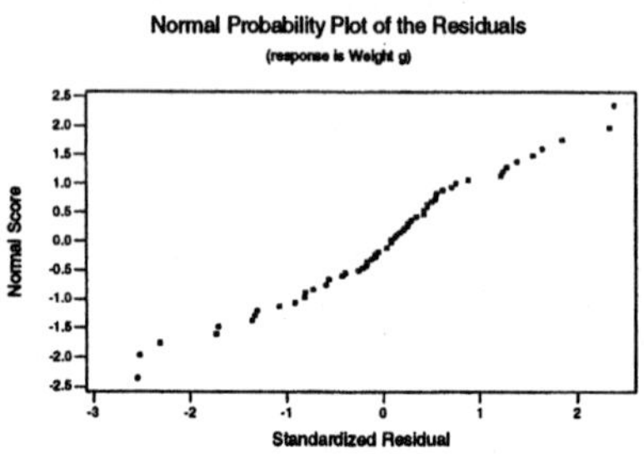

A plot of the standardized residuals against the factor gossypol follows.

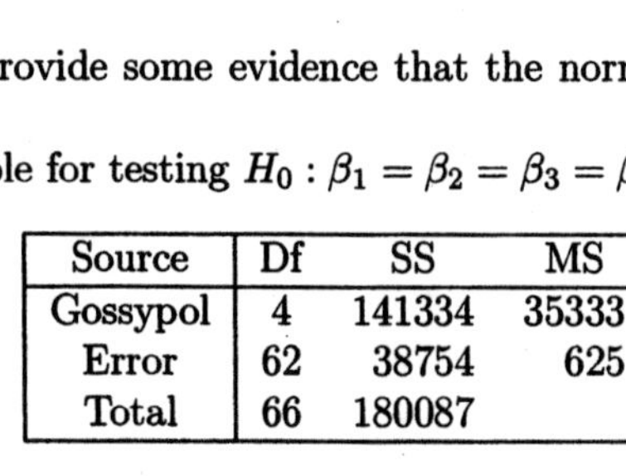

Again, these plots provide some evidence that the normality assumption may not be appropriate.
(c) The ANOVA table for testing $H_0 : \beta_1 = \beta_2 = \beta_3 = \beta_4 = \beta_5$ follows.

Source	Df	SS	MS
Gossypol	4	141334	35333
Error	62	38754	625
Total	66	180087	

The F statistic for testing H_0 is given by $F = 35333/625 = 56.533$, and since $F \sim F(4, 62)$ under H_0, the P-value equals $P(F > 56.533) = 0.000$. Therefore,

we have strong evidence against the null hypothesis of no difference amongst the mean level of the response given different amount of gossypol.
(d) The 0.95-confidence intervals for the difference between the means are given in the following table.

Family error rate = 0.279
Individual error rate = 0.0500

Critical value = 1.999

Intervals for (column level mean) - (row level mean)

	0.00	0.04	0.07	0.10
0.04	-14.75 24.40			
0.07	28.10 66.27	21.50 63.23		
0.10	84.60 119.42	77.85 116.53	35.98 73.67	
0.13	85.34 124.49	78.78 121.40	36.87 78.59	-16.44 22.24

Only the mean at the 0.00% level does not differ from the mean at the 0.04% level and the mean at the 0.10% level does not differ from the mean at the 0.13% level at the 5% significant level, as both intervals include the value 0.

10.4.5
(a) A side-by-side boxplot of the data follows.

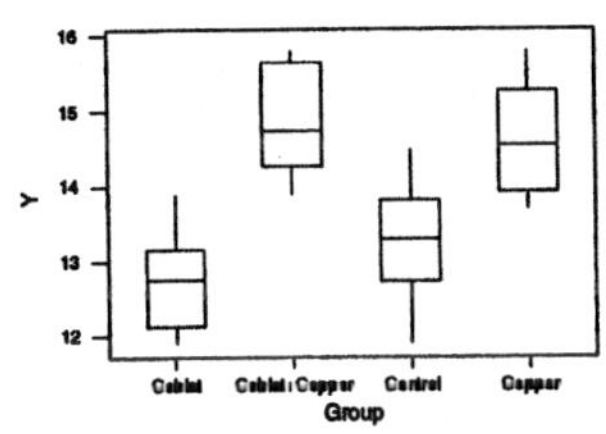

(b) A plot of the standardized residuals against the predictor follows.

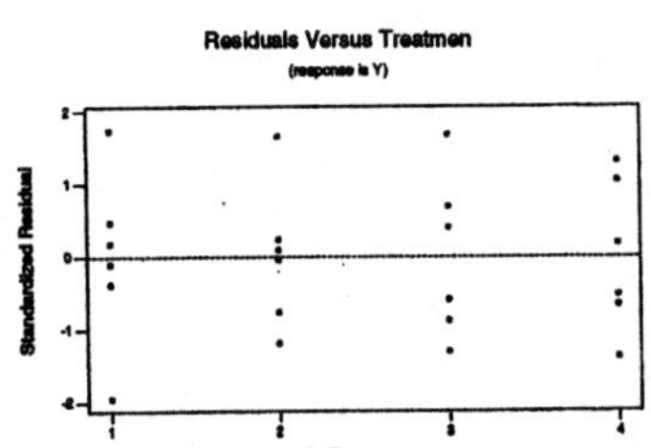

A normal probability plot of the standardized residuals follows.

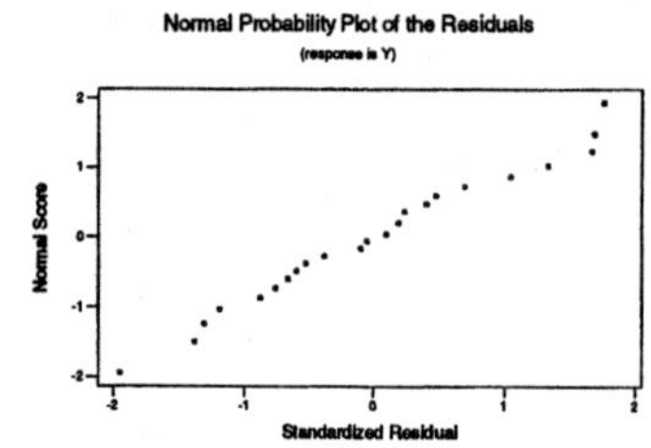

Both plots look reasonable, indicating no concerns about the correctness of the model assumptions.
(c) The ANOVA table for testing $H_0 : \beta_1 = \beta_2 = \beta_3 = \beta_4$ follows.

Source	Df	SS	MS
Treatment	3	19.241	6.414
Error	20	11.788	0.589
Total	23	31.030	

The F statistic for testing H_0 is given by $F = 6.414/0.589 = 10.89$ and, since $F \sim F(3, 20)$ under H_0, the P-value equals $P(F > 10.89) = .00019$. Therefore, we have strong evidence against the null hypothesis of no difference among the conditional means of Y given the predictor.
(d) The 0.95-confidence intervals for the difference between the means are given in the following table.

```
Family error rate = 0.192
Individual error rate = 0.0500

Critical value = 2.086

Intervals for (column level mean) - (row level mean)

              1          2          3

  2     -0.3913
         1.4580

  3     -2.2746    -2.8080
        -0.4254    -0.9587

  4     -2.5246    -3.0580    -1.1746
        -0.6754    -1.2087     0.6746
```

The mean response for the control treatment does not differ from the mean response given the Cobalt treatment and the mean response for the Copper treatment does not differ from the mean response for the Cobalt+Copper treatment at the 5% level, since both intervals include the value 0. All other mean differences are judged to be nonzero at the 5% level.

10.4.6

(a) A side-by-side boxplot of the data follows.

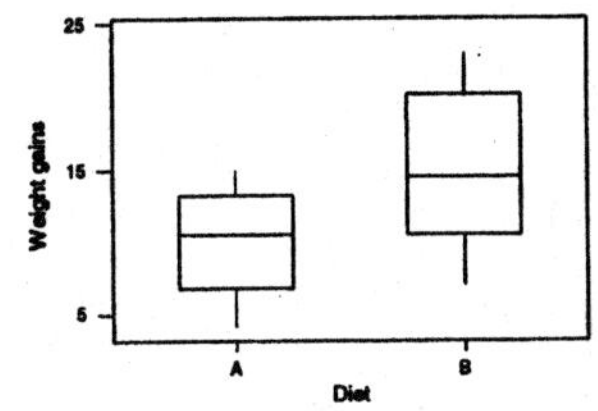

(b) A plot of the standardized residuals against Diet follows.

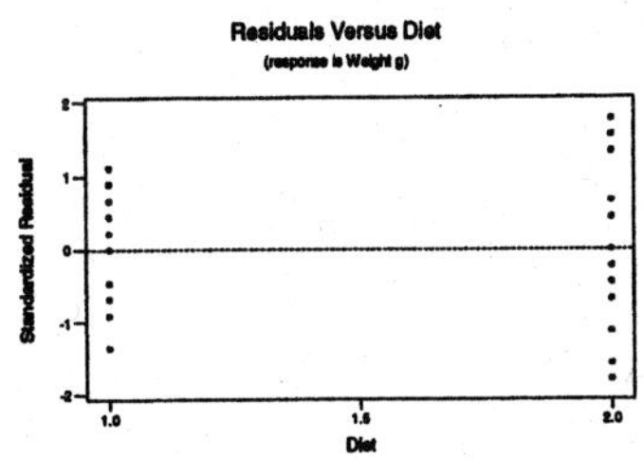

A normal probability plot of the standardized residuals follows.

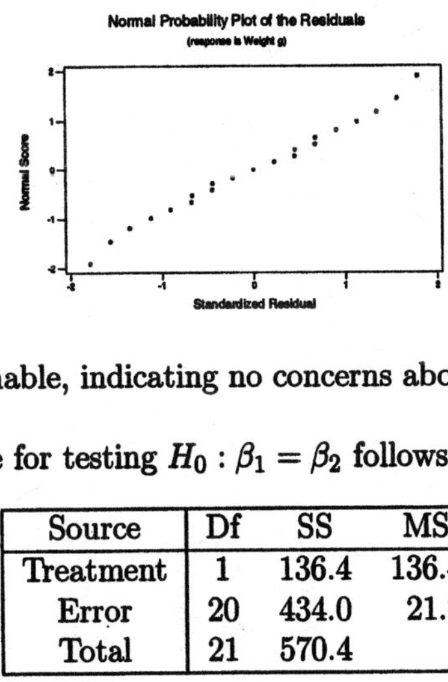

Both plots look reasonable, indicating no concerns about the correctness of the model assumptions.
(c) The ANOVA table for testing $H_0 : \beta_1 = \beta_2$ follows.

Source	Df	SS	MS
Treatment	1	136.4	136.4
Error	20	434.0	21.7
Total	21	570.4	

The F statistic for testing H_0 is given by $F = 136.4/21.7 = 6.2857$ and, since $F \sim F(1, 20)$ under H_0, the P-value equals $P(F > 6.2857) = .02091$. Therefore, we have evidence against the null hypothesis of no difference among the conditional means of Y given Diet at the 5% level but not at 1%.
(d) A .95-confidence interval for the difference between the means follows.

$$\beta_1 - \beta_2 \in (10.0 - 15.0) \pm 4.658 \left(\frac{1}{10} + \frac{1}{12}\right)^{1/2} 2.086 = (-9.1604, -.83961)$$

Note that this does not include the value 0 and therefore supports our conclusion from part (c).

10.4.7

(a) A side-by-side boxplot of the data follows.

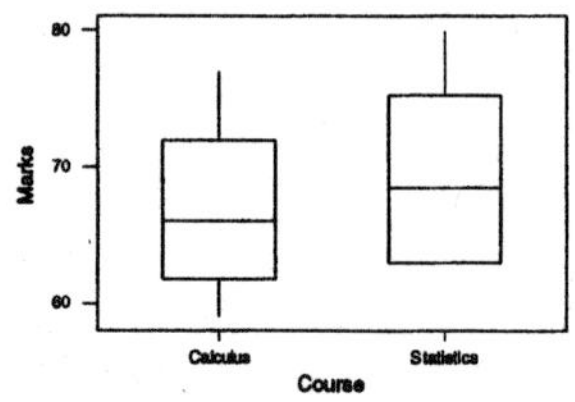

(b) Treating the marks as separate samples, the ANOVA table for testing any difference between the mean mark in Calculus and the mean mark in Statistics follows.

Source	Df	SS	MS
Course	1	36.45	36.45
Error	18	685.30	38.07
Total	19	721.75	

The F statistic for testing $H_0 : \beta_1 = \beta_2$ is given by $F = 36.45/38.07 = .95745$ and, since $F \sim F(1, 19)$ under H_0, the P-value equals $P(F > .95745) = .3408$. Therefore, we don't have any evidence against the null hypothesis of no difference among the conditional means of Y given Course.

A plot of the standardized residuals against Course follows.

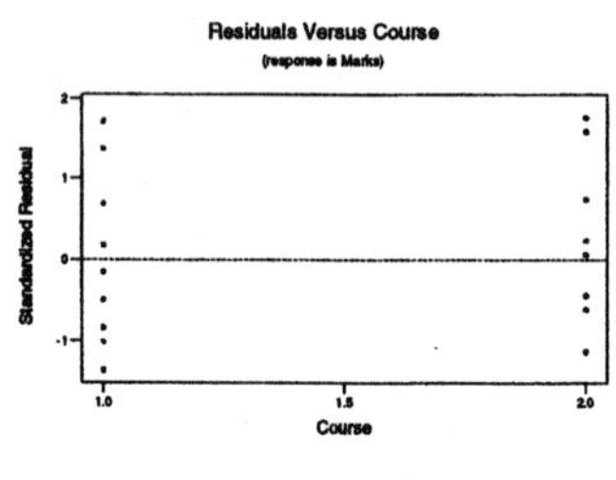

A normal probability plot of the standardized residuals follows.

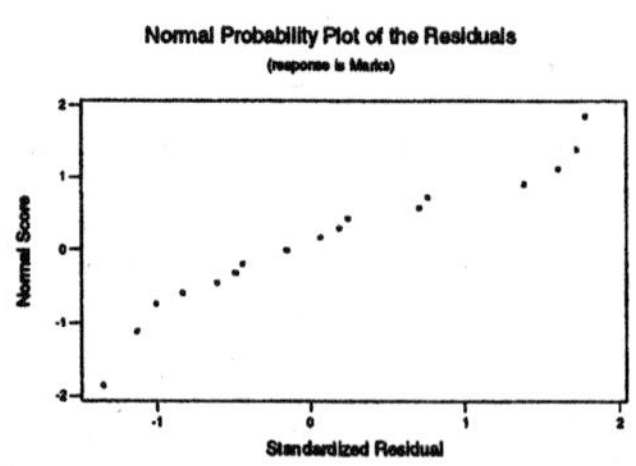

Both plots look reasonable, indicating no concerns about the correctness of the model assumptions.
(c) Treating this data as repeated measures, the mean difference between the mark in Calculus and the mark in Statistics is given by $\bar{d} = -2.7$ with standard deviation $s = 2.00250$. The P-value for testing $H_0 : \mu_1 = \mu_2$, since $T \sim t(9)$ under H_0, the P-value is given by $P\left(|T| > \left|-2.7/\left(2.00250/\sqrt{10}\right)\right|\right) = 2P(T > 4.2637) = .0021$, so we have strong evidence against the null. Hence we conclude that there is a difference between the mean mark in Calculus and the mean mark in Statistics. A normal probability plot of the data follows and this doesn't indicate any reason to doubt model assumptions.

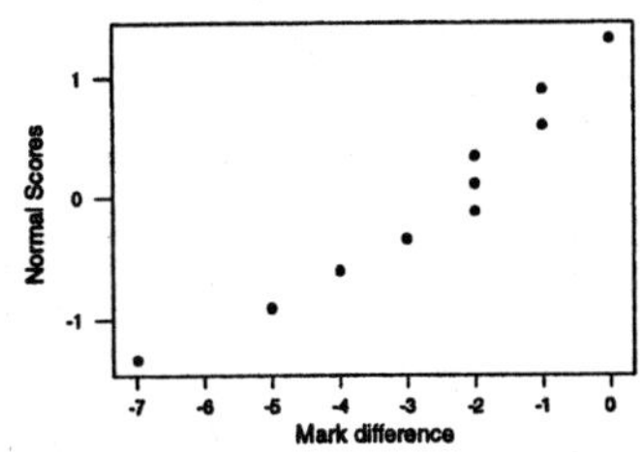

(d) The estimate of the correlation between the Calculus and Statistics marks is given by the sample correlation coefficient $r_{xy} = 0.944155$.

10.4.8
(a) A side-by-side boxplot of the data follows.

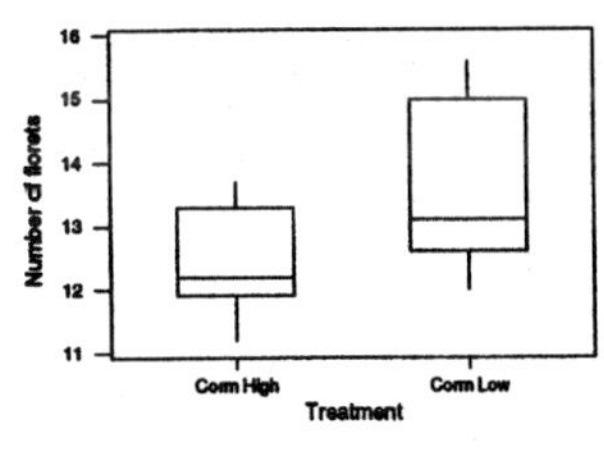

(b) Treating the Corm High and Corm Low measurements as separate samples, the ANOVA table for testing any difference between the population means follows.

Source	Df	SS	MS
Treatment	1	5.040	5.040
Error	12	16.014	1.335
Total	13	21.054	

The F statistic for testing $H_0 : \beta_1 = \beta_2$ is given by $F = 5.040/1.335 = 3.7753$ and, since $F \sim F(1, 19)$ under H_0, the P-value equals $P(F > 3.7753) = .07585$.

Therefore, we don't have substantial evidence against the null hypothesis of no difference amongst the conditional means of Y given Corm level.

A plot of the standardized residuals against Treatment follows.

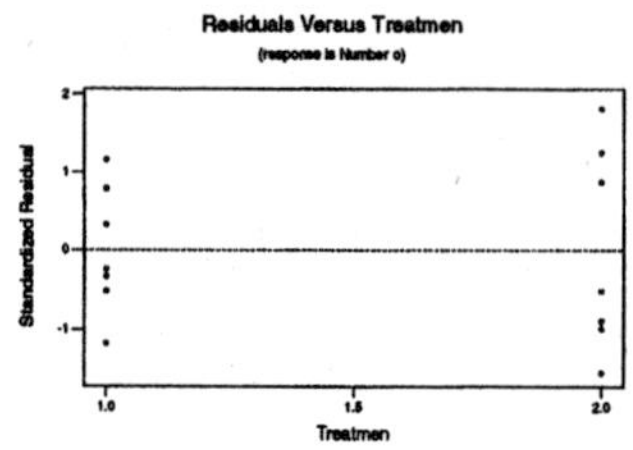

A normal probability plot of the standardized residuals follows.

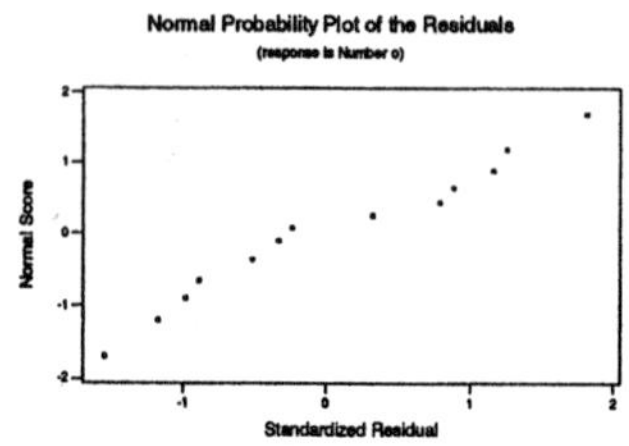

Both plots look reasonable, indicating no concerns about the correctness of the model assumptions.

(c) Treating this data as repeated measures, the mean difference between the number of florets in plots with Corm High and the number of florets in plots with Corm Low is given by $\bar{d} = -1.2$ with standard deviation $s = 1.395$. The P-value for testing $H_0 : \mu_1 = \mu_2$, since $T \sim t(6)$ under H_0, equals $P\left(|T| > \left|-1.2/\left(1.395/\sqrt{7}\right)\right|\right) = 2P\left(T > 2.2759\right) = .06316$, so we don't have substantial evidence against the null. Hence, we conclude that there is no difference between the mean number of florets in plots with Corm High and the mean number of florets in plots with Corm Low.

A normal probability plot of the data follows and this reveals no evidence of model incorrectness.

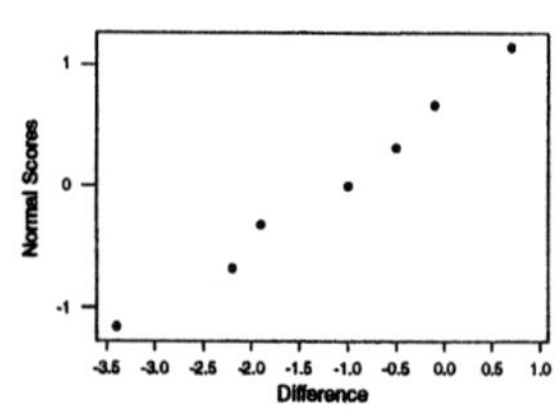

(d) The estimate of the correlation between the Calculus and Statistics marks is given by the sample correlation coefficient $r_{xy} = 0.301949$.

Problems

10.4.9 To prove this we express the sum of squares as follows

$$\sum_{i=1}^{a}\sum_{j=1}^{n_i}(y_{ij}-\beta_i)^2=\sum_{i=1}^{a}\sum_{j=1}^{n_i}(y_{ij}-\bar{y}_i+\bar{y}_i-\beta_i)^2$$
$$=\sum_{i=1}^{a}\sum_{j=1}^{n_i}(y_{ij}-\bar{y}_i)^2+2\sum_{i=1}^{a}\sum_{j=1}^{n_i}(y_{ij}-\bar{y}_i)(\bar{y}_i-\beta_i)+\sum_{i=1}^{a}n_i(\bar{y}_i-\beta_i)^2.$$

First, note that the second term is equal to 0 since

$$\sum_{j=1}^{n_i}(y_{ij}-\bar{y}_i)(\bar{y}_i-\beta_i)=(\bar{y}_i-\beta_i)\left(\sum_{j=1}^{n_i}y_{ij}-n_i\bar{y}_i\right)$$
$$=(\bar{y}_i-\beta_i)\left(\sum_{j=1}^{n_i}y_{ij}-n_i\sum_{j=1}^{n_i}\frac{y_{ij}}{n_i}\right)=0.$$

So the above sum of squares is minimized as a function of β_i if and only if the second term is equal to 0, and if and only if $\bar{y}_i-\beta_i=0$, if and only if $\beta_i=\bar{y}_i$.

10.4.10 To prove this we express the sum of squares as follows.

$$\sum_{i=1}^{a}\sum_{j=1}^{n_i}(y_{ij}-\bar{y})^2=\sum_{i=1}^{a}\sum_{j=1}^{n_i}(y_{ij}-\bar{y}_i+\bar{y}_i-\bar{y})^2$$
$$=\sum_{i=1}^{a}\sum_{j=1}^{n_i}(y_{ij}-\bar{y}_i)^2+2\sum_{i=1}^{a}\sum_{j=1}^{n_i}(y_{ij}-\bar{y}_i)(\bar{y}_i-\bar{y})+\sum_{i=1}^{a}n_i(\bar{y}_i-\bar{y})^2.$$

Note that the second term is equal to 0 since

$$\sum_{i=1}^{a}\sum_{j=1}^{n_i}(y_{ij}-\bar{y}_i)(\bar{y}_i-\bar{y})=\sum_{i=1}^{a}\left((\bar{y}_i-\bar{y})\left(\sum_{j=1}^{n_i}y_{ij}-n_i\bar{y}_i\right)\right)$$
$$=\sum_{i=1}^{a}\left((\bar{y}_i-\bar{y})\left(\sum_{j=1}^{n_i}y_{ij}-n_i\sum_{j=1}^{n_i}\frac{y_{ij}}{n_i}\right)\right)=0$$

10.4.11 If an interaction exists between the two factors, then the b response curves are not parallel and therefore cannot be horizontal, i.e., there must be effect due to both factors.

10.4.12 By assumption we have $Y_{ij}\sim N\left(\beta_i,\sigma^2\right)$ and these are all independent. Therefore, we have that $\bar{Y}_i\sim N\left(\beta_i,\sigma^2/n_i\right)$. Further, $\text{Cov}(Y_{ij},\bar{Y}_i)=\text{Cov}\left(Y_{ij},\sum_{k=1}^{n_i}\frac{Y_{ik}}{n_i}\right)=\sigma^2/n_i$. Therefore, by Theorem 4.6.1 we have that $Y_{ij}-\bar{Y}_i\sim N\left(0,\sigma^2\left(1-1/n_i\right)\right)$ as claimed.

10.4.13 By assumption we have $Y_{ijk} \sim N(\beta_{ij}, \sigma^2)$ and these are all independent. Then we have that $\bar{Y}_{ij} \sim N(\beta_{ij}, \sigma^2/n_{ij})$. Further, $\text{Cov}(Y_{ijk}, \bar{Y}_{ij}) = \text{Cov}\left(Y_{ijk}, \sum_{l=1}^{n_{ij}} \frac{y_{ijl}}{n_{ij}}\right) = \sigma^2/n_{ij}$. Therefore, by Theorem 4.6.1 we have $Y_{ijk} - \bar{Y}_{ij} \sim N(0, \sigma^2(1 - 1/n_{ij}))$.

10.4.14 First, recall that $s^2 = \frac{1}{N-ab}\sum_{i=1}^{a}\sum_{j=1}^{b}\sum_{k=1}^{n_{ij}}(y_{ijk} - \bar{y}_{ij})^2$. Now if $n_{ij} = 1$ then $\bar{y}_{ij} = y_{ijk}$ for all i and j. Hence, $y_{ijk} - \bar{y}_{ij} = 0$, which establishes that $s^2 = 0$ as claimed.

10.4.15 By looking at various plots of the residuals, for example, a normal probability plot of the standardized residuals.

Computer Problems

10.4.16 Controlling a family error rate of 0.0455, the 0.95-confidence intervals for the difference between the means are given in the following table. It required a 0.01 individual error rate.

```
Family error rate = 0.0455
Individual error rate = 0.0100

Critical value = 2.845

Intervals for (column level mean) - (row level mean)

             Coblat   Coblat+C   Control

Coblat+C    -3.3944
            -0.8723

Control     -1.7944    0.3389
             0.7277    2.8611

Copper      -3.1444   -1.0111    -2.6111
            -0.6223    1.5111    -0.0889
```

10.4.17
(a) A side-by-side boxplot of the data by treatment (using the coding $3(i-1)+j$) follows.

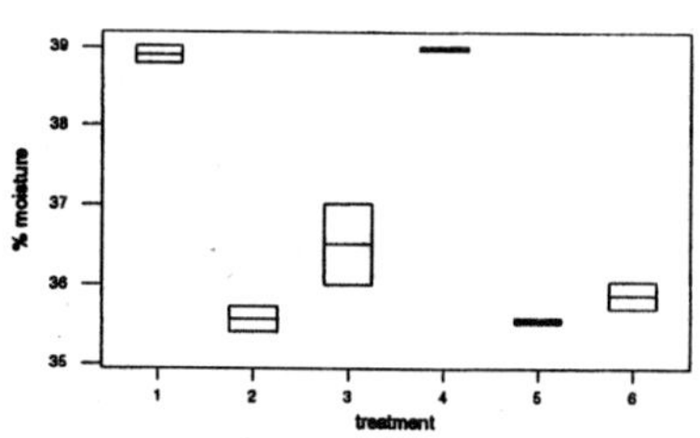

(b) A table of the cell means is given follows.

	Lot 1	Lot 2	Lot 3
Cheese 1	38.905	35.575	36.510
Cheese 2	38.985	35.550	35.870

(c) A normal probability plot of the standardized residuals follows.

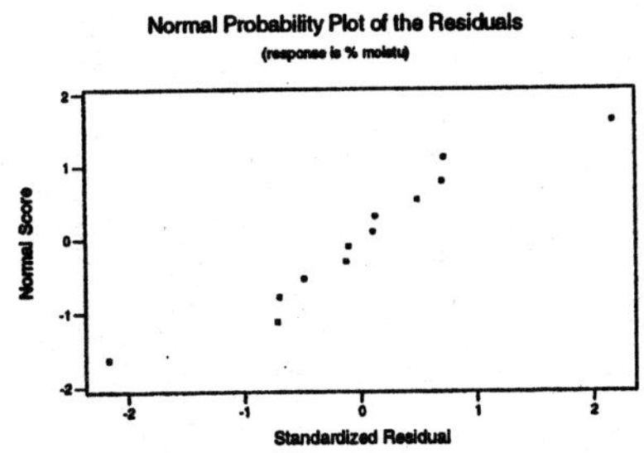

A plot of the standardized residuals against each of the treatment combination (using the coding $3(i-1)+j$) follows.

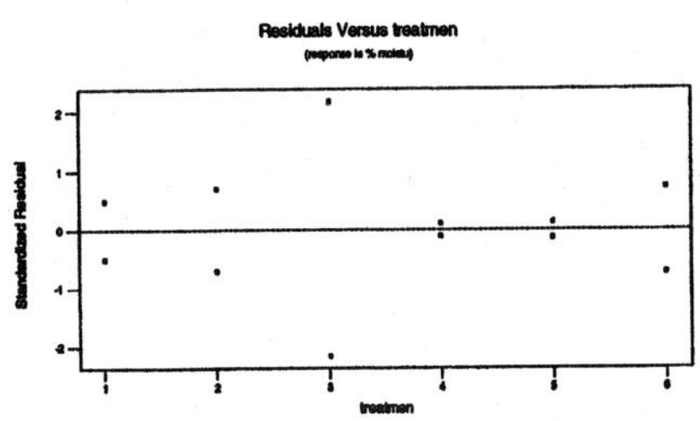

Both plots looks reasonable, so indicating no serious concerns about the correctness of the model assumptions.

(d) The ANOVA table for testing all relevant hypotheses follows.

Source	Df	SS	MS
Cheese	1	0.114	0.114
Lot	2	25.900	12.950
Interaction	2	0.303	0.151
Error	6	0.662	0.110
Total	11	26.979	

The F statistic for testing H_0 : no interaction between cheese and lot, is given by $F = 0.151/0.110 = 1.3727$ and, since $F \sim F(2,6)$ under H_0, the P-value equals $P(F > 1.3727) = .32293$. Therefore, we don't have evidence against the null hypothesis of no interaction effect.

We can then proceed to calculate the P-value for testing H_0 : no effect due to cheese. This is given by $P(F > 0.114/0.110 = 1.0364) = .34794$, since $F \sim F(1,6)$ under H_0. Therefore, we don't have any evidence against the null hypothesis of no effect due to Cheese.

The P-value for testing H_0 : no effect due to lot, since $F \sim F(2,6)$ under H_0, is given by $P(F > 12.950/0.110 = 117.73) = .00002$. Therefore, we have strong evidence against the null hypothesis of no effect due to Lot.

(e) Since we conclude that only the factor Lot has a significant effect on the response, we calculate the following table of means.

	Lot 1	Lot 2	Lot 3
Mean	38.945	35.563	36.190

The corresponding response curve follows.

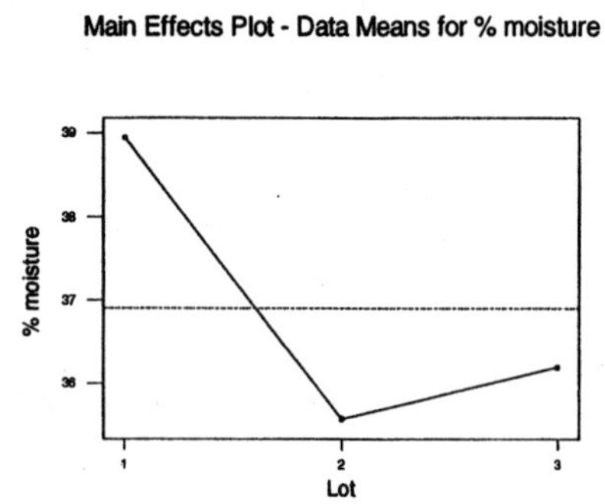

The .95-confidence intervals for the difference between the means are given in the following table.

```
Family error rate = 0.113
Individual error rate = 0.0500

Critical value = 2.262

Intervals for (column level mean) - (row level mean)

              1          2

    2    2.8288
         3.9362

    3    2.2013    -1.1812
         3.3087    -0.0738
```

As we can see, all the confidence intervals above do not include the value 0, indicating differences between all the means at the 5% level.

(f) Our result here agrees with the result of Exercise 10.4.4 as in both case no significant effect due to cheese was found although we do find an effect due to Lot.

10.4.18

(a) A side-by-side boxplot of the data by treatment (using the coding $3(i-1)+j$ with $A = i, B = j$) follows.

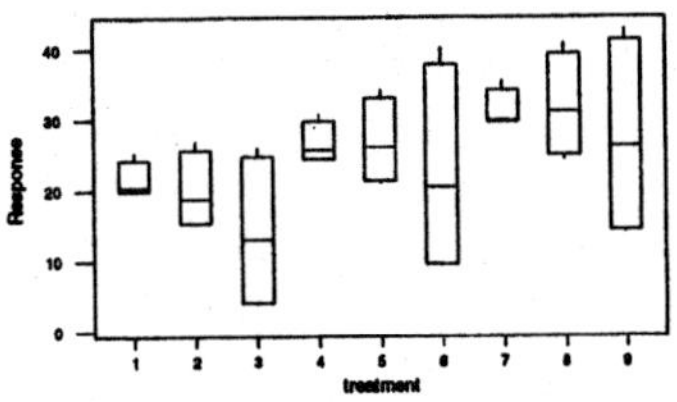

(b) A table of the cell means is given follows.

	$A=1$	$A=2$	$A=3$
$B=1$	21.58	26.64	31.23
$B=2$	20.00	26.83	31.72
$B=3$	14.02	22.59	27.26

(c) A normal probability plot of the standardized residuals follows.

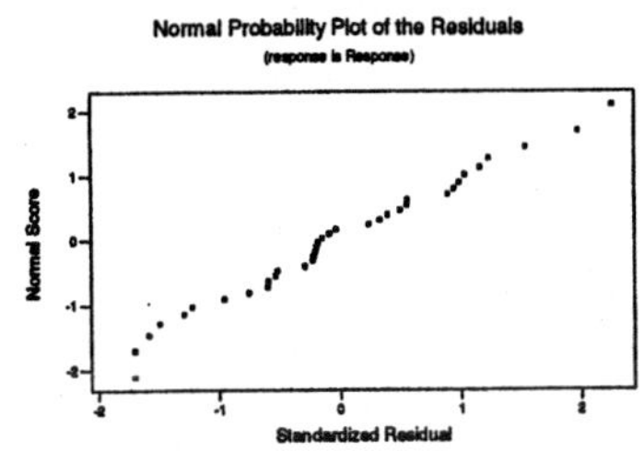

A plot of the standardized residuals against each of the treatment combination follows.

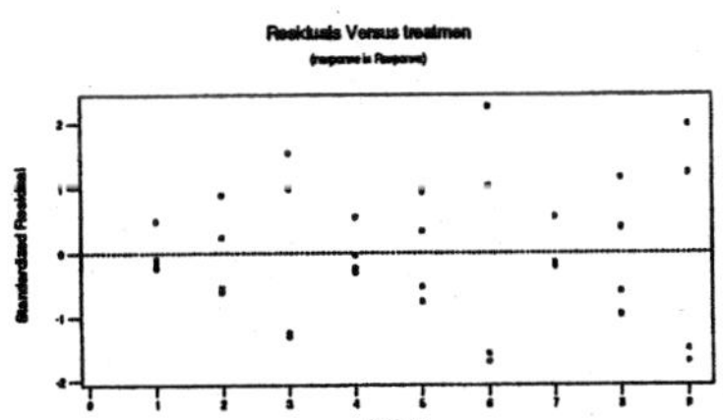

Both plots indicate a possible problem with the model assumptions, but nothing severe.

(d) The ANOVA table for testing all relevant hypotheses follows.

Source	Df	SS	MS
A	2	807.2	403.6
B	2	204.2	102.1
$A \times B$	4	17.0	4.2
Error	27	2158.0	79.9
Total	35	3186.3	

The F statistic for testing H_0 : no interaction between Cheese and Lot, is given by $F = 4.2/79.9 = 0.0526$ and, since $F \sim F(4, 27)$ under H_0, the relevant P-value equals $P(F > 0.0526) = .99451$. Therefore, we don't have any evidence against the null hypothesis of no interaction effect.

We can then proceed to calculate the P-value for testing H_0 : no effect due to A and, since $F \sim F(2, 27)$ under H_0, this is given by $P(F > 403.6/79.9) = .01369$. Therefore, we have some evidence against the null hypothesis of no effect due to A.

We can also test H_0 : no effect due to B and, since $F \sim F(2, 27)$ under H_0, the P-value equals $P(F(2, 6) > 102.1/79.9) = .29497$. Therefore, we have enough no evidence against the null hypothesis of no effect due to B.

(e) Since we conclude that only factor A has a significant effect on the response we calculate the following table of means.

	$A = 1$	$A = 2$	$A = 3$
Mean	18.53	25.35	30.07

A plot of the corresponding response curve follows.

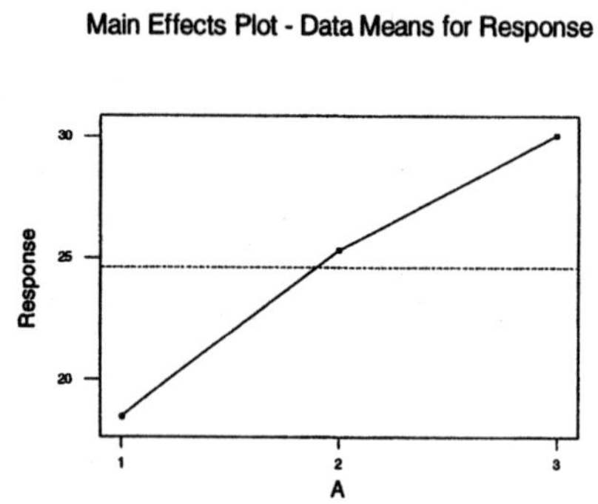

The 0.95-confidence intervals for the difference between the means are given in the following table.

```
Family error rate = 0.120
Individual error rate = 0.0500

Critical value = 2.035

Intervals for (column level mean) - (row level mean)

              1          2

   2    -13.872
          0.237

   3    -18.589    -11.772
         -4.481      2.337
```

As we can see, at the 5% significance level, only the means of the first and third groups do not differ.

10.4.19

(a) A side-by-side boxplot of the data by treatment (using the coding $3(i-1)+j$ with $A = i, B = j$) follows.

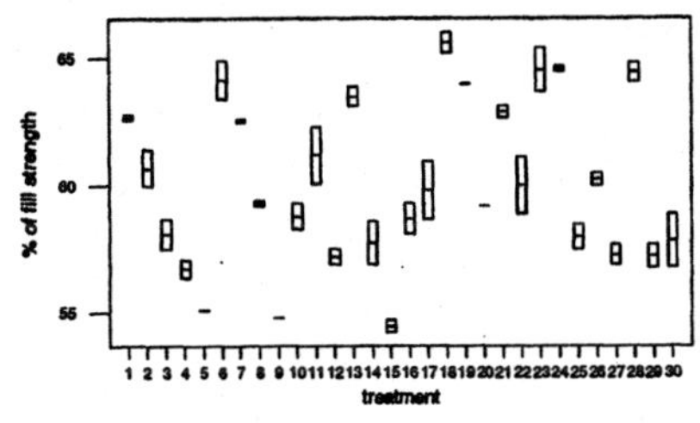

(b) A table of the cell means follows.

	Cask 1	Cask 2	Cask 3
Batch 1	62.70	61.20	62.90
Batch 2	60.70	57.20	60.00
Batch 3	58.10	63.50	64.55
Batch 4	56.75	57.75	64.60
Batch 5	55.10	54.45	58.00
Batch 6	64.15	58.70	60.25
Batch 7	62.55	59.85	57.30
Batch 8	59.30	65.60	64.45
Batch 9	54.80	64.00	57.25
Batch 10	58.80	59.20	57.85

(c) A normal probability plot of the standardized residuals follows.

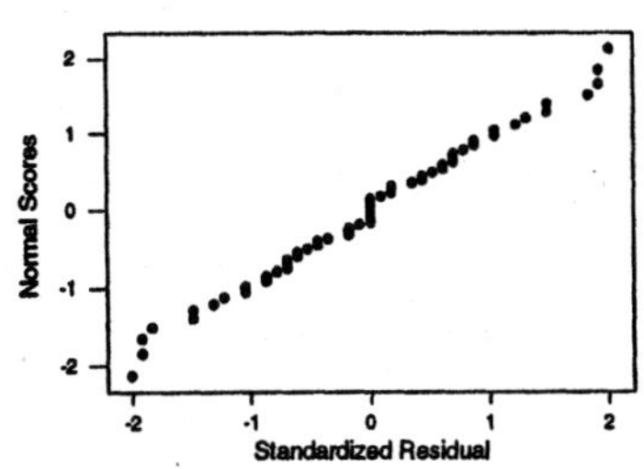

A plot of the standardized residuals against each of the treatment combination follows.

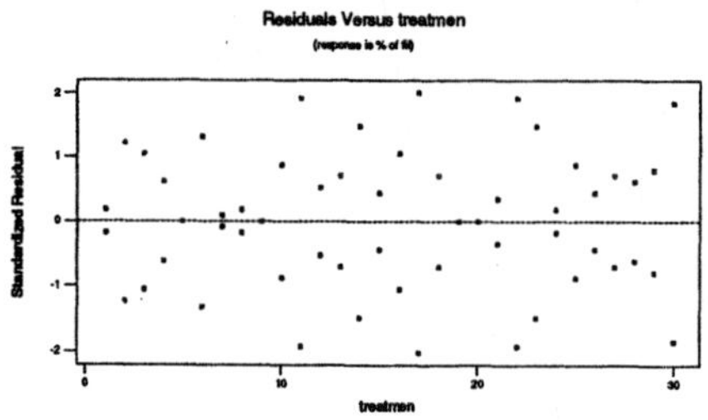

Both plots looks reasonable, indicating no concerns about the correctness of the model assumptions.
(d) The ANOVA table for testing all relevant hypothesis follows.

Source	Df	SS	MS
Cask	2	20.425	10.213
Batch	9	249.328	27.703
Interaction	18	328.841	18.269
Error	30	19.555	0.652
Total	59	618.150	

The F statistic for testing H_0 : no interaction between Cask and Batch, is given by $F = 18.269/0.652 = 28.02$ and, since $F \sim F(18, 30)$ under H_0, the P-value equals $P(F > 28.02) = .0000$. Therefore, we have strong evidence against the null hypothesis of no interaction. It is clear now that both predictors, Cask and Batch, will have a significant effect on the response. There is no need to test for an effect due to either Cask or Batch.
(e) Since we conclude that Cask and Batch interact, the appropriate table of means is given in part (b). The plot of the response curves follows.

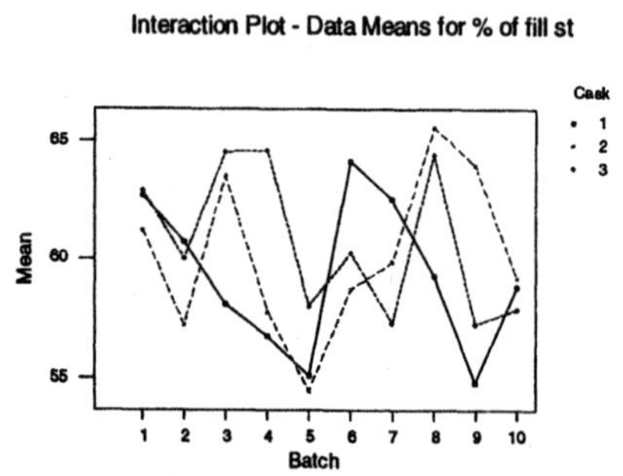

10.4.20
(a) First, since there is only one observation in each combination of the two factors, Fertilizer and Plot of Land, we have to assume that no interaction exists between the two in order to be able to detect an effect due to fertilizer.

The ANOVA table for testing no effect due to fertilizer follows.

Source	Df	SS	MS
Fertilizer	4	0.7308	0.1827
Block	2	0.1086	0.0543
Error	8	0.3384	0.0423
Total	14	1.1778	

The F statistic for testing H_0 : no effect due to fertilizer, is given by $F = 0.1827/0.0423 = 4.3191$ and, since $F \sim F(4,8)$ under H_0, the P-value equals $P(F(4,8) > 4.3191) = .03747$. Therefore, we have some evidence against the null hypothesis of no effect due to Fertilizer.

(b) As mentioned in part (a), since $n_{ij} = 1$, we assume, in addition to the usual assumptions, that there is no interaction between Fertilizer and Plot of Land.

(c) This would increase the number of degrees of freedom available for error, as we would need only 1 degree of freedom to estimate the effect (slope) of Fertilizer. So we would have 11 degrees of freedom for error and thus make our comparisons more accurate.

(d) Using Minitab we fit this model obtaining the following ANOVA table.

Analysis of Variance for Response, using Adjusted SS for Tests

Source	DF	Seq SS	MS	F	P
A	1	0.55981	0.55981	12.09	0.005
B	2	0.10864	0.05432	1.17	0.345
Error	11	0.50939	0.04631		
Total	14	1.17784			

From this we can see that there is strong evidence of an effect due to A.

To check the validity of this model we provide a normal probability plot of the standardized residuals below.

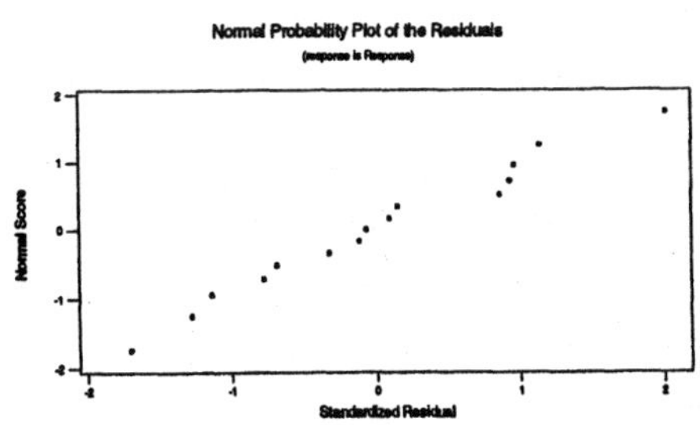

A plot of the standardized residuals against each of the treatment combination is given below.

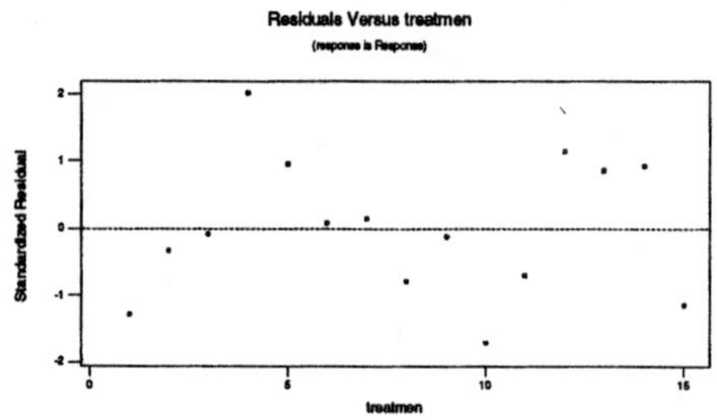

Both plots looks reasonable, indicating no serious concerns about the correctness of the model assumptions.

10.5 Categorical Response and Quantitative Predictors

Exercises

10.5.1 We have $\int_{-\infty}^{\infty} f(x)\,dx = \int_0^{\infty} \frac{e^{-x}}{(1+e^{-x})^2}dx = \frac{1}{1+e^{-x}}\Big|_{-\infty}^{\infty} = 1$. Hence, f is indeed a density function. The distribution function is then given by $F(x) = \int_{-\infty}^{x} \frac{e^{-t}}{(1+e^{-t})^2}dt = \frac{1}{1+e^{-t}}\Big|_{-\infty}^{x} = \frac{1}{1+e^{-x}}$ as claimed. Let $p = P(X \le x) = F(x)$, then $p \in [0,1]$ and we have $p = (1+e^{-x})^{-1}$, which implies $F^{-1}(p) = x = \ln(p/(1-p))$ as claimed.

10.5.2 Let $p = P(Y = 1\,|\,x)$. The log odds at $X = x$ is then given by (10.5.1) as follows.

$$\ln\left(\frac{p}{1-p}\right) = \ln\left(\left(\frac{\exp\{\beta_1+\beta_2 x\}}{1+\exp\{\beta_1+\beta_2 x\}}\right) / \left(\frac{1}{1+\exp\{\beta_1+\beta_2 x\}}\right)\right)$$
$$= \ln(\exp\{\beta_1+\beta_2 x\}) = \beta_1+\beta_2 x$$

as claimed.

Computer Exercises

10.5.3 The results should be the same as presented in Example 10.5.1.

10.5.4
(a) Fitting the model using Minitab leads to the estimates given in the following table.

Coefficient	Estimate	Std. Error	Z	P-value
β_1	-1.1850	0.9338	-1.27	0.204
β_2	0.9436	0.3966	2.38	0.017
β_3	0.0597	0.1462	0.41	0.683

(b) The Chi-squared statistic for testing the validity of the model is then equal to 4.66204 with P-value given by $P\left(\chi^2(8) > 4.66204\right) = .79301$. Therefore, we have no evidence that the model is incorrect.
(c) The P-value for testing $H_0 : \beta_3 = 0$ is 0.638, so we don't have any evidence against the null hypothesis.
(d) Since the null hypothesis $H_0 : \beta_3 = 0$ is not rejected, we dropped the quadratic term and refit the model. This leads to the estimates given in the following table.

Coefficient	Estimate	Std. Error	Z	P-value
β_1	-1.0063	0.8150	-1.23	0.217
β_2	0.9969	0.4163	2.39	0.017

The P-value for testing $H_0 : \beta_2 = 0$ is 0.017, so we have some evidence against the null hypothesis and we conclude that there is a linear effect.
(e) The plot of $P(Y = 1 \mid X = x)$ as a function of x using the estimates found above follows.

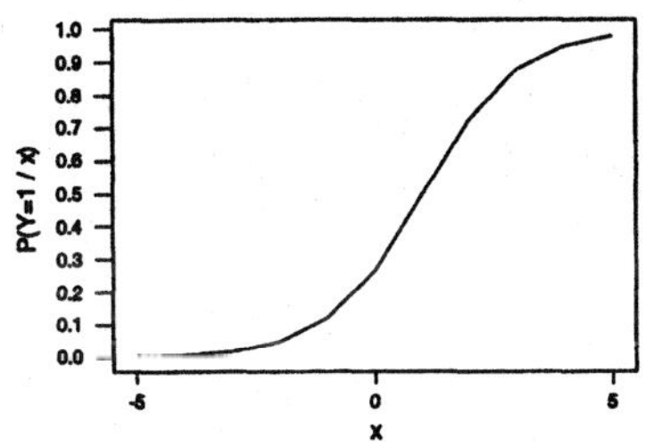

Problem

10.5.5 The cell count (number of successes) is an observation from a Binomial$(m(x_2, x_3), P(Y = 1 \mid X_2 = x_2, X_3 = x_3))$ distribution. Let $p_{(x_2,x_3)}(\theta)$ $= P(Y = 1 \mid X_2 = x_2, X_3 = x_3)$, where $\theta = (\beta_2, \beta_3)$. Then by Theorem 9.1.2 we have that

$$X^2 = \sum_{(x_2,x_3)} \frac{\left(s(x_2,x_3) - m(x_2,x_3)\hat{p}(x_2,x_3)\right)^2}{m(x_2,x_3)\hat{p}(x_2,x_3)} \xrightarrow{D} \chi^2(k - 1 - \dim\Omega)$$

where k is the number of combination of (x_2, x_3) and $\dim\Omega = 2$. Hence, (10.5.3) is the correct form of the Chi-squared goodness of fit test statistic.

Chapter 11

Advanced Topic—Stochastic Processes

11.1 Simple Random Walk

Exercises

11.1.1
(a) 0.
(b) 0.
(c) 1/3.
(d) 2/3.
(e) 0.
(f) $2(1/3)(2/3) = 4/9$.
(g) 0.
(h) $(1/3)(1/3) = 1/9$.
(i) 0.
(j) 0.
(k) $\binom{20}{12}(1/3)^{12}(2/3)^8 \doteq 0.00925$.
(l) 0.
(m) $\binom{20}{9}(1/3)^9(2/3)^{11} \doteq 0.0987$.

11.1.2
(a) $P(X_1 = 6,\ X_2 = 5) = (2/5)(3/5) = 6/25$.
(b) $P(X_1 = 4,\ X_2 = 5) = (3/5)(2/5) = 6/25$.
(c) $P(X_2 = 5) = \binom{2}{1}(2/5)(3/5) = 12/25$.
(d) By the law of total probability, $P(X_2 = 5) = P(X_1 = 6,\ X_2 = 5) + P(X_1 = 4,\ X_2 = 5)$.

11.1.3
(a) $P(X_1 = X_3 = 8) = P(X_1 = 8,\ X_2 = 7,\ X_3 = 8) + P(X_1 = 8,\ X_2 = 9,\ X_3 = 8) = (1/6)(5/6)(1/6) + (1/6)(1/6)(5/6) = 10/216 = 5/108$.
(b) $P(X_1 = 6,\ X_3 = 8) = P(X_1 = 6,\ X_2 = 7,\ X_3 = 8) = (5/6)(1/6)(1/6) =$

5/216 .
(c) $P(X_3 = 8) = \binom{3}{2}(1/6)^2(5/6)^1 = 15/216 = 5/72$.
(d) By the law of total probability, $P(X_3 = 8) = P(X_1 = 6,\ X_3 = 8) + P(X_1 = 8,\ X_3 = 8)$.

11.1.4
(a) Here $E(X_n) = a + n(2p-1) = 1000 - n(0.02)$. Hence, $E(X_0) = 1000$; $E(X_1) = 999.98$; $E(X_2) = 999.96$; $E(X_{10}) = 999.80$; $E(X_{20}) = 999.60$; $E(X_{100}) = 998$; $E(X_{1000}) = 980$.
(b) If $E(X_n) < 0$, then $1000 - n(0.02) < 0$, i.e., $n > 1000/(0.02) = 50,000$.

11.1.5
(a) Here $P(\tau_c < \tau_0) \doteq 0.89819$. That is, if you start with \$9 and repeatedly make \$1 bets having probability 0.499 of winning each bet, then the probability you will reach \$10 before going broke is equal to 0.89819.
(b) Here $P(\tau_c < \tau_0) \doteq 0.881065$.
(c) Here $P(\tau_c < \tau_0) \doteq 0.664169$.
(d) Here $P(\tau_c < \tau_0) \doteq 0.0183155$.
(e) Here $P(\tau_c < \tau_0) \doteq 4 \times 10^{-18}$.
(f) Here $P(\tau_c < \tau_0) \doteq 2 \times 10^{-174}$.

11.1.6 If $p = 0.4$, then $P(\tau_0 < \infty) = 1$. If $p = 0.6$, then $P(\tau_0 < \infty) = ((1-p)/p)^a = (0.4/0.6)^{10} \doteq 0.0173415$, i.e., less than 2%. That is, if we start with \$10 and repeatedly make bets with probability 0.4 of winning each bet, then we will eventually go broke with certainty. However, if the probability of winning each bet is 0.6, then there is less than 2% chance of eventually going broke.

Problem

11.1.7
(a) This is *equivalent* to having \$5, and trying to reach \$50, by making bets of just \$1 each time — since we can count things in units of \$2 instead of \$1. Hence, the desired probability is $(1-((1-p)/p)^5)/(1-((1-p)/p)^{50})$.
(b) If $p = 0.4$, then this equals approximately 1.034×10^{-8}.
(c) The corresponding probability with \$1 bets is $(1-((1-p)/p)^{10})/(1-((1-p)/p)^{100}) \doteq 1.39 \times 10^{-16}$. Hence, we have a larger probability of reaching our goal if we bet \$2 each time, rather than \$1.
(d) The corresponding probability with \$10 bets is $(1-((1-p)/p)^1)/(1-((1-p)/p)^{10}) \doteq 0.00882$. Hence, the probability of success *increases* if we bet \$10 each time, rather than \$1 or \$2.

Challenge

11.1.8 Fix a and c and let $f(x) = (1-x^a)/(1-x^c)$. We wish to compute $\lim_{p\to 1/2} f((1-p)/p)$. Since $\lim_{p\to 1/2}((1-p)/p) = 1$, the desired limit is equal to $\lim_{x\to 1} f(x)$ (if it exists). But from L'Hospital's rule, $\lim_{x\to 1} f(x) = \lim_{x\to 1}(\frac{d}{dx}(1-x^a)/\frac{d}{dx}(1-x^c)) = \lim_{x\to 1}(ax^{a-1})/(cx^{c-1}) = a/c$, as desired.

11.2 Markov Chains

Exercises

11.2.1
(a) $P(X_0 = 1) = \mu_1 = 0.7$.
(b) $P(X_0 = 2) = \mu_2 = 0.1$.
(c) $P(X_0 = 3) = \mu_3 = 0.2$.
(d) $P(X_1 = 2 \,|\, X_0 = 1) = p_{12} = 1/4$.
(e) $P(X_3 = 2 \,|\, X_2 = 1) = p_{12} = 1/4$.
(f) $P(X_1 = 2 \,|\, X_0 = 2) = p_{22} = 1/2$.
(g) $P(X_1 = 2) = \sum_i \mu_i p_{i2} = (0.7)(1/4) + (0.1)(1/2) + (0.2)(3/8) = 0.3$.

11.2.2
(a) $P(X_0 = \text{high}) = \mu_{\text{high}} = 1/3$.
(b)$P(X_0 = \text{low}) = \mu_{\text{low}} = 2/3$.
(c) $P(X_1 = \text{high} \,|\, X_0 = \text{high}) = p_{\text{high,high}} = 1/4$.
(d) $P(X_3 = \text{high} \,|\, X_2 = \text{high}) = p_{\text{high,high}} = 1/4$.
(e) $P(X_1 = \text{high}) = \sum_i \mu_i p_{i,\text{high}} = (1/3)(1/4) + (2/3)(1/6) = 7/36$.

11.2.3
(a) $P_0(X_2 = 0) = \sum_i p_{0i} p_{i0} = (0.2)(0.2) + (0.8)(0.3) = 0.28$. $P_0(X_2 = 1) = \sum_i p_{0i} p_{i1} = (0.2)(0.8) + (0.8)(0.7) = 0.72$. $P_1(X_2 = 0) = \sum_i p_{1i} p_{i0} = (0.3)(0.2)+(0.7)(0.3) = 0.27$. $P_1(X_2 = 1) = \sum_i p_{1i} p_{i1} = (0.3)(0.8)+(0.7)(0.7) = 0.73$.
(b) $P_0(X_3 = 1) = \sum_{i,j} p_{0i} p_{ij} p_{j1} = (0.2)(0.2)(0.8)+(0.2)(0.8)(0.7)+(0.8)(0.3)(0.8)+(0.8)(0.7)(0.7) = 0.728$.

11.2.4
(a) We need $\pi_0(0.2) + \pi_1(0.3) = \pi_0$ and $\pi_0(0.8) + \pi_1(0.7) = \pi_1$, where $\pi_1 = (8/3)\pi_0$, where $\pi_0 = 3/11$, and $\pi_1 = 8/11$.
(b) Since $p_{ij} > 0$ for all i and j, the chain is irreducible and aperiodic, so $\lim_{n\to\infty} P_0(X_n = 0) = \pi_0 = 3/11$.
(c) Similarly, $\lim_{n\to\infty} P_1(X_n = 0) = \pi_0 = 3/11$.

11.2.5
(a) $P_2(X_1 = 1) = p_{21} = 1/2$.
(b) $P_2(X_1 = 2) = p_{22} = 0$.
(c) $P_2(X_1 = 3) = p_{23} = 1/2$.
(d) $P_2(X_2 = 1) = \sum_i p_{2i} p_{i3} = (1/2)(1) + (1/2)(0) = 1/2$.
(e) $P_2(X_2 = 2) = \sum_i p_{2i} p_{i3} = (1/2)(0) + (1/2)(1/5) = 1/10$.
(f) $P_2(X_2 = 3) = \sum_i p_{2i} p_{i3} = (1/2)(0) + (1/2)(4/5) = 2/5$.
(g) $P_2(X_3 = 3) = \sum_{ij} p_{2i} p_{ij} p_{j3} = (1/2)(0)+(1/2)(1/5)(1/2)+(1/2)(4/5)(4/5) = 37/100$.
(h) $P_2(X_3 = 1) = \sum_{ij} p_{2i} p_{ij} p_{j1} = (1/2)(1)(1) + (1/2)(1/5)(1/2) = 11/20$.
(i) $P_2(X_1 = 7) = 0$.
(j) $P_2(X_2 = 7) = 0$.
(k) $P_2(X_3 = 7) = 0$.

(l) $\max_n P_2(X_n = 7) = 0$ since it is impossible to get from 2 to 7 in any number of steps.
(m) No, since e.g. it is impossible to get from 2 to 7 in any number of steps.

11.2.6
(a) Here $p_{ij} > 0$ for all $i$and j, so the chain is irreducible and aperiodic.
(b) Here $p_{ij} > 0$ for all $i$and j, so the chain is irreducible and aperiodic.
(c) Here $p_{12}p_{21} > 0$ and $p_{22} > 0$, so the chain is irreducible. Also, $p_{12}p_{21} > 0$ and $p_{12}p_{22}p_{21} > 0$ and $p_{22} > 0$, so the chain is aperiodic.
(d) Since $p_{i2} > 0$ for all i, and $p_{2j} > 0$ for all j, the chain is irreducible. Also, since $p_{ii}^{(2)} > 0$ and $p_{ii}^{(3)} > 0$ for all i, the chain is aperiodic.
(e) This chain is irreducible since for all i and j, $p_{ij}^{(n)} > 0$ for some $n \le 3$. However, the chain is *not* aperiodic since $p_{ii}^{(n)} > 0$ only when n is a multiple of 3.
(f) This chain is irreducible since for all i and j, $p_{ij}^{(n)} > 0$ for some $n \le 3$. Also, since $p_{ii}^{(3)} > 0$ and $p_{ii}^{(4)} > 0$ for all i, the chain is aperiodic.

11.2.7 This chain is doubly stochastic, i.e., has $\sum_i p_{ij} = 1$ for all j. Hence, as in Example 11.2.15, we must have the uniform distribution ($\pi_1 = \pi_2 = \pi_3 = \pi_4 = 1/4$) as a stationary distribution.

11.2.8
(a) By moving clockwise one step at a time, we see that for all i and j, we have $p_{ij}^{(n)} > 0$ for some $n \le d$. Hence, the chain is irreducible.
(b) Since $p_{ii} > 0$ for all i, each state has period 1, so the chain is aperiodic.
(c) If i and j are two or more apart, then $p_{ij} = p_{ji} = 0$. If i and j are one apart, then $\pi_i p_{ij} = (1/d)(1/3) = 1/3d$ and $\pi_j p_{ji} = (1/d)(1/3) = 1/3d$. Hence, the chain is reversible with respect to $\{\pi_i\}$.

11.2.9
(a) By either increasing or decreasing one step at a time, we see that for all i and j, we have $p_{ij}^{(n)} > 0$for some $n \le d$. Hence, the chain is irreducible.
(b) The chain can only move from even numbers to odd, and from odd numbers to even. Hence, each state has period 2.
(c) If i and j are two or more apart, then $p_{ij} = p_{ji} = 0$. If $j = i + 1$, then $\pi_i p_{ij} = (1/2^d)\binom{d}{i}((d-i)/d) = (1/2^d)(d!/i!(d-i)!)((d-i)/d) = (1/2^d)((d-1)!/i!(d-i-1)!)$, while $\pi_j p_{ji} = (1/2^d)\binom{d}{i+1}((i+1)/d) = (1/2^d)(d!/(i+1)!(d-i-1)!)((i+1)/d) = (1/2^d)((d-1)!/i!(d-i-1)!)$. Hence, the chain is reversible with respect to $\{\pi_i\}$.

11.2.10
(a) This chain is irreducible since for all i and j, $p_{ij}^{(n)} > 0$ for some $n \le 3$.
(b) Since $p_{ii}^{(3)} > 0$ and $p_{ii}^{(5)} > 0$ for all i, the chain is aperiodic.
(c) We need $\pi_3(1/2) = \pi_1$, $\pi_1(1) + \pi_3(1/2) = \pi_2$, and $\pi_2(1) = \pi_3$. Hence, $\pi_1 = 1/4$ and $\pi_2 = \pi_3 = 1/2$.
(d) We have $\lim_{n\to\infty} P_1(X_n = 2) = \pi_2 = 1/2$. Hence, $P_1(X_{500} = 2) \approx 1/2$.

11.2.11
(a) This chain is irreducible since for all i and j, $p_{ij}^{(n)} > 0$ for some $n \leq 3$.
(b) Since $p_{ii}^{(3)} > 0$ and $p_{ii}^{(5)} > 0$ for all i, the chain is aperiodic.
(c) We need $\pi_3(1/2) = \pi_1$, $\pi_1(1/2) + \pi_3(1/2) = \pi_2$, and $\pi_1(1/2) + \pi_2(1) = \pi_3$. Hence, $\pi_1 = 2/9$, $\pi_2 = 3/9$, and $\pi_3 = 4/9$.
(d) We have $\lim_{n\to\infty} P_1(X_n = 2) = \pi_2 = 3/9 = 1/3$. Hence, $P_1(X_{500} = 2) \approx 1/3$.

Problems

11.2.12 For reversibility, we need $\pi_1 p_{12} = \pi_2 p_{21}$, so $\pi_2 = (p_{12}/p_{21})\pi_1 = ((4/5)/(1/5))\pi_1 = 4\pi_1$. Then $\pi_3 = (p_{23}/p_{32})\pi_2 = ((4/5)/(1/5))\pi_2 = 4\pi_2 = 4^2\pi_1$. Then $\pi_4 = (p_{34}/p_{43})\pi_3 = ((4/5)/(1/5))\pi_3 = 4\pi_3 = 4^3\pi_1$. Then $\pi_5 = (p_{45}/p_{54})\pi_4 = ((4/5)/(1/5))\pi_4 = 4\pi_4 = 4^4\pi_1$. Hence, since $1+4+4^2+4^3+4^4 = 341$, we have $\pi_1 = 1/341$, $\pi_2 = 4/341$, $\pi_3 = 4^2/341$, $\pi_4 = 4^3/341$, and $\pi_5 = 4^4/341$.

11.2.13 We know that this example is irreducible and aperiodic, with $\pi_i = 1/d = 1/100$for all i. Hence, $\lim_{n\to\infty} P_0(X_n = 55) = 1/100$. Hence, for large n (such as the number of seconds in a month), $P_0(X_n = 55) \approx 1/100$.

11.2.14 We use induction on n. The case $n = 1$ follows by definition. Assuming the theorem is true for some n, then $P_i(X_{n+1} = j) = \sum_k P_i(X_n = k,\ X_{n+1} = j) = \sum_k P_i(X_n = k)\, p_{kj} = \sum_k \sum_{i_1,\ldots,i_{n-1}} p_{ii_1} p_{i_1 i_2} \cdots p_{i_{n-1}k}\, p_{kj}$. The result follows by replacing the dummy variable k by i_n.

11.2.15 If $j = 0$, then $p_{ij} > 0$ only for $i = 1$ when $p_{10} = 1/d$. Hence, $\sum_{i\in S} \pi_i p_{ij} = \pi_1(1/d) = (1/2^d)\binom{d}{1}(1/d) = (1/2^d)(d)(1/d) = (1/2^d) = \pi_0$. If $j = d$, then $p_{ij} > 0$ only for $i = d-1$ when $p_{d-1,d} = 1/d$. Hence, $\sum_{i\in S} \pi_i p_{ij} = \pi_1(1/d) = (1/2^d)\binom{d}{d-1}(1/d) = (1/2^d)(d)(1/d) = (1/2^d) = \pi_d$.

11.3 Markov Chain Monte Carlo

Exercises

11.3.1 First, choose any initial value X_0. Then, given $X_n = i$, let $Y_{n+1} = i+1$ or $i-1$, with probability 1/2 each. Let $j = Y_{n+1}$ and let $\alpha_{ij} = \min(1,\ \pi_j/\pi_i) = \min(1,\ e^{-(j-13)^4+(i-13)^4})$. Then let $X_{n+1} = j$ with probability α_{ij}, otherwise let $X_{n+1} = i$ with probability $1 - \alpha_{ij}$.

11.3.2 First, choose any initial value X_0. Then, given $X_n = i$, let $Y_{n+1} = i+1$ with probability 5/8 or $Y_{n+1} = i - 1$ with probability 3/8. Let $j = Y_{n+1}$ and let $\alpha_{ij} = \min(1,\ \pi_j q_{ji}/\pi_i q_{ij}) = \min(1,\ (i+7.5)^{-8}(3/8)/(i+6.5)^{-8}(5/8))$ if $j = i+1$ or $\alpha_{ij} = \min(1,\ \pi_j q_{ji}/\pi_i q_{ij}) = \min(1,\ (i+5.5)^{-8}(5/8)/(i+6.5)^{-8}(3/8))$ if $j = i - 1$. Then let $X_{n+1} = j$ with probability α_{ij}, otherwise let $X_{n+1} = i$ with probability $1 - \alpha_{ij}$.

11.3.3 First, choose any initial value X_0. Then, given $X_n = i$, let $Y_{n+1} = i+1$ with probability 7/9 or $Y_{n+1} = i - 1$ with probability 2/9. Let $j = Y_{n+1}$ and let $\alpha_{ij} = \min(1,\ \pi_j q_{ji}/\pi_i q_{ij}) = \min(1,\ e^{-j^4-j^6-j^8}(2/9)/e^{-i^4-i^6-i^8}(7/9))$ if $j = i+1$ or $\alpha_{ij} = \min(1,\ \pi_j q_{ji}/\pi_i q_{ij}) = \min(1,\ e^{-j^4-j^6-j^8}(7/9)/e^{-i^4-i^6-i^8}(2/9))$ if $j = i-1$. Then let $X_{n+1} = j$ with probability α_{ij}, otherwise let $X_{n+1} = i$ with probability $1 - \alpha_{ij}$.

11.3.4 Let $\{Z_n\}$be i.i.d. $\sim N(0,1)$. First, choose any initial value X_0. Then, given $X_n = x$, let $Y_{n+1} = X_n + Z_{n+1}$. Let $y = Y_{n+1}$ and let $\alpha_{xy} = \min(1,\ f(y)/f(x)) = \min(1,\ e^{-y^4-y^6-y^8+x^4+x^6+x^8})$. Then let $X_{n+1} = y$ with probability α_{xy}, otherwise let $X_{n+1} = x$ with probability $1 - \alpha_{xy}$.

11.3.5 Let $\{Z_n\}$ be i.i.d. $\sim N(0,1)$. First, choose any initial value X_0. Then, given $X_n = x$, let $Y_{n+1} = X_n + \sqrt{10}\, Z_{n+1}$. Let $y = Y_{n+1}$ and let $\alpha_{xy} = \min(1,\ f(y)/f(x)) = \min(1,\ e^{-y^4-y^6-y^8+x^4+x^6+x^8})$. Then let $X_{n+1} = y$ with probability α_{xy}, otherwise let $X_{n+1} = x$ with probability $1 - \alpha_{xy}$.

Problem

11.3.6 First, choose any initial value X_0. Then, given $X_n = (i_1, i_2)$, choose Y_{n+1} so that $P(Y_{n+1} = (i_1, j)) = 2^{-j}$ for $j = 1, 2, 3, \ldots$. Then, given $Y_{n+1} = (i_1, j)$, choose Z_{n+1} so that $P(Z_{n+1} = (k, j)) = 2^{-k}$ for $k = 1, 2, 3, \ldots$. Then set $X_{n+1} = Z_{n+1} = (k, j)$.

11.4 Martingales

Exercises

11.4.1 Here $E(X_{n+1} \mid X_n) = (3/8)(X_n - 4) + (5/8)(X_n + C) = X_n + (5C - 12)/8$. This equals X_n if $C = 12/5$.

11.4.2 Here $E(X_{n+1} \mid X_n) = (p)(X_n + 7) + (1-p)(X_n - 2) = X_n + (9p - 2)$. This equals X_n if $p = 2/9$.

11.4.3 Here $E(X_{n+1} \mid X_n) = (p)(2X_n) + (1-p)(X_n/2) = X_n(2p + (1-p)/2) = X_n((3p/2) + (1/2))$. This equals X_n if $(3p/2) + (1/2) = 1$, i.e. if $p = 1/3$.

11.4.4 Let $p = P(X_n = 14)$. Then $E(X_n) = (0.1)(8) + (0.9 - p)(12) + (p)(14) = 2p + 11.6$. But $\{X_n\}$ is a martingale, so we know that $E(X_n) = X_0 = 14$. Hence, $2p + 11.6 = 14$, so $p = 1.2$.

11.4.5 Let $p = P(X_n = 4)$. Then $E(X_n) = (2/3)(1-p)(3) + (1/3)(1-p)(4) + (p)(6) = (8p + 10)/3$. But $\{X_n\}$ is a martingale, so we know that $E(X_n) = X_0 = 5$. Hence, $(8p + 10)/3 = 5$, so $p = 5/8$.

11.4.6
(a) Let X_n be the number of pennies at time n. Then $\{X_n\}$ is a martingale. Hence, $E(X_{20}) = X_0 = 175$.
(b) Let p be the probability you have 200 pennies when you stop. Then let T be the time at which you stop. Then $E(X_T) = (p)(200) + (1-p)(100) = 100 + 100p$. But $\{X_n\}$ is a bounded martingale, so $E(X_T) = X_0 = 175$. Hence, $100 + 100p = 175$, so $p = 3/4$.

11.4.7
(a) Here $E(X_{n+1} \mid X_n) = (1/4)(3X_n) + (3/4)(X_n/3) = X_n$, so $\{X_n\}$ is a martingale.
(b) T is non-negative integer-valued and does not look into the future, so it is a stopping time.
(c) Since $\{X_n\}_{n \le T}$ is bounded between 1 and 81, we have $E(X_T) = X_0 = 27$.
(d) Let $p = P(X_T = 1)$. Then $E(X_T) = (p)(1) + (1-p)(81) = 81 - 80p$. Hence, $81 - 80p = 27$, so $p = 54/80 = 27/40$.

Problems

11.4.8 Since T_2 happens *later* than T_1, it also does not look into the future, so it also must be a stopping time. However, since T_3 happens *earlier*, it is possible that T_3 looks into the future, so T_3 may not be a stopping time.

11.4.9
(a) Since $\{T_3 \le n\} = \{T_1 \le n\} \cup \{T_2 \le n\}$ and T_1 and T_2 are stopping times, then $\{T_3 \le n\}$ is also a function of $X_0, X_1, \ldots, X_n$ alone, so that T_3 is also a stopping time.
(b) Since $\{T_4 \le n\} = \{T_1 \le n\} \cap \{T_2 \le n\}$ and T_1 and T_2 are stopping times, then $\{T_4 \le n\}$ is also a function of $X_0, X_1, \ldots, X_n$ alone, so that T_4 is also a stopping time.

11.5 Brownian Motion

Exercises

11.5.1
(a) $P(Y_1^{(1)} = 1) = 1/2$.
(b) $P(Y_1^{(2)} = 1) = 0$.
(c) $P(Y_1^{(2)} = \sqrt{2}) = P(Y_{2/2}^{(2)} = 2/\sqrt{2}) = (1/2)(1/2) = 1/4$.
(d) We have $P(Y_1^{(M)} \ge 1) = P(Y_{M/M}^{(M)} \ge \sqrt{M}/\sqrt{M})$. Hence, $P(Y_1^{(1)} \ge 1) = 1/2$. Also, $P(Y_1^{(2)} \ge 1) = (1/2)(1/2) = 1/4$. Also, $P(Y_1^{(3)} \ge 1) = (1/2)(1/2)(1/2) + (1/2)(1/2)(1/2) + (1/2)(1/2)(1/2) = 3/8$. Also, $P(Y_1^{(4)} \ge 1) = (1/2)(1/2)(1/2)(1/2) + (1/2)(1/2)(1/2)(1/2) + (1/2)(1/2)(1/2)(1/2) + (1/2)(1/2)(1/2)(1/2) + (1/2)(1/2)(1/2)(1/2) = 5/16$.

11.5.2 Since $B_1 \sim N(0,1)$, $P(B_1 \ge 1) = P(B_1 \le -1) = \Phi(-1) \doteq 0.1587$.

11.5.3
(a) $P(B_2 \ge 1) = P((1/\sqrt{2})B_2 \ge 1/\sqrt{2}) = \Phi(-1/\sqrt{2}) \doteq 0.2397$.
(b) $P(B_3 \le -4) = P((1/\sqrt{3})B_2 \le -4/\sqrt{3}) = \Phi(-4/\sqrt{3}) \doteq 0.0105$.
(c) $P(B_9 - B_5 \le 2.4) = P(B_4 \le 2.4) = P((1/2)B_4 \le 2.4/2) = \Phi(2.4/2) = 1 - \Phi(-2.4/2) \doteq 0.8849$.
(d) $P(B_{26} - B_{11} > 9.8) = P(B_{15} > 9.8) = P((1/\sqrt{15})B_{15} > 9.8/\sqrt{15}) = \Phi(-9.8/\sqrt{15}) \doteq 0.0057$.

(e) $P(B_{26.3} \leq -6) = P((1/\sqrt{26.3})B_{26.3} \leq -6/\sqrt{26.3}) = \Phi(-6/\sqrt{26.3}) \doteq 0.1210$.
(f) $P(B_{26.3} \leq 0) = P((1/\sqrt{26.3})B_{26.3} \leq 0/\sqrt{26.3}) = \Phi(0/\sqrt{26.3}) = 1/2$.

11.5.4
(a) $P(B_2 \geq 1,\ B_5 - B_2 \geq 2) = P(B_2 \geq 1)\,P(B_5 - B_2 \geq 2) = P((1/\sqrt{2})B_2 \geq 1/\sqrt{2})\,P((1/\sqrt{3})(B_5 - B_2) \geq 2/\sqrt{3}) = \Phi(-1/\sqrt{2})\,\Phi(-2/\sqrt{3}) \doteq 0.02975$.
(b) $P(B_5 < -2,\ B_{13}-B_5 \geq 4) = P(B_5 < -2)\,P(B_{13}-B_5 \geq 4) = P((1/\sqrt{5})B_5 < -2/\sqrt{5})\,P((1/\sqrt{8})(B_{13} - B_5) \geq 4/\sqrt{8}) = \Phi(-2/\sqrt{5})\,\Phi(-4/\sqrt{8}) \doteq 0.01459$.
(c) $P(B_{8.4} > 3.2,\ B_{18.6} - B_{8.4} \geq 0.9) = P(B_{8.4} > 3.2)\,P(B_{18.6} - B_{8.4} \geq 0.9) = P((1/\sqrt{8.4})B_{8.4} > 3.2/\sqrt{8.4})\,P((1/\sqrt{10.2})(B_{18.6} - B_{8.4}) \geq 0.9/\sqrt{10.2}) = \Phi(-3.2/\sqrt{8.4})\,\Phi(-0.9/\sqrt{10.2}) \doteq 0.05243$.

11.5.5 $E(B_{13}B_8) = \min(13, 8) = 8$.

11.5.6
(a) Since $B_{17} - B_{14} \sim N(0,3)$, $E((B_{17} - B_{14})^2) = 3 + 0^2 = 3$.
(b) $E((B_{17}-B_{14})^2) = E(B_{17}^2)-2E(B_{17}B_{14})+E(B_{14}^2) = 17+0^2-2\min(17,14)+14+0^2 = 17-2\cdot 14+14 = 3$.

11.5.7
(a) Let $p = P(\text{hit } -5 \text{ before } 15)$ and let T be the first time we hit either. Then $0 = E(B_T) = p(-5) + (1-p)(15) = 15 - 20p$, so that $p = 15/20 = 3/4$.
(b) Let $p = P(\text{hit } -15 \text{ before } 5)$ and let T be the first time we hit either. Then $0 = E(B_T) = p(-15) + (1-p)(5) = 5 - 20p$, so that $p = 5/20 = 1/4$.
(c) The answer in (a) is larger because -5 is closer to $B_0 = 0$ than 15 is, while -15 is farther than 5 is.
(d) Let $p = P(\text{hit } 15 \text{ before } -5)$ and let T be the first time we hit either. Then $0 = E(B_T) = p(15) + (1-p)(-5) = -5 + 20p$, so that $p = 5/20 = 1/4$.
(e) We have $3/4 + 1/4 = 1$, which it must since the events in parts (a) and (d) are complimentary events.

11.5.8
(a) $E(X_7) = E(5 + 3\cdot 7 + 2B_7) = 5 + 3\cdot 7 + 2\cdot 0 = 26$.
(b) $\mathrm{Var}(X_{8.1}) = \mathrm{Var}(5 + 3\cdot 8.1 + 2B_{8.1}) = 4\,\mathrm{Var}(B_{8.1}) = 4\cdot 8.1 = 32.4$.
(c) $P(X_{2.5} < 12) = P(5 + 3\cdot 2.5 + 2B_{2.5} < 12) = P(B_{2.5} < -1/4) = P((1/\sqrt{2.5})B_{2.5} < -1/4\sqrt{2.5}) = \Phi(-1/4\sqrt{2.5}) \doteq 0.4372$.
(d) $P(X_{17} > 50) = P(5+3\cdot 17+2B_{17} > 50) = P(B_{17} > -3) = P((1/\sqrt{17})B_{17} > -3\sqrt{17}) = \Phi(-3/\sqrt{17}) \doteq 0.2334$.

11.5.9 $E(X_3X_5) = E((10 - 1.5\cdot 3 + 4B_3)(10 - 1.5\cdot 5 + 4B_5)) = E((5.5 + 4B_3)(2.5 + 4B_5)) = E(5.5\cdot 2.5 + 4\cdot 2.5\cdot B_3 + 4\cdot 5.5\cdot B_5 + 4\cdot 4\cdot B_3\cdot B_5) = 5.5\cdot 2.5 + 4\cdot 2.5\cdot 0 + 4\cdot 5.5\cdot 0 + 4\cdot 4\cdot \min(3,5) = 61.75$.

11.5.10
(a) $P(X_8 > 500) = P(400+0\cdot 8+9B_8 > 500) = P(B_8 > 100/9) = P((1/\sqrt{8})B_8 > 100/9\sqrt{8}) = \Phi(-100/9\sqrt{8}) \doteq 0.00004276$.
(b) $P(X_8 > 500) = P(400+5\cdot 8+9B_8 > 500) = P(B_8 > 60/9) = P((1/\sqrt{8})B_8 > 60/9\sqrt{8}) = \Phi(-60/9\sqrt{8}) \doteq 0.009211$.

(c) $P(X_8 > 500) = P(400{+}10{\cdot}8{+}9B_8 > 500) = P(B_8 > 20/9) = P((1/\sqrt{8})B_8 > 20/9\sqrt{8}) = \Phi(-20/9\sqrt{8}) \doteq 0.2160$.
(d) $P(X_8 > 500) = P(400 + 20 \cdot 8 + 9B_8 > 500) = P(B_8 > -60/9) = P((1/\sqrt{8})B_8 > -60/9\sqrt{8}) = 1 - \Phi(-60/9\sqrt{8}) \doteq 0.9908$.

11.5.11
(a) $P(X_{10} > 250) = P(200 + 3 \cdot 10 + 1B_{10} > 250) = P(B_{10} > 20/1) = P((1/\sqrt{10})B_{10} > 20/1\sqrt{10}) = \Phi(-20/1\sqrt{10}) \doteq 1.27 \times 10^{-10}$.
(b) $P(X_{10} > 250) = P(200 + 3 \cdot 10 + 4B_{10} > 250) = P(B_{10} > 20/4) = P((1/\sqrt{10})B_{10} > 20/4\sqrt{10}) = \Phi(-20/4\sqrt{10}) \doteq 0.05692$.
(c) $P(X_{10} > 250) = P(200 + 3 \cdot 10 + 10B_{10} > 250) = P(B_{10} > 20/10) = P((1/\sqrt{10})B_{10} > 20/10\sqrt{10}) = \Phi(-20/10\sqrt{10}) \doteq 0.2635$.
(d) $P(X_{10} > 250) = P(200 + 3 \cdot 10 + 100B_{10} > 250) = P(B_{10} > 20/100) = P((1/\sqrt{10})B_{10} > 20/100\sqrt{10}) = \Phi(-20/100\sqrt{10}) \doteq 0.4748$.

Problems

11.5.12 We have that $E(X) = E(2B_3 - 7B_5) = 2 \cdot 0 - 7 \cdot 0 = 0$. Also, $E(X^2) = E((2B_3 - 7B_5)^2) = E(4B_3^2 + 49B_5^2 - 28B_3B_5) = 4 \cdot 3 + 49 \cdot 5 - 28\min(3,5) = 173$. Hence, $\text{Var}(X) = 173$.

11.5.13 We have that $P(B_t < x) = P((1/\sqrt{t})B_t < x/\sqrt{t}) = \Phi(x/\sqrt{5})$, while $P(B_t > -x) = P((1/\sqrt{t})B_t > -x/\sqrt{t}) = 1 - \Phi(-x/\sqrt{5}) = \Phi(x/\sqrt{5}) = P(B_t < x)$.

Challenges

11.5.14 Let $g(x,y) = f_{B_s B_t}(x,y)$ be the joint density function of B_s and $B_t - B_s$. Since $B_s \sim N(0,s)$ and $B_t - B_s \sim N(0, t-s)$ and they are independent, then $g(x,y) = (1/2\pi\sqrt{s(t-s)})e^{-x^2/2s}e^{-y^2/2(t-s)}$. Then let $h(x,y) = f_{B_s,B_t}(x,y)$ be the joint density function of B_s and B_t. Then by the two-dimensional change-of-variable theorem $h(x,y) = g(x, y-x)$
$= (1/2\pi\sqrt{s(t-s)})e^{-x^2/2s}e^{-(y-x)^2/2(t-s)}$. Then the conditional density of B_s given B_t is equal to

$$
\begin{aligned}
&h(x|y) = h(x,y)/f_{B_t}(y)\\
&= (1/2\pi\sqrt{s(t-s)})e^{-x^2/2s}e^{-(y-x)^2/2(t-s)} \,/\, (1/\sqrt{2\pi t})e^{-y^2/2t}\\
&= \sqrt{t/2\pi s(t-s)}e^{-x^2/2s}e^{-y^2/2(t-s)}e^{+y^2/2t} = \sqrt{t/2\pi s(t-s)}e^{-(tx-sy)^2/2st(t-s)}\\
&= \sqrt{1/2\pi[s(t-s)/t]}e^{-(x-sy/t)^2/2[s(t-s)/t]}.
\end{aligned}
$$

We conclude that, conditional on $B_t = y$, the conditional distribution of B_s is normally distributed with mean sy/t and variance $s(t-s)/t$. Hence, choosing $Z \sim N(sy/t,\ s(t-s)/t)$, we have $P(B_s \le x \mid B_t = y) = P(Z \le x) = P((Z - sy/t)/\sqrt{s(t-s)/t} \le (x - sy/t)/\sqrt{s(t-s)/t}) = \Phi((x - sy/t)/\sqrt{s(t-s)/t})$.

11.5.15
(a) $\lim_{h\searrow 0} |(f(t+h) - f(t))^2/h| \le \lim_{h\searrow 0} |(Kh)^2/h| = \lim_{h\searrow 0} |Kh| = 0$, so $\lim_{h\searrow 0}(f(t+h) - f(t))^2/h = 0$.

(b) $\lim_{h\searrow 0} E((B_{t+h} - B_t)^2/h) = \lim_{h\searrow 0} h/h = 1 \neq 0$.
(c) They imply that Brownian motion is not always a Lipschitz function. (In fact, it never is.)
(d) This implies that Brownian motion is not always a differentiable function. (In fact, it never is.)

11.6 Poisson Processes

Exercises

11.6.1
(a) $N_2 \sim \text{Poisson}(14)$, so $P(N_2 = 13) = e^{-14}14^{13}/13! \doteq 0.1060$.
(b) $P(N_5 = 3) = e^{-35}35^3/3! \doteq 4.5 \times 10^{-12}$.
(c) $P(N_6 = 20) = e^{-42}42^{20}/20! \doteq 6.9 \times 10^{-5}$.
(d) $P(N_{50} = 340) = e^{-350}350^{340}/340! \doteq 0.01873$.
(e) We have that $P(N_2 = 13,\ N_5 = 3) = 0$ since we always have $N_5 \geq N_2$.
(f) We have that $P(N_2 = 13,\ N_5 = 20) = P(N_2 = 13,\ N_5 - N_2 = 7) = P(N_2 = 13)\,P(N_5 - N_2 = 7) = (e^{-14}14^{13}/13!)\,(e^{-21}21^7/7!) \doteq 2.9 \times 10^{-5}$.
(g) $P(N_2 = 13,\ N_5 = 3,\ N_6 = 20) = 0$ since we always have $N_5 \geq N_2$.

11.6.2 We have that $P(N_{1/2} = 6) = e^{-3/2}(3/2)^6/6! \doteq 0.00353$. Also, $P(N_{0.3} = 5) = e^{-0.9}(0.9)^5/5! \doteq 0.00200$.

11.6.3 We have that $P(N_2 = 6) = e^{-2/3}(2/3)^6/6! \doteq 6.26 \times 10^{-5}$. Also, $P(N_3 = 5) = e^{-3/3}(3/3)^5/5! \doteq 0.00307$.

11.6.4 We have that $P(N_2 = 6,\ N_3 = 5) = 0$ since we always have $N_3 \geq N_2$.

11.6.5 $P(N_{2.6} = 2\,|\,N_{2.9} = 2) = P(N_{2.6} = 2,\ N_{2.9} = 2)/F(N_{2.9} = 2) = P(N_{2.6} = 2,\ N_{2.9} - N_{2.6} = 0)/F(N_{2.9} = 2) = P(N_{2.6} = 2)\,P(N_{2.9} - N_{2.6} = 0)/F(N_{2.9} = 2) = (e^{-2.6a}2.6^2/2!)(e^{-0.3a}0.3^0/0!)/(e^{-2.9a}2.9^2/2!) = (2.6/2.9)^2 \doteq 0.8038$

11.6.6
(a) $P(N_6 = 5\,|\,N_9 = 5) = P(N_6 = 5,\ N_9 = 5)\,/\,F(N_9 = 5) = P(N_6 = 5,\ N_9 - N_6 = 0)\,/\,F(N_9 = 5) = P(N_6 = 5)\,P(N_9 - N_6 = 0)\,/\,F(N_9 = 5) = (e^{-6/3}(6/3)^5/5!)(e^{-3/3}(3/3)^0/0!)/(e^{-9/3}(9/3)^5/5!) \doteq 0.1317$.
(b) $P(N_6 = 5\,|\,N_9 = 7) = P(N_6 = 5,\ N_9 = 7)\,/\,F(N_9 = 7) = P(N_6 = 5,\ N_9 - N_6 = 2)\,/\,F(N_9 = 7) = P(N_6 = 5)\,P(N_9 - N_6 = 2)\,/\,F(N_9 = 7) = (e^{-6/3}(6/3)^5/5!)(e^{-3/3}(3/3)^2/2!)/(e^{-9/3}(9/3)^7/7!) \doteq 0.6145$.
(c) We have that $P(N_9 = 5\,|\,N_6 = 7) = 0$ since we always have $N_9 \geq N_6$.
(d) $P(N_9 = 7\,|\,N_6 = 7) = P(N_9 - N_6 = 0\,|\,N_6 = 7) = P(N_9 - N_6 = 0) = e^{-3/3}(3/3)^0/0! = 1/e \doteq 0.3679$.
(e) $P(N_9 = 12\,|\,N_6 = 7) = P(N_9 - N_6 = 5\,|\,N_6 = 7) = P(N_9 - N_6 = 5) = e^{-3/3}(3/3)^5/5! \doteq 0.00307$.

Problems

11.6.7

(a) $P(N_s = j \,|\, N_t = j) = P(N_s = j,\ N_t = j) \,/\, P(N_t = j) = P(N_s = j,\ N_t - N_s = 0) \,/\, P(N_t = j) = P(N_s = j)\, P(N_t - N_s = 0) \,/\, P(N_t = j) = (e^{-as}(as)^j/j!)(e^{-a(t-s)}(as)^0/0!)/(e^{-at}(at)^j/j!) = (s/t)^j$.

(b) No, the answer does not depend on a. Intuitively, once we condition on knowing that we have exactly j events between time 0 and time t, then we no longer care what was the intensity which produced them.

11.6.8

(a) $P(N_s = 1 \,|\, N_t = 1) = (s/t)^1 = s/t$.

(b) This says that $P(T_1 \le s \,|\, N_t = 1) = s/t$. It follows that, conditional on $N_t = 1$, the distribution of T_1 is *uniform* on the interval $[0, t]$.

Appendix: Selected Answers

Chapter 1

Section 1.2

1.2.1 (a) 5/6. (b) 1. (c) 1/2.

1.2.4 No.

1.2.8 No.

1.2.10 Yes.

Section 1.3

1.3.1 (a) 0.9. (b) 0.1 is the smallest possible value of $P(\{1\})$.

1.3.4 The maximum is 25% and the minimum is 15%.

Section 1.4

1.4.1 (a) 1/1679616. (b) 1/279936. (c) 1/209952.

1.4.6 $P(\text{sum} \geq 4) = 1 - 1/221 - 8/663 = 652/663$.

1.4.10 5/12.

Section 1.5

1.5.1 (a) 0.75. (b) The conditional probability equals 0.762.

1.5.4 $P(\text{five Spades} \mid \text{at least 4 Spades}) \doteq 0.044$.

1.5.9 (a) A and B are not independent. (b) A and C are independent. (c) A and D are independent. (d) C and D are independent. (e) A and C and D are not all independent.

1.5.10 We have from the Exercise 1.4.11 solution that $P(\text{all red}) = 5/4488$, while $P(\text{all blue}) = 35/816$. Hence, $P(\text{all red} \mid \text{all same color}) = P(\text{all red}) \,/\, P(\text{all same color}) = (5/4488)/[(5/4488) + (35/816)] = 2/79 \doteq 0.025$.

Section 1.6

1.6.2 $\lim_{n\to\infty} P([1/4,\ 1 - e^{-n}]) = 3/4$.

1.6.3 $\lim_{n\to\infty} P(A_n) = P(A) = P(S) = 1$.

Chapter 2

Section 2.1

2.1.1(a) $\min_{s\in S} X(s) = X(1) = 1$. (b) $\max_{s\in S} X(s)$ does not exist. (c) $\min_{s\in S} Y(s)$ does not exist. (d) $\max_{s\in S} Y(s) = Y(1) = 1$.

2.1.4 $Z(1) = 3$, $Z(2) = 20$, $Z(3) = 87$, $Z(4) = 1032$, $Z(5) = 15,635$, $Z(6) = 279,948$.

2.1.7 No.

Section 2.2

2.2.1 X must equal 0, 1, or 2, and $P(X = 0) = 1/4$, $P(X = 1) = 1/2$, $P(X = x) = 0$ for $x \neq 0, 1, 2$.

2.2.5 (a) $P(X = 1) = .3, P(X = 2) = .2, P(X = 3) = .5$, and $P(X = x) = 0$ for all $x \notin \{1, 2, 3\}$. (b) $P(Y = 1) = .3, P(Y = 2) = .2, P(Y = 3) = .5$, and $P(Y = y) = 0$ for all $y \notin \{1, 2, 3\}$. (c) $P(W = 2) = 0.09$, $P(W = 3) = 0.12$, $P(W = 4) = 0.34$, $P(W = 5) = 0.2$, $P(W = 6) = 0.25$, and $P(W = w) = 0$ for all other choices of w.

2.2.8 Note that each number $w \in \{0, 1, \ldots, 99\}$ can occur and $P(W = w) = 1/100$.

Section 2.3

2.3.1 $p_Y(2) = 1/36$, $p_Y(3) = 2/36$, $p_Y(4) = 3/36$, $p_Y(5) = 4/36$, $p_Y(6) = 5/36$, $p_Y(7) = 6/36$, $p_Y(8) = 5/36$, $p_Y(9) = 4/36$, $p_Y(10) = 3/36$, $p_Y(11) = 2/36$, $p_Y(12) = 1/36$, and $p_Y(y) = 0$ otherwise.

2.3.5 $p_W(1) = 1/36$, $p_W(2) = 2/36$, $p_W(3) = 2/36$, $p_W(4) = 2/36 + 1/36 = 3/36$, $p_W(5) = 2/36$, $p_W(6) = 2/36 + 2/36 = 4/36$, $p_W(8) = 2/36$, $p_W(9) = 1/36$, $p_W(10) = 2/36$, $p_W(12) = 2/36+2/36 = 4/36$, $p_W(15) = 2/36$, $p_W(16) = 1/36$, $p_W(18) = 2/36$, $p_W(20) = 2/36$, $p_W(24) = 2/36$, $p_W(25) = 1/36$, $p_W(30) = 2/36$, $p_W(36) = 1/36$, and $p_W(w) = 0$ otherwise.

2.3.10 $P(X^2 \le 15) = 369/625$.

2.3.15 (a) $\binom{10}{3}(.35)^3(.65)^7$. (b) $(.35)(.65)^9$. (c) $\binom{9}{1}(.35)^2(.65)^8$.

2.3.24 $Z \sim$ Binomial$(n_1 + n_2, p)$.

2.3.28 $Z \sim$ Negative Binomial$(r + s, \theta)$.

Section 2.4

2.4.1 (a) $P(U \le 0) = 0$. (b) $P(U = 1/2) = 0$. (c) $P(U < -1/3) = 0$.
(d) $P(U \le 2/3) = 2/3$. (f) $P(U < 1) = 1$. (g) $P(U \le 17) = 1$.

2.4.4 (a) $c = 2$. (b) $c = n + 1$. (c) $c = 3/\left(\sqrt{24}\right)$. (d) $c = 1$.

2.4.11 We have that $f(x) \ge 0$ for every x, and putting $u = x^\alpha$, $du = \alpha x^{\alpha-1}\,dx$, we have $\int_0^\infty \alpha x^{\alpha-1} e^{-x^\alpha}\,dx = \int_0^\infty e^{-u}\,du = -e^{-u}|_0^\infty = 1$.

Section 2.5

2.5.2 $F_X(x) = 0$ when $x < 1, F_X(x) = 1/6$ when $1 \le x < 4, F_X(x) = 2/6$ when $4 \le x < 9, F_X(x) = 3/6$ when $9 \le x < 16, F_X(x) = 4/6$ when $16 \le x < 25, F_X(x) = 5/6$ when $25 \le x < 36$, and$F_X(x) = 1$ when $36 \le x$.

2.5.3(a) No. (c) Yes. (g) No.

2.5.7 $\lim_{n\to\infty} |F(2n) - F(n)| = 0$.

2.5.11 $F(x) = (1 + e^{-x})^{-1}$.

2.5.15 $F(x) = \frac{1}{2}e^{x}$ when $x \le 0$ and $F(x) = \frac{1}{2} + \frac{1}{2}(1 - e^{-x})$ when $x > 0$.

Section 2.6

2.6.3 Let $h(x) = cx + d$. Then $Y = h(X)$ and h is strictly increasing, so $f_Y(y) = f_X(h^{-1}(y)) / |h'(h^{-1}(y))| = e^{-[y-d-c\mu]^2/2c^2\sigma^2}/c\sigma\sqrt{2\pi}$.

2.6.13 $Y \sim$ Weibull(α/β).

2.6.15 $Y \sim$ Exponential(1).

Section 2.7

2.7.1 $F_{X,Y}(x, y) = 0$ when $\min[x,\ (y+2)/4] < 0, F_{X,Y}(x, y) = 1/3$ when $0 \le \min[x,\ (y+2)/4] < 1$, and $F_{X,Y}(x, y) = 1$ when $\min[x,\ (y+2)/4] \ge 1$.

2.7.3 (a) $p_X(2) = p_X(3) = p_X(-3) = p_X(-2) = p_X(17) = 1/5$, with $p_X(x) = 0$ otherwise (d) $P(Y = X) = 0$ since this never occurs.

2.7.4 (a) $C = 4$, and $P(X \le 0.8,\ Y \le 0.6) \doteq 0.0863$.

2.7.6 $F_{X,Y}(x, y) = 1 - e^{-\lambda \min(x,\ y^{1/3})}$ for $x, y > 0$, otherwise equals 0.

2.7.11 (a) $C = 2$. (b) We have that $f_X(x) = 2e^{-2x}$so $X \sim$ Exponential(2) and $f_Y(y) = 2e^{-y}(1 - e^{-y})$ for $y > 0$.

Section 2.8

2.8.1 (a) $p_X(-2) = 1/4, p_X(9) = 1/4, p_X(13) = 1/2$ otherwise $p_X(x) = 0$. (b) $p_Y(3) = 2/3, p_Y(5) = 1/3$, otherwise $p_Y(y) = 0$. (c) Yes.

2.8.5 (a) $P(Y = 4 \mid X = 9) = 1/6$. (c) $P(Y = 0 \mid X = -4) = 0$.

2.8.7 (a) $C = 4$ and X and Y are not independent since $f_{Y|X}(y \mid x) \ne f_Y(y)$.

2.8.12 Since X and Y are independent, $P(X = 1 \mid Y = 5) = P(X = 1) = 1/3$.

2.8.19 $P(X_2 = f_2 \mid X_1 = f_1) = \binom{n-f_1}{f_2}\left(\frac{\theta_2}{1-\theta_1}\right)^{f_2}\left(1 - \frac{\theta_2}{1-\theta_1}\right)^{n-f_1-f_2}$, so $X_2 \mid X_1 = f_1 \sim$ Binomial$(n - f_1, \theta_2/(1-\theta_1))$.

2.8.22 The distribution function of $X_{(3)}$, for $0 < x < 1$, is given by $P(X_{(3)} \le x) = 10x^3(1-x)^2 + 5x^4(1-x) + x^5 = 10x^3 - 15x^4 + 6x^5$, so $f(x) = 30x^2 - 60x^3 + 30x^4 = 30x^2(x-1)^2$. This is the Beta(3,3) density.

Section 2.9

2.9.2 (a) $f_{X,Y}(x, y) = e^{-x}$ for $x \ge 0$ and $1 \le y \le 4$, otherwise $f_{X,Y}(x, y) = 0$. (b) $h(x, y) = (x + y,\ x - y)$. (c) $h^{-1}(z, w) = ((z + w)/2,\ (z - w)/2)$. (d) Here $J(x, y) = \frac{\partial h_1}{\partial x}\frac{\partial h_2}{\partial y} - \frac{\partial h_2}{\partial x}\frac{\partial h_1}{\partial y} = |(1)(-1) - (1)(1)| = 2$, so $f_{Z,W}(z, w) = f_{X,Y}(h^{-1}(z, w)) / |J(h^{-1}(z, w))| = f_{X,Y}((z+w)/2,\ (z-w)/2)/2$, which equals

$e^{-(z+w)/2}$ for $(z+w)/2 \geq 0$ and $1 \leq (z-w)/2 \leq 4$, i.e., for $z \geq 1$ and $\max(-z,\, z-8) \leq w \leq z-2$, otherwise $f_{Z,W}(z,w) = 0$.

2.9.6 (a) $p_{Z,W}(5,5) = 1/7$, $p_{Z,W}(8,2) = 1/7$, $p_{Z,W}(9,1) = 1/7$, $p_{Z,W}(8,0) = 3/7$, $p_{Z,W}(12,4) = 1/7$, and $p_{Z,W}(z,w) = 0$ otherwise.

Section 2.10

2.10.2 (a) $F^{-1}(t) = t$, so $X = U$ (b) $F^{-1}(t) = \sqrt{t}$, so $X = \sqrt{U}$.

2.10.6 $x = F^{-1}(u) = \ln(u/(1-u))$ for $0 \leq u \leq 1$.

2.10.10 $x = F^{-1}(u) = \ln(2u)$ and, for $1/2 \leq u \leq 1$, $x = F^{-1}(u) = -\ln 2(1-u)$.

Chapter 3

Section 3.1

3.1.1 (a) $E(X) = 8/7$. (b) $E(X) = 1$. (c) $E(X) = 8$.

3.1.3 (a) $E(X) \doteq -14.4$. (d)$E(Y^2) \doteq 123.3$.

3.1.7 $E(XY) = 30$.

3.1.9 $E(X) = 6$.

3.1.12 (b) $E(Y - X) = E(Y) - E(X) = -(1-\theta)^{101}(1/\theta + 100)$.

Section 3.2

3.2.1 (a) $C = 1/4$, $E(X) = 7$. (b) $C = 1/16$, $E(X) \doteq 7.04$. (c) $C = 5/3093$ $E(X) \doteq -4.19$.

3.2.4 (a) $E(X) = 57/70$. (b) $E(Y) = 157/280$.

3.2.8 $E(Y + Z) = 49/36$.

3.2.15 $E(X) = \int_0^1 x \frac{\Gamma(a+b)}{\Gamma(a)\Gamma(b)} x^{a-1}(1-x)^{b-1}\, dx = \frac{\Gamma(a+b)}{\Gamma(a)\Gamma(b)} \int_0^1 x^a (1-x)^{b-1}\, dx = \frac{\Gamma(a+b)}{\Gamma(a)\Gamma(b)} \frac{\Gamma(a+1)\Gamma(b)}{\Gamma(a+b+1)} = \frac{\Gamma(a+b)}{\Gamma(a)\Gamma(b)} \frac{a\Gamma(a)\Gamma(b)}{(a+b)\Gamma(a+b)} = \frac{a}{a+b}$.

Section 3.3

3.3.1 (a) $\mathrm{Cov}(X,Y) = 2/3$. (b) $\mathrm{Var}(X) = 2$, $\mathrm{Var}(Y) = 32/9$. (c) $\mathrm{Corr}(X,Y) = 1/4$.

3.3.4 $E(X) = 63/80$, $E(Y) = 61/72$, $\mathrm{Var}(X) = 191/6400$,
$\mathrm{Var}(Y) = 3253/181440$, $\mathrm{Cov}(X,Y) = -1/1920$.

3.3.10 $\mathrm{Corr}(X,Y) = sgn(c)$, where $sgn(c) = 1$ for $c > 0$, $sgn(c) = 0$ for $c = 0$, and $sgn(c) = -1$ for $c < 0$, (a) $\lim_{c \searrow 0} \mathrm{Cov}(X,Y) = \lim_{c \searrow 0} c = 0$, (b)$\lim_{c \nearrow 0} \mathrm{Cov}(X,Y) = \lim_{c \nearrow 0} c = 0$.

3.2.17 $\mathrm{Var}(X) = 2$.

Section 3.4

3.4.1 (a) $r_Z(t) = t/(2-t)$ (b) $E(Z) = 2$, $E(Z^2) = 6$.

3.4.4 $r_Y(t) = t^4 r_X(t^3)$.

3.4.11 $r_X(t) = \theta^r (1 - t(1-\theta))^{-r}$ provided $|t(1-\theta)| < 1$.

3.4.15 Write $Z = \mu + \sigma X$, where $X \sim \text{Normal}(0,1)$. Then $m_Z(s) = E(e^{sZ}) = E(e^{s(\mu+\sigma X)}) = e^{s\mu} + E(e^{s\sigma X}) = e^{s\mu} + m_X(\sigma s) = e^{s\mu} + e^{(\sigma s)^2/2} = e^{s\mu} + e^{\sigma^2 s^2/2}$.

3.4.25 $m_{S_N}(s) = r_N(m_{X_1}(s)) = \theta / \left(1 - \frac{\lambda}{\lambda - s}(1-\theta)\right), m'_{S_N}(0) = (1-\theta)/(\lambda\theta)$.

Section 3.5

3.5.1 (a) $E(X \mid Y = 3) = 5/2$. (b) $E(Y \mid X = 3) = 22/3$. (c) $E(X \mid Y = 2) = 5/2$, $E(X \mid Y = 17) = 3$. (d) $E(Y \mid X = 2) = 5/2$.

3.5.5 $E(\text{earnings} \mid Y = \text{"takes course"}) = \$2700$, $E(X \mid Y = \text{"doesn't take course"}) = \$2100$, $E(\text{earnings}) = \$2340$

3.5.14 (a) $\text{Var}(X) \doteq 0.244967$. (b) $\text{Var}(E(X \mid Y)) \doteq 0.0000002196$. (c) $\text{Var}(X \mid Y) = (8/49)(49 + 8064Y^4 + 98304Y^8)/(3 + 256Y^4)^2$. (d) $E(\text{Var}(X \mid Y)) \doteq 0.244967$.

Section 3.6

3.6.1 $P(Z \geq 7) \leq 3/7$.

3.6.6 Largest value is $\text{Cov}(Y, Z) \doteq 38.73$ and smallest is $\text{Cov}(Y, Z) \doteq -38.73$. Therefore, r_{XY} is as stated.

3.6.11 $P(X \notin (\bar{x} - 2\hat{s}, \bar{x} + 2\hat{s})) \leq \frac{\hat{s}_X^2}{(2\hat{s}_X)^2} = \frac{1}{4}$, so the largest possible proportion is 1/4.

Section 3.7

3.7.1 $E(X_1) = 3$, $E(X_2) = 0$, and $E(Y) = (1/5)E(X_1) + (4/5)E(X_2) = 3/5$.

Chapter 4

Section 4.1

4.1.2 If Z is the sample mean, then $P(Z = 1) = 1/36$, $P(Z = 1.5) = 2/36$, $P(Z = 2) = 3/36$, $P(Z = 2.5) = 4/36$, $P(Z = 3) = 5/36$, $P(Z = 3.5) = 6/36$, $P(Z = 4) = 5/36$, $P(Z = 4.5) = 4/36$, $P(Z = 5) = 3/36$, $P(Z = 5.5) = 2/36$, and $P(Z = 6) = 1/36$.

4.1.4 If Z is the sample mean, then $P(Z = 0) = \frac{N}{N+M}\frac{N-1}{N+M-1}$, $P(Z = 0.5) = 2\frac{N}{N+M}\frac{M}{N+M-1}$, and $P(Z = 1) = \frac{M}{N+M}\frac{M-1}{N+M-1}$.

4.1.8 $Y \sim \text{Poisson}(n\lambda)$.

Section 4.2

4.2.2 For any $\epsilon > 0$, $P(|X_n - 0| \geq \epsilon) = P(Y^n \geq \epsilon) = P(Y \geq \epsilon^{1/n}) = 1 - \epsilon^{1/n} \to 0$ as $n \to \infty$, so $X_n \to 0$ in probability.

Section 4.3

4.3.1 Note that $Z_n = Z$ unless $7 \leq U < 7 + 1/n^2$. Hence, if $U < 7$, then $Z_n = Z$ for all n, so of course $Z_n \to Z$. Also, if $U > 7$, then $Z_n = Z$ whenever $1/n^2 < 7 - U$, i.e., $n > 1/\sqrt{7 - U}$, so again $Z_n \to Z$. Hence, $P(Z_n \to Z) \geq P(U \neq 7) = 1 - P(U = 7) = 1 - 0 = 1$, i.e., $Z_n \to Z$ with probability 1.

4.3.5 $P(X_n \to X \text{ and } Y_n \to Y) \geq 1 - P(X_n \not\to X) - P(Y_n \not\to Y)$.

4.3.12 This is false.

Section 4.4

4.4.1 Here $\lim_{n\to\infty} P(X_n = i) = 1/3 = P(X = i)$ for $i = 1, 2, 3$, so $\lim_{n\to\infty} P(X_n \le x) = P(X \le x)$ for all x, so $X_n \to X$ in distribution.

4.4.5$P(S \le 540) \approx 0.6915$.

4.4.16 $P(M_n \le m) = P(\frac{\sqrt{n}(M_n-0)}{\sqrt{15}} \le \frac{\sqrt{n}(m-0)}{\sqrt{15}}) \approx P(Z \le \frac{\sqrt{n}(m-0)}{\sqrt{15}})$, where $Z \sim N(0,1)$.

Section 4.5

4.5.1 This integral equals $\sqrt{2\pi}E(\cos^2(Z))$, where $Z \sim N(0,1)$.

4.5.4 $n \ge 9\,(10)\,/(.1)^2 = 9000.0$.

4.5.11 $n \ge 9/\left(4\delta^2\right)$.

Section 4.6

4.6.1 (a) $U \sim N(44, 629)$. $V \sim N(-18 - 8C, 144 + 25C^2)$. (b) U and V are independent if and only if $C = -24/125$.

4.6.3 $C_1 = 1/\sqrt{5}$. $C_2 = -3$. $C_3 = 1/\sqrt{2}$. $C_4 = 7$. $C_5 = 2$.

4.6.11 We see that $f_Z(-z) = \Gamma((n+1)/2)(1+(-z)^2/n)^{-(n+1)/2}/\Gamma(n/2)\sqrt{\pi n} = \Gamma((n+1)/2)(1+z^2/n)^{-(n+1)/2}/\Gamma(n/2)\sqrt{\pi n} = f_Z(z)$. Then using the substitution $s = -t$, we have $P(Z < -x) = \int_{-\infty}^{-x} f_Z(t)\,dt = -\int_x^{\infty} f_Z(-s)\,(-ds) = \int_x^{\infty} f_Z(s)\,ds = P(Z > x)$.

Chapter 5

Section 5.1

5.1.1 The mean survival times for the control group and the treatment group are 93.2 days and 356.2 days, respectively. As we can see, there is a big difference between the two means, which might suggest that the treatment is indeed effective, but we can't base our conclusions about the effectiveness of the treatment based only on these numbers. We have to consider sampling variability as well.

5.1.3 For those who are still alive their survival times will be longer than the recorded values, so these data values are incomplete.

Section 5.2

5.2.1 In Example 5.2.1 the mode is given by 0. In Example 5.2.2 the mode of this density is 1. In both cases the mode is at the extreme left end of the distribution and so doesn't seem like a very good predictor.

5.2.3 The density is given by $.5\,(2\pi)^{-1/2}\exp\{-\,(x+4)^2/2\}+$ $.5\,(2\pi)^{-1/2}\exp\{-\,(x-4)^2/2\}$ for $-\infty < x < \infty$.

5.2.8 Suppose that $X \sim$ Beta(a,b). We have that $E(X) = a/(a+b)$ with $E((X - a/(a+b))^2) = ab/(a+b+1)(a+b)^2$. The mode is given by $(a-1)/(a+b-2)$ and $E((X-(a-1)/(a+b-2))^2) = ab/(a+b+1)(a+b)^2 + (a/(a+b) - (a-1)/(a+b-2))^2 \ge E((X - a/(a+b))^2)$. Therefore, the mean is a better predictor.

Section 5.3

5.3.1 The statistical model for a single response consists of three probability functions $\{f_1, f_2, f_3\}$, where f_1 is the probability function for the Bernoulli(1/2) distribution, f_1 is the probability function for the Bernoulli(1/3) distribution, and f_3 is the probability function for the Bernoulli(2/3) distribution. Then $(x_1, x_2, \ldots, x_5)$ is a sample from one of these Bernoulli(θ) distributions.

5.3.6 The first quartile of the Uniform$[0, \beta]$ distribution is $c = 0.25\beta$. Since c is a 1-1 transformation of β, we can parameterize this model by the first quartile.

5.3.9 $\sum_{i=1}^n (x_i - \mu)^2 = \sum_{i=1}^n (x_i - \bar{x} + \bar{x} - \mu)^2 = \sum_{i=1}^n ((x_i - \bar{x})^2 + 2(x_i - \bar{x})(\bar{x} - \mu) + (\mu - \bar{x})^2) = \sum_{i=1}^n (x_i - \bar{x})^2 - 2(\mu - \bar{x}) \sum_{i=1}^n (x_i - \bar{x}) + \sum_{i=1}^n (\mu - \bar{x})^2 = \sum_{i=1}^n (x_i - \bar{x})^2 + n(\mu - \bar{x})^2$ since $\sum_{i=1}^n (x_i - \bar{x}) = \sum_{i=1}^n x_i - n\bar{x} = 0$.

5.3.11 The first quartile of a $N(\mu, \sigma^2)$ distribution is $c = \mu + \sigma z_{.25}$, where $z_{.25}$ is the first quartile of the $N(0,1)$ distribution, i.e., $\Phi(z_{.25}) = .25$. But we see from this that several different values of (μ, σ^2) can give the same first quartile, e.g., $(\mu, \sigma^2) = (0, 1)$ and $(\mu, \sigma^2) = (z_{.25}/2, 1/4)$ both give rise to normal distributions whose first quartile equals $z_{.25}$. Therefore, we cannot parameterize this model by the first quartile.

Section 5.4

5.4.1 $F_X(x) = 0$ when $x < 1$, $F_X(x) = 4/10$ when $1 \le x < 2$, $F_X(x) = 7/10$ when $2 \le x < 3$, $F_X(x) = 9/10$ when $3 \le x < 4$, $F_X(x) = 1$ when $4 \le x$, $f_X(1) = 4/10, f_X(2) = 3/10, f_X(3) = 2/10, f_X(4) = 1/10$, and $\mu_X = 2, \sigma_X^2 = 1$.

5.4.4 (a) $f_X(0) = a/N, f_X(1) = (N-a)/N$. This is a Bernoulli$((N-a)/N)$ distribution. (b) $P\left(\hat{f}_X(0) = f_X(0)\right) = P\left(n\hat{f}_X(0) = nf_X(0)\right)$ $= \binom{a}{n - nf_X(0)}\binom{N-a}{nf_X(0)}/\binom{N}{n}$ since $n\hat{f}_X(0) \sim$ Hypergeometric(N, a, n). (c) We have that $n\hat{f}_X(0) \sim$ Binomial$(n, a/N)$ so $P\left(\hat{f}_X(0) = f_X(0)\right)$ $= P\left(n\hat{f}_X(0) = nf_X(0)\right) = P$(number of 0's in the sample equals $nf_X(0)$) $=$ $\binom{n}{nf_X(0)} (a/N)^{nf_X(0)} (1 - a/N)^{n - nf_X(0)}$.

5.4.9 (a) $f_X(0) = a/N, f_X(1) = b/N, f_X(2) = (N - a - b)/N$. (b) Assuming f_1, f_2, f_3 are nonnegative integers summing to n (otherwise probability is 0), the probability is $\binom{a}{f_0}\binom{b}{f_1}\binom{N-a-b}{f_2}/\binom{N}{n}$. (c) The probability that $\hat{f}_X(0) = f_0, \hat{f}_X(1) = f_1$ and $\hat{f}_X(2) = f_2$ is $\binom{a}{f_0\, f_1\, f_2} (a/N)^{f_0} (b/N)^{f_1} ((N - a - b)/N)^{f_2}$.

5.4.12 When $f_X(0) = a/N$ is unknown we estimate it by $\hat{f}_X(0)$. Now $N = a/f_X(0)$, so we can estimate N by setting $\hat{N} = a/\hat{f}_X(0)$, provided $\hat{f}_X(0) \ne 0$.

Section 5.5

5.5.1 (a) $\hat{f}_X(0) = .2667, \hat{f}_X(1) = .2, \hat{f}_X(2) = .2667$, and $\hat{f}_X(3) = \hat{f}_X(4) = .1333$. (b) $\hat{F}_X(0) = .2667, \hat{F}_X(1) = .4667, \hat{F}_X(2) = .7333, \hat{F}_X(3) = .8667$, and $\hat{F}_X(4) = 1.000$. (d) The mean $\bar{x} = 15$ and the variance $s^2 = 1.952$. (e) The median is 2 and the $IQR = 3$. According to the 1.5 IQR rule, there are no outliers.

5.5.6 The distribution is skewed to the right, so we choose the median as a measure of location and the IQR as a measure of spread.

5.5.7 $\psi(\mu) = x_{0.25} = \mu + \sigma_0 z_{0.25}$, where $z_{0.25}$ satisfies $\Phi(z_{0.25}) = .25$.

5.5.13 $\psi(\theta) = 2\theta(1-\theta)$.

Chapter 6

Section 6.1

6.1.1 The appropriate statistical model is the Binomial(n, θ), where $\theta \in \Omega = [0,1]$ is the probability of having this antibody in the blood (we can also think of θ as the unknown proportion of the population who have this antibody in their blood). The likelihood function is given by $L(\theta \,|\, s) = \binom{n}{s}\theta^s(1-\theta)^{n-s}$, where s is the number of people whose result was positive. The likelihood function for $n = 10$ people and $s = 3$ is given by $L(\theta \,|\, 3) = \binom{10}{3}\theta^3(1-\theta)^7$.

6.1.3 The likelihood function is given by $L(\theta \,|\, x_1, ..., x_{20}) = \theta^{20}\exp(-(20\bar{x})\,\theta)$. By the factorization theorem (Theorem 6.1.1) $\bar{x}$ is a sufficient statistic, so we only need to observe its value to obtain a representative likelihood. The likelihood function when $\bar{x} = 5.2$ is given by $L(\theta \,|\, x_1, ..., x_{20}) = \theta^{20}\exp(-20\,(5.2)\,\theta)$.

6.1.7 The likelihood function is given by $L(\theta \,|\, x_1, ..., x_n) = \theta^{n\bar{x}}e^{-n\theta}/\prod_{i=1}^n x_i!$. By the factorization theorem $\bar{x}$ is a sufficient statistic. If we differentiate $\ln L(\theta \,|\, x_1, ..., x_n) = -\ln \prod x_i! + n\bar{x}\ln\theta - n\theta$, we get $(\ln L(\theta \,|\, x_1, ..., x_n))' = n\bar{x}/\theta - n$ and setting this equal to 0 gives the solution $\theta = \bar{x}$. Therefore, we can obtain $\bar{x}$ from the likelihood and we conclude that it is a minimal sufficient statistic.

6.1.13 The likelihood function is given by $L(\theta \,|\, x_1, ..., x_n) = \theta^{-n}I_{[x_{(n)},\infty)}(\theta)$ when $\theta > 0$. By the factorization theorem $x_{(n)}$ is a sufficient statistic. Now notice that the likelihood function is 0 to the left of $x_{(n)}$ and positive to the right. So given the likelihood, we can determine $x_{(n)}$ and it is minimal sufficient.

Section 6.2

6.2.1 The MLEs are $\hat{\theta}(1) = a, \hat{\theta}(2) = b$, $\hat{\theta}(3) = b$, and $\hat{\theta}(4) = a$.

6.2.4 The likelihood function is $L(\theta \,|\, x_1, ..., x_n) = e^{-n\theta}\theta^{n\bar{x}}$, the log-likelihood function is $l(\theta \,|\, x_1, ..., x_n) = -n\theta + n\bar{x}\ln\theta$ and the score function is given by $S(\theta \,|\, x_1, ..., x_n) = -n + n\bar{x}/\theta$. Solving the score equation gives $\hat{\theta}(x_1, ..., x_n) = \bar{x}$. Note that since $\bar{x} \geq 0$, we have $\left.\frac{\partial S(\theta \,|\, x_1,...,x_n)}{\partial\theta}\right|_{\theta=\bar{x}} = -\left.\frac{n\bar{x}}{\theta^2}\right|_{\theta=\bar{x}} = -\frac{n}{\bar{x}} < 0$. So $\bar{x}$ is the MLE.

6.2.9 The likelihood function is $L(\alpha \,|\, x_1, ..., x_n) = \alpha^n\left(\prod_{i=1}^n(1+x_i)\right)^{-(\alpha+1)}$, the log-likelihood function is $l(\alpha \,|\, x_1, ..., x_n) = n\ln\alpha - (\alpha+1)\sum_{i=1}^n \ln(1+x_i)$, and the score function is $S(\alpha \,|\, x_1, ..., x_n) = \frac{n}{\alpha} - \sum_{i=1}^n \ln(1+x_i)$. Solving the score equation gives $\hat{\alpha}(x_1, ..., x_n) = n/\sum_{i=1}^n \ln(1+x_i)$. Note also that $\frac{\partial}{\partial\alpha}S(\alpha \,|\, x_1, ..., x_n) = -\frac{n}{\alpha^2} < 0$ for every α, so $\hat{\alpha}$ is the MLE.

6.2.13 The log-likelihood function is given by $l(\mu \,|\, x_1, ..., x_n) = -n\,(\bar{x} - \mu)^2/2$, and as a function of μ, its graph is a concave parabola and its maximum value occurs at $\bar{x}$. So if $\bar{x} \geq 0$, this is the MLE. If $\bar{x} < 0$, however, the maximum occurs at 0 and this is the MLE.

Section 6.3

6.3.1 This is a two sided z-test with the z statistic equal to -0.54 and the P-value is equal to 0.592. So there is no evidence against H_0. A .95-confidence interval for the unknown μ is $(4.442, 5.318)$. Note that the confidence interval contains the value 5, which confirms our conclusion using the above test.

6.3.4 This is a two sided t-test with the t statistic equal to 9.12 and (using the Student(3) distribution) the P-value equals 0.452, so we don't have evidence against the null hypothesis.

6.3.8 The P-value equals 0.32, so we conclude that there is no evidence against H_0. A .90-confidence interval for θ is given by $(0.559832, 0.680168)$ which includes the value 0.65, so agrees with the result of the above test.

6.3.19 The form of the power function associated with the above hypothesis assessment procedure is given by $\beta(\mu) = 1 - \Phi\left(\frac{\mu_0 - \mu}{\sigma_0/\sqrt{n}} + z_{1-\alpha}\right)$.

Section 6.4

6.4.1 An approximate .95-confidence interval for μ_3 is given by $(26.027, 151.373)$.

6.4.4 Let $\psi(\mu) = \exp(\mu)$ then $\psi'(\mu) = \exp(\mu)$. By the delta theorem (6.4.1) an approximate γ-confidence interval for $\psi(\mu)$ is given by $(-5.6975, 42.046)$.

6.4.13 For a random variable with this distribution, we have that $E_{\hat{F}}(X^i) = \sum_{j=1}^{n} x_{(j)}^i(\hat{F}(x_{(j)}) - \hat{F}(x_{(j-1)}))$ where we take $x_{(0)} = -\infty$. Now $\hat{F}(x_{(j)}) - \hat{F}(x_{(j-1)}) = 1/n$ since all the $x_{(j)}$ are distinct.

Section 6.5

6.5.1 The Fisher information is $nI(\sigma^2) = n/2\sigma^4$.

6.5.3 The Fisher information is $nI(\alpha) = n/\alpha^2$.

6.5.5 An approximate .90-confidence interval is given by $\frac{2}{\bar{x}} \pm \frac{1}{\sqrt{2n}}\left(\frac{2}{\bar{x}}\right) z_{.95} = (9.5413 \times 10^{-4}, 1.5045 \times 10^{-3})$.

Chapter 7

Section 7.1

7.1.1 $\pi(1 \,|\, s=1) = 3/16, \pi(2 \,|\, s=1) = 4/16, \pi(3 \,|\, s=1) = 9/16, \pi(1 \,|\, s=2) = 3/14, \pi(2 \,|\, s=2) = 4/7$, and $\pi(3 \,|\, s=2) = 3/14$.

7.1.4 The posterior is a Gamma$(n\bar{x} + a, n + \beta)$ distribution.

7.1.13 $m(x_1, ...x_n) = \frac{\Gamma(\alpha+\beta)}{\Gamma(\alpha)\Gamma(\beta)} \frac{\Gamma(n\bar{x}+\alpha)\Gamma(n(1-\bar{x})+\beta)}{\Gamma(\alpha+\beta+n)}$ for $(x_1, ...x_n) \in \{0,1\}^n$.

Section 7.2

7.2.1 $E(\theta^m \,|\, x_1, ..., x_n) = \frac{\Gamma(\alpha+\beta+n)\Gamma(n\bar{x}+\alpha+m)}{\Gamma(n\bar{x}+\alpha)\Gamma(\alpha+\beta+n+m)}$.

7.2.3 $E\left(1/\sigma^2 \mid x_1, ..., x_n\right) = (\alpha_0 + n/2) / \beta_x$, where β_x is given by (7.1.8) and the posterior mode is $1/\hat{\sigma}^2 = (\alpha_0 + n/2 - 1) / \beta_x$.

7.2.12 (a) The posterior mode is $\hat{\mu} = 64.053$. A .95-credible interval for μ is given by $64.053 \pm \sqrt{9/19} z_{0.975} = (62.704, 65.402)$. This interval has length equal to 2.698 and the margin of error is less than 1.5 marks, which is quite small, so we conclude that the estimate is quite accurate. (b) Based on the .95-credible interval, we can't reject $H_0 : \mu = 65$ at the 5% level since 65 falls inside the interval. (c) The posterior probability of the null hypothesis above is .3885. (d) The Bayes factor in favor of $H_0 : \mu = 65$ is given by .6353.

Section 7.3

7.3.7 As we increase the Monte Carlo sample size N, the interval that contains the exact value of the posterior expectation with virtual certainty becomes shorter and shorter. But for a given sample size n for the data, the posterior expectation will not be equal to the true value of $1/(\alpha + 1)$, so this interval will inevitably exclude the true value.

7.3.9 $F_{Y|X}^{-1}(u) = \left(\left(1 - x^2\right) u\right)^{1/2}$ for $0 < u < 1$. Therefore, we can generate Y given $X = x$ by generating $U \sim$ Uniform[0, 1] and putting $Y = \left(\left(1 - x^2\right) U\right)^{1/2}$, $F_{X|Y}^{-1}(u) = \left(y^2 u\right)^{1/2} = y u^{1/2}$ for $0 < u < 1$. Therefore, we can generate X given $Y = y$ by generating $U \sim$ Uniform[0, 1] and putting $X = yU^{1/2}$. So we select x_0. Then we generate $Y \sim f_{Y|X}(\cdot \mid x_0)$, using the above algorithm, obtaining y_1. Next we generate $X \sim f_{X|Y}(\cdot \mid y_1)$, using the above algorithm, obtaining x_1. Then we generate $Y \sim f_{Y|X}(\cdot \mid x_1)$, using the above algorithm, obtaining y_2, etc. We can generate exactly from this distribution as follows: $F_Y^{-1}(u) = u^{1/4}$ for $0 < u < 1$, so we can generate $Y \sim F_Y$ by generating $U \sim$ Uniform[0, 1] and putting $y = U^{1/4}$. Then we use the above algorithm to generate $X \sim f_{X|Y}(\cdot \mid y)$.

Section 7.4

7.4.1 The posterior density is $\propto \lambda^{n+\alpha-1} \exp\left[-\lambda \left(\ln\left(\prod (1 + x_i)\right) + \beta\right)\right]$ and we recognize this as being proportional to the Gamma$(n + \alpha, \ln\left(\prod (1 + x_i)\right) + \beta)$ density. Hence, this is a conjugate family.

7.4.3 (a) $m_1(1, 1, 3) = 59/1728, m_2(1, 1, 3) = 43/1296$ and the maximum value of the prior predictive is obtained when $\tau = 1$. (b) $\pi_1(a \mid 1, 1, 3) = 32/59, \pi_1(b \mid 1, 1, 3) = 27/59$.

7.4.7 Jeffreys' prior for this model is $\sqrt{n}\theta^{-1/2}(1 - \theta)^{-1/2}$. The posterior density of θ is then proportional to $\theta^{n\bar{x}-1/2}(1 - \theta)^{n(1-\bar{x})-1/2}$, which we recognize as the unnormalized density of a Beta$(n\bar{x} + 1/2, n(1 - \bar{x}) + 1/2)$ distribution.

7.4.11 Jeffreys' prior is given by $1/\sigma^2$.

Chapter 8

Section 8.1

8.1.1 We have that $L(1 \mid \cdot) = (3/2) L(2 \mid \cdot)$, so by Section 6.1.1 T is a sufficient statistic. Then, given $T = 1$, the conditional distributions of s are given by

$f_a(1\,|\,T=1) = 2/3, f_a(2\,|\,T=1) = 1/3, f_a(3\,|\,T=1) = 0, f_a(4\,|\,T=1) = 0,$ and $f_b(1\,|\,T=1) = 2/3, f_b(2\,|\,T=1) = 1/3, f_b(3\,|\,T=1) = 0, f_b(4\,|\,T=1) = 0$, and we see that these are the same (i.e., independent of θ). When $T = 3$ the conditional distributions of s are degenerate at $s = 3$ and when $T = 4$ the conditional distributions are degenerate at $s = 4$. These are also independent of θ.

8.1.4 $E(\bar{x} + \sigma_0 z_{.25}) = E(\bar{x}) + \sigma_0 z_{.25} = \mu + \sigma_0 z_{.25}$. Since $\bar{x}$ is complete, this implies that $\bar{x} + \sigma_0 z_{.25}$ is UMVU.

8.1.14 Suppose that c is a function such that $E_\theta(c(U)) = 0$ for every θ. Then $E_\theta(c(h(T))) = 0$ for every θ and the completeness of T implies that $P_\theta(\{s : c(h(T(s))) = 0\}) = 1$ for every θ. Now suppose u is such that $c(u) \neq 0$. Then $P_\theta(U = u) = P_\theta(h(T) = u) = P_\theta\left(T = h^{-1}(u)\right) =$
$P_\theta\left(\{s : T(s) = h^{-1}(u)\}\right) = 0$ since $c(h(T(s)) = c(u)$ for s in $\left\{s : T(s) = h^{-1}(u)\right\}$. This implies that U is complete.

Section 8.2

8.2.1 The ratio $f_b(s)/f_a(s)$ has the following distribution when $\theta = a$: $P_a(f_b(s)/f_a(s) = 3/2) = P_a(\{1,2\}) = 1/2, P_a(f_b(s)/f_a(s) = 2) = P_a(\{3\}) = 1/12$, and $P_a(f_b(s)/f_a(s) = 1/5) = P_a(\{4\}) = 5/12$. When $\alpha = .1$, using (8.2.4) and (8.2.5), we have that $c_0 = 3/2$ and $\gamma = ((1/10) - (1/12))\,/\,(1/2) = 1/30$. The power of the test is $P_b(\{3\}) + P_b(\{1,2\})\,/30 = 1/6 + (3/4)/30 = 23/120$. When $\alpha = .05$ we have that $c_0 = 2$ and $\gamma = ((1/20) - 0)\,/\,(1/12) = 3/5$. The power of the test is $P_b(\{3\})\,(3/5) = (1/6)\,(3/5) = 1/10$.

8.2.3 By (8.2.6) the optimal size .01 test is of the form (using $z_{.99} = 2.3263$) $\varphi_0(\bar{x}) = 1$ when $\bar{x} \geq 2.0404$ and is 0 otherwise.

8.2.7 With $x_\alpha(\beta_0)$ denoting the αth quantile of the Gamma$(n\alpha_0, \beta_0)$ distribution, the UMP size α test for $H_0 : \beta \leq \beta_0$ versus $H_a : \beta > \beta_0$ is to reject whenever $n\bar{x} \leq x_\alpha(\beta_0)$.

Section 8.3

8.3.1 The posterior distribution of θ is given by $\Pi(\theta = 1\,|\,2) = 2/5, \Pi(\theta = 2\,|\,2) = 3/5$, so $\Pi(\theta = 2\,|\,2) > \Pi(\theta = 1\,|\,2)$. We accept $H_0 : \theta = 2$.

8.3.3 The Bayes rule is given by the posterior mean $(1/\tau_0^2 + n/\sigma_0^2)^{-1}(\mu_0/\tau_0^2 + n\bar{x}/\sigma_0^2)$ and this converges to $\bar{x}$ as $\tau_0 \to \infty$.

8.3.6 The Bayes rule is given by the posterior mean of $1/\beta$, namely $(n\bar{x} + v_0)\,/\,(n\alpha_0 + \tau_0 - 1)$, which converges with probability 1 to$(\alpha_0/\beta)\,/\alpha_0 = \beta^{-1}$ as $n \to \infty$.

Section 8.4

8.4.1 The model is given by the collection of probability functions $\left\{\theta^{n\bar{x}}(1-\theta)^{n-n\bar{x}} : \theta \in [0,1]\right\}$ on the set of all sequences $(x_1, \ldots, x_n)$ of 0's and 1's. The action space is $\mathcal{A} = [0,1]$, the correct action function is $A(\theta) = \theta$, and the loss function is $L(\theta, a) = (\theta - a)^2$. The risk function for T is given by $R_T(\theta) = E_\theta\left((\theta - \bar{x})^2\right) = \text{Var}_\theta(\bar{x}) = \theta(1-\theta)\,/n$.

8.4.4 The model is given by the collection of probability functions $\left\{\theta^{n\bar{x}}(1-\theta)^{n-n\bar{x}} : \theta \in [0,1]\right\}$ on the set of all sequences $(x_1,\ldots,x_n)$ of 0's and 1's. The action space is $\mathcal{A} = \{H_0, H_a\}$, where $H_0 : \theta = 1/2$, the correct action function is $A(\theta) = H_0$ when $\theta = 1/2$, and $A(\theta) = H_a$ when $\theta \neq 1/2$. The loss function is $L(\theta,a) = 0$ when $\theta = 1/2, a = H_0$ or $\theta \neq 1/2, a = H_a$ and $L(\theta,a) = 1$ when $\theta = 1/2, a = H_a$ or $\theta \neq 1/2, a = H_0$. The test function $\varphi(n\bar{x}) = 0$ when $n\bar{x} \notin \{0,1,n-1,n\}$ and $\varphi(n\bar{x}) = 1$ when $n\bar{x} \in \{0,1,n-1,n\}$ has risk function given by $R_\varphi(\theta) = \binom{n}{0}(1-\theta)^n + \binom{n}{1}\theta(1-\theta)^{n-1} + \binom{n}{n-1}\theta^{n-1}(1-\theta) + \binom{n}{n}\theta^n$.

8.4.8 Suppose we have that $\delta(s,\cdot)$ is degenerate at $d(s)$ for each s. Then, clearly, $d : S \to \mathcal{A}$. Now suppose we have $d : S \to \mathcal{A}$ and define $\delta(s,B) = 1$ when $d(s) \in B$ and $\delta(s,B) = 0$ otherwise. Then $\delta(s,\mathcal{A}) = 1$, and if $B_1, B_2, \ldots$ are mutually disjoint subsets of $\mathcal{A}$, then $d(s) \in B_i$ for one i (and only one) if and only if $d(s) \in \cup_{j=1}^{\infty} B_j$, so $\delta\left(s, \cup_{j=1}^{\infty} B_j\right) = \sum_{j=1}^{\infty}\delta(s,B_j)$. Therefore, $\delta(s,\cdot)$ is a probability measure for each s, and δ is a decision function. Now, using the fact that $\delta(s,\cdot)$ is a discrete probability measure degenerate at $d(s)$, we have that $R_\delta(\theta) = E_\theta\left(E_{\delta(s,\cdot)}(L(\theta,a))\right) = E_\theta(\delta(s,\{d(s)\})(L(\theta,d(s)))) = E_\theta(L(\theta,d(s)))$ since $\delta(s,\{d(s)\}) = 1$.

Chapter 9

Section 9.1

9.1.1 $D(r) = \left(\sigma_0^2\right)^{-1}\sum_{i=1}^{n}(x_i-\bar{x})^2 = 22.761$. Now $D(R) \sim \chi^2(19)$ distribution, so the P-value is then given by $P(D(R) > 22.761) = .248$, which doesn't indicate any evidence against the model being correct.

9.1.5 The Chi-squared statistic is equal to 3.50 and the P-value is given by $(X^2 \sim \chi^2(4))\ P\left(X^2 \geq 3.5\right) = 0.4779$. Therefore, we have no evidence against the Uniform model being correct.

9.1.11 We have $E(a_1Y_1 + \cdots + a_kY_k) = a_1\mu_1 + \cdots + a_k\mu_k$ and so $E(Y_i) = \mu_i$ by taking $a_i = 1$ and $a_j = 0$ whenever $j \neq i$. By Theorem 3.3.3 (b) we have $\text{Var}(a_1Y_1 + \cdots + a_kY_k) = a_1^2\,\text{Var}(Y_1) + \cdots + a_k^2\,\text{Var}(Y_k) + 2\sum_{i<j} a_ia_j\,\text{Cov}(Y_i,Y_j) = \sum_{i=1}^{k}\sum_{j=1}^{k} a_ia_j\sigma_{ij}$. Putting $a_i = 1$ and $a_j = 0$ whenever $i \neq j$, we obtain $\text{Var}(Y_i) = \sigma_{ii}$ and this implies that $Y_i \sim N(\mu_i,\sigma_{ii})$. Putting $a_i = a_j = 1$ and $a_l = 0$ whenever $l \notin \{i,j\}$, we obtain $\text{Var}(Y_i + Y_j) = \sigma_{ii} + \sigma_{jj} + 2\sigma_{ij} = \text{Var}(Y_i) + \text{Var}(Y_j) + 2\,\text{Cov}(Y_i,Y_j)$. This immediately implies that $\text{Cov}(Y_i,Y_i) = \sigma_{ij}$.

Section 9.2

9.2.1 (a) The probability of obtaining $s = 2$ from f_1 is 1/3, which is a reasonable value, so we have no evidence against the model $\{f_1, f_2\}$. (b) The prior predictive M distribution is given by $m(1) = 1/3, m(2) = 1/10, m(3) = 17/30$. So the probability of a data set occurring with probability as small or smaller than $m(2)$ is 1/10, so the observation 2 is not very surprising. Accordingly, there is no evidence of a prior-data conflict. (c) The prior predictive M now is given by $m(1) = 1/3, m(2) = 1/300, m(3) = 199/300$. So the probability of a data set occurring with probability as small or smaller than $m(2)$ is 1/300, so the observation 2 is surprising. Accordingly, there is some evidence of a prior-data conflict.

9.2.4 First, by Corollary 4.6.1 we have $\bar{X} \sim N(\mu, \sigma_0^2/n)$. Then we can write $\bar{X}$ as $\bar{X} = \mu + Z/\sqrt{n}$, where $Z \sim N(0, \sigma_0^2)$ is independent of $\mu \sim N(\mu_0, \tau_0^2)$. Hence, by Theorem 4.6.1 we have that the prior predictive distribution of $\bar{X}$ is the $N(\mu_0, \tau_0^2 + \sigma_0^2/n)$ distribution.

Chapter 10

Section 10.1

10.1.1 From the definitions we know that if the conditional distribution of Y given X does not change as we change X, then X and Y are unrelated. Then for any x_1, x_2 (that occur with positive probability), and y we have $P(Y = y \mid X = x_1) = P(Y = y \mid X = x_2)$. Hence, $P(X = x_1, Y = y) / P(X = x_1) = P(X = x_2, Y = y) / P(X = x_2)$, so $P(X = x_1, Y = y) = P(X = x_2, Y = y) P(X = x_1) / P(X = x_2)$. Summing this over x_1 gives $P(Y = y) = P(X = x_2, Y = y) / P(X = x_2)$, so we must have $P(X = x_2, Y = y) = P(X = x_2) P(Y = y)$ and this implies that X and Y are statistically independent. Now on the other hand, if X and Y are statistically independent, then for all x and y we have $P(Y = y \mid X = x) = P(Y = y)$, so the conditional distribution of Y given X does not change as we change X and therefore X and Y are unrelated.

10.1.3 The conditional distribution of Y given $X = x$ does change as we change x, so we conclude that X and Y are related.

10.1.7 If the conditional distribution of life-length given various smoking habits changes, then we can conclude that these two variables are related. However, since we can't assign the value of smoking habit (perhaps different amount of smoking), and considering there may be many other confounding variables that should be take into account, e.g., exercise habits, eating habits, sleeping habits, etc., we can't conclude that this relationship is a cause-effect relationship.

10.1.13 (a) The response variable could be whether or not they have watched the particular program. Predictor variables might be the number of members in each household, whether or not they received the brochure, number of televisions in the house, number of children, etc. (b) We cannot claim for a cause-effect relationship since for many of the predictor variables it would be impossible to assign the different values it can take.

10.1.19 U and V are related.

Section 10.2

10.2.1 The Chi-squared statistic is equal to $X_0^2 = 5.7143$ and, with $X^2 \sim \chi^2(2)$, the P-value equals $P(X^2 > 5.7143) = .05743$. Therefore, we don't have evidence against the null hypothesis of no difference in the distributions of thunderstorms between the two years (at least at the .05 level).

10.2.5

(a) First, note that the predictor variable, X (gender), is not random. The Chi-squared statistic is equal to $X_0^2 = 10.4674$ and, with $X^2 \sim \chi^2(4)$, the P-value equals $P(X^2 > 10.4674) = .03325$. Therefore, we have some evidence against the null hypothesis of no relationship between hair color and gender. (b) The appropriate bar plots are the two conditional distributions. (c) The standardized residuals are given in the following table. They all look reasonable, so nothing stands out as an explanation of why the model of independence doesn't fit. Overall, it looks like a large sample size has detected a small difference.

	$Y=$ fair	$Y=$ red	$Y=$ medium	$Y=$ dark	$Y=$ jet black
$X=$ m	-1.07303	0.20785	1.05934	-0.63250	1.73407
$X=$ f	1.16452	-0.22557	-1.14966	0.68642	-1.88191

10.2.16 $E\left(X_1^{l_1}\cdots X_k^{l_k}\right) = \frac{\Gamma(\alpha_1+\cdots+\alpha_k)}{\Gamma(\alpha_1)\cdots\Gamma(\alpha_k)}\frac{\Gamma(\alpha_1+l_1)\cdots\Gamma(\alpha_k+l_k)}{\Gamma(\alpha_1+\cdots+\alpha_k+l_1+\cdots+l_k)}$.

Section 10.3

10.3.2 Since $\bar{x} \in [0,\theta] \subset [0,\infty)$ with probability 1, we have that $\bar{x}$ is the least squares estimate of the mean $\theta/2 \in [0,\infty)$.

10.3.4 (b) The least squares estimates of β_1 and β_2 are given by $b_2 = 2.1024$ and $b_1 = \bar{y} = -0.00091$, so the least-squares line is given by $y = -0.00091 + 2.1024x$. (e) Both graphs indicate that the normal simple linear regression model is reasonable. (f) A .95-confidence interval for the intercept is $(-1.0533, 1.0515)$ and a .95-confidence interval for the slope is $(1.7696, 2.4352)$. (g) The F statistic for testing $H_0 : \beta_2 = 0$ is given by $F = 204.28$ and, since $F \sim F(1,9)$ under H_0, the P-value is given by $P(F > 204.28) = .000$, so we reject the null hypothesis of no effect between X and Y. (h) The proportion of the observed variation in the response that is being explained by changes in the predictor is given by the coefficient of determination $R^2 = .9578$. (i) The prediction is given by $y = -0.00091$. This is an interpolation because 0.0 is in the range of observed X values. The standard error of this prediction is 0.46515. (j) The prediction is given by $y = 12.613$. This is an extrapolation because 6 is not in the range of observed X values. The standard error of this prediction is 0.99763. (k) The prediction is given by $y = 42.047$. This is an extrapolation because 12 is not in the range of observed X values. The standard error of this prediction is 2.978 4. The standard errors get larger as we move away from the observed X values.

10.3.12 Since $\sum_{i=1}^n (x_i - \bar{x})^2 = 0$, we must have $(x_i - \bar{x})^2 = 0$, so $x_i = \bar{x}$ for every i and all the x_i are equal to the same value, say x. Then we need to estimate the conditional mean of Y at $X = x$ based on a sample $(y_1, \ldots, y_n)$ from this distribution. The model says that this conditional mean is of the form $E(Y \mid X = x) = \beta_1 + \beta_2 x$, where $\beta_1, \beta_2 \in R^1$. Therefore, $E(Y \mid X = x)$ can be any value in R^1 and the least squares estimate is given by the sample average $\bar{y}$.

10.3.16

(a) Putting $b = \sum_{i=1}^n x_i y_i / \sum_{i=1}^n x_i^2$, we have that $\sum_{i=1}^n (y_i - \beta x_i)^2 =$ $\sum_{i=1}^n (y_i - bx_i)^2 + (b-\beta)^2 \sum_{i=1}^n x_i^2$ since $\sum_{i=1}^n (y_i - bx_i)\, x_i =$

$\sum_{i=1}^n x_i y_i - b \sum_{i=1}^n x_i^2 = 0$, and this is clearly minimized, as a function of β, by b. (b) $E(B \mid X_1 = x_1, \ldots, X_n = x_n) = \sum_{i=1}^n x_i (\beta x_i) / \sum_{i=1}^n x_i^2 = \beta$ and $\text{Var}(B \mid X_1 = x_1, \ldots, X_n = x_n) = \sigma^2 / \sum_{i=1}^n x_i^2$. (c) $E(S^2 \mid X_1 = x_1, \ldots, X_n = x_n) = (n-1)^{-1} \sum_{i=1}^n E\left((Y_i - Bx_i)^2 \mid X_1 = x_1, \ldots, X_n = x_n\right)$ and $E\left((Y_i - Bx_i)^2 \mid X_1 = x_1, \ldots, X_n = x_n\right) = \sigma^2 - \sigma^2 x_i^2 / \sum_{i=1}^n x_i^2 = \sigma^2 (1 - x_i^2 / \sum_{i=1}^n x_i^2)$, so $E\left(S^2 \mid X_1 = x_1, \ldots, X_n = x_n\right) = \sigma^2$. (d) We have that $\sum_{i=1}^n y_i^2 = \sum_{i=1}^n (y_i - bx_i)^2 + b^2 \sum_{i=1}^n x_i^2$. Here we have that $\sum_{i=1}^n (y_i - bx_i)^2$ is the error sum of squares and $b^2 \sum_{i=1}^n x_i^2$ is the regression sum of squares. The coefficient of determination is then given by $R^2 = b^2 \sum_{i=1}^n x_i^2 / \sum_{i=1}^n y_i^2$ and this is the proportion of the total variation observed in Y (as measured by $\sum_{i=1}^n y_i^2$) due to changes in X. (e) Since B is a linear combination of independent normal variables, we have that B is normally distributed with mean given by (part (b)) β and variance (part (b)) given by $\sigma^2 / \sum_{i=1}^n x_i^2$. (f) We have that $(B - \beta) / \sigma \left(\sum_{i=1}^n x_i^2\right)^{-1/2} \sim N(0,1)$ independent of $(n-1)S^2/\sigma^2 \sim \chi^2(n-1)$, so $(B - \beta) / S \left(\sum_{i=1}^n x_i^2\right)^{-1/2} \sim t(n-1)$. Now there is no relationship between X and Y if and only if $\beta = 0$ so we test $H_0 : \beta = 0$ by computing the P-value $P(|T| > |b / s \left(\sum_{i=1}^n x_i^2\right)^{-1/2}|)$, where $T \sim t(n-1)$. (g) We have that $y_i = bx_i + (y_i - bx_i)$, and when the model is correct $y_i - bx_i$ is a value from a distribution with mean 0 and variance (see part (b)) $\sigma^2 \left(1 - x_i^2 / \sum_{i=1}^n x_i^2\right)$. Therefore, the ith standardized residual is given by $(y_i - bx_i) / s \left(1 - x_i^2 / \sum_{i=1}^n x_i^2\right)^{1/2}$. We can plot these in residual plots and normal probability plots to see if they look like samples from the $N(0,1)$ distribution.

Section 10.4

10.4.1 (c) The F statistic for testing H_0 is given by $F = 2.18/2.09 = 1.0431$ and, since $F \sim F(2,9)$ under H_0, we have P-value $P(F > 1.0431) = .39135$. Therefore, we don't have evidence against the null hypothesis of no difference among the conditional means of Y given X. (d) Since we didn't find any relationship between Y and X, there is no need to calculate these confidence intervals.

10.4.5 (c) The F statistic for testing H_0 is given by $F = 6.414/0.589 = 10.89$ and, since $F \sim F(3,20)$ under H_0, the P-value equals $P(F > 10.89) = .00019$. Therefore, we have strong evidence against the null hypothesis of no difference among the conditional means of Y given the predictor.

10.4.11 If an interaction exists between the two factors, then the b response curves are not parallel, so cannot be horizontal, i.e., there must be effect due to both factors.

10.4.16 An individual error rate of .01 gives a family error rate of 0.0455.

10.4.17 (d) The F statistic for testing H_0 : no interaction between Cheese and Lot, is given by $F = 0.151/0.110 = 1.3727$ and, since $F \sim F(2,6)$ under H_0, the P-value equals $P(F > 1.3727) = .32293$. Therefore, we don't have evidence against the null hypothesis of no interaction effect. We can then proceed to calculate the P-value for testing H_0 : no effect due to Cheese. Since $F \sim F(1,6)$

under H_0, this is given by $P(F > 0.114/0.110 = 1.0364) = .34794$. Therefore, we don't have any evidence against the null hypothesis of no effect due to Cheese. The P-value for testing H_0 : no effect due to Lot, since $F \sim F(2, 6)$ under H_0, is given by $P(F > 12.950/0.110 = 117.73) = .00002$. Therefore, we have strong evidence against the null hypothesis of no effect due to Lot.

Section 10.5

10.5.4 (b) The Chi-squared statistic for testing the validity of the model is then equal to 4.66204 with P-value given by $P(\chi^2(8) > 4.66204) = .79301$. Therefore, we have no evidence that the model is incorrect. (c) The P-value for testing $H_0 : \beta_3 = 0$ is 0.638, so we don't have any evidence against the null hypothesis.

Chapter 11

Section 11.1

11.1.1 (a) 0. (b) 0. (f) 4/9. (k) 0.00925.

11.1.5 (a) $P(\tau_c < \tau_0) \doteq 0.89819$. (b) Here $P(\tau_c < \tau_0) \doteq 0.881065$. (d) Here $P(\tau_c < \tau_0) \doteq 0.0183155$.

Section 11.2

11.2.1 (a) $P(X_0 = 1) = \mu_1 = 0.7$. (b) $P(X_0 = 2) = \mu_2 = 0.1$. (d) $P(X_1 = 2 \,|\, X_0 = 1) = p_{12} = 1/4$

11.2.5 (a) $P_2(X_1 = 1) = p_{21} = 1/2$. (b) $P_2(X_1 = 2) = p_{22} = 0$. (g) $P_2(X_3 = 3) = 37/100$.

11.2.10 (a) This chain is irreducible. (b) The chain is aperiodic. (c) $\pi_1 = 1/4$ and $\pi_2 = \pi_3 = 1/2$. (d) $\lim_{n\to\infty} P_1(X_n = 2) = \pi_2 = 1/2$. Hence, $P_1(X_{500} = 2) \approx 1/2$.

Section 11.3

11.3.1 First, choose any initial value X_0. Then, given $X_n = i$, let $Y_{n+1} = i + 1$ or $i - 1$ with probability 1/2 each. Let $j = Y_{n+1}$ and let $\alpha_{ij} = \min(1,\ \pi_j/\pi_i) = \min(1,\ e^{-(j-13)^4+(i-13)^4})$. Then let $X_{n+1} = j$ with probability α_{ij}, otherwise let $X_{n+1} = i$ with probability $1 - \alpha_{ij}$.

Section 11.4

11.4.1 $C = 12/5$.

11.4.4 $P(X_n = 14) = 1.2$.

11.4.7 (a) $\{X_n\}$ is a martingale. (b) T is a stopping time. (c) $E(X_T) = 27$. (d) $P(X_T = 1) = 27/40$.

Section 11.5

11.5.1 (a) $P(Y_1^{(1)} = 1) = 1/2$. (c) $P(Y_1^{(2)} = \sqrt{2}) = 1/4$.

11.5.5 $E(B_{13}B_8) = 8$.

11.5.10 (a) $P(X_8 > 500) \doteq 0.00004276$.

Section 11.6

11.6.1 (a) $N_2 \sim \text{Poisson}(14)$, so $P(N_2 = 13) \doteq 0.1060$. (b) $P(N_5 = 3) \doteq 4.5 \times 10^{-12}$. (f) $P(N_2 = 13,\ N_5 = 20) \doteq 2.9 \times 10^{-5}$.
11.6.6 (a) $P(N_6 = 5 \mid N_9 = 5) \doteq 0.1317$.